Advances in Spatial Science

Springer
Berlin
Heidelberg
New York
Barcelona
Budapest
Hong Kong
London
Milan
Paris
Santa Clara
Singapore
Tokyo

Titles in the Series

C. S. Bertuglia,
M. M. Fischer and *G. Preto* (Eds.)
Technological Change,
Economic Development and Space
XVI, 354 pages. 1995. ISBN 3-540-59288-1

H. Coccossis and *P. Nijkamp* (Eds.)
Overcoming Isolation
VIII, 272 pages. 1995. ISBN 3-540-59423-X

L. Anselin and *R. J.G.M. Florax* (Eds.)
New Directions in Spatial
Econometrics
XIX, 420 pages. 1955. ISBN 3-540-60020-5

H. Eskelinen and *F. Snickars* (Eds.)
Competitive European Peripheries
VIII, 271 pages. 1995. ISBN 3-540-60211-9

J. C.J.M. van den Bergh, P. Nijkamp
and *P. Rietveld* (Eds.)
Recent Advances in Spatial
Equilibrium Modelling
VIII, 392 pages. 1996. ISBN 3-540-60708-0

P. Nijkamp, G. Pepping
and *D. Banister* (Eds.)
Telematics and Transport Behaviour
XII, 227 pages. 1996. ISBN 3-540-60919-9

D. F. Batten and
C. Karlsson (Eds.)
Infrastructure and the
Complexity of Economic Development
VIII, 298 pages. 1996. ISBN 3-540-61333-1

T. Puu
Mathematical Location and
Land Use Theory
IX, 294 pages. 1997. ISBN 3-540-61819-8

Y. Leung
Intelligent Spatial Decision
Support Systems
XV, 470 pages. 1997. ISBN 3-540-62518-6

C. S. Bertuglia, S. Lombardo
and *P. Nijkamp* (Eds.)
Innovative Behaviour in Space
and Time
X, 437 pages. 1997. ISBN 3-540-62542-9

A. Nagurney and *S. Siokos*
Financial Networks
XVI, 492 pages. 1997. ISBN 3-540-63116-X

M. M. Fischer and *A. Getis* (Eds.)
Recent Developments in Spatial Analysis
X, 434 pages. 1997. ISBN 3-540-63180-1

Richard H. M. Emmerink

Information and Pricing in Road Transportation

With 85 Figures
and 16 Tables

Springer

Dr. Richard H. M. Emmerink
Free University
Department of Spatial Economics
De Boelelaan 1105
NL-1081 HV Amsterdam, The Netherlands

ISBN 3-540-64088-6 Springer-Verlag Berlin Heidelberg New York

Cataloging-in-Data applied for
Die Deutsche Bibliothek - CIP-Einheitsaufnahme
Emmerink, Richard H. M.:
Information and pricing in road transportation : with 16 tables / Richard H. M. Emmerink. -
Berlin; Heidelberg; New York; Barcelona; Budapest; Hong Kong; London; Milan; Paris;
Santa Clara; Singapore; Tokyo : Springer, 1998
(Advances in spatial science
ISBN 3-540-64088-6

Printed in Germany

Hardcoverdesign: Erich Kirchner, Heidelberg
SPIN 10655433 42/2202-5 4 3 2 1 0 - Printed on acid-free paper

PREFACE

This monologue in transport economics is based on my last four years of research at the Free University in Amsterdam. It addresses two instruments that might resolve part of the traffic congestion problem in metropolitan areas around the world: (1) the provision of traffic information and (2) road pricing.

From 1992 until 1996 I have been working on this monologue at the department of Spatial Economics at the Free University in Amsterdam. Here, I greatly benefited from the stimulating research environment created by Professor Peter Nijkamp and Professor Piet Rietveld. I also learned a lot by doing joint research at this department with Erik Verhoef. Many of the chapters are based on co-authored work with Peter Nijkamp, Piet Rietveld and Erik Verhoef.

During these four years I spent about half a year at the Centre for Transport Studies at Imperial College, London, where I worked with Professor Kay Axhausen. Kay Axhausen has always been very enthusiastic about my research ideas and provided me with many constructive comments. Our cooperation culminated in a stay at Leopold-Franzens-Universitaet, Innsbruck, in early 1996.

In the final stage of this monologue I have benefited from an anonymous referee who provided me with many constructive and detailed comments that have definitely improved the quality.

In addition, I would like to thank various publishers for giving me permission to use material that have appeared earlier in their journals.

Richard Emmerink

Amsterdam, December 1997

TABLE OF CONTENTS

PART I INTRODUCTION

LIST OF FIGURES

LIST OF TABLES

PART I

INTRODUCTION

1 INTRODUCTION

1.1 PROLOGUE

Human activities require movement of persons and/or goods from one place to another. In order for persons to engage in activities such as work, school, shopping, leisure, and socialising, trips are frequently needed since the locations of the activities usually do not coincide. Furthermore, both production and consumption goods go through different stages of treatment - for example from raw materials, to intermediate goods, to consumption goods -that require transportation. Dependent on the characteristics and requirements of an activity and the persons and goods involved, transportation can take place via the air, the sea, or the land. The present book is concerned with the latter, transportation on land using road infrastructure.

In the past, transportation on land was carried out either by foot or on horseback. At the end of the 18*th* century the steam engine enabled more efficient transportation by train. The invention of the combustion engine at the end of the 19*th* century made motorised transport by means of automobiles practical. As the automobile was a relatively fast and comfortable mode of transport, it became increasingly popular, and began to replace the traditional modes. However, it was not until the 1950s that the price of the car became affordable for many households. Since then, the rate of growth in car ownership has been immense. For example, in 1950 the number of cars and vans per capita in Great Britain was equal to 0.045. This figure increased to 0.213 in 1970 and to 0.374 in 1990 (Button, 1992a, p. 58).

In order to facilitate this increase in automobile ownership, governments realised that the existing road infrastructure had to be dramatically improved. However, as the rate of growth of car ownership was so fast, supply of new road infrastructure fell behind. The capacity of the existing road infrastructure became too small, particularly at specific sites during specific times of the day, the so-called peak-hours. A new type of transportation problem entered society: traffic congestion *caused by automobile travel.*[1] Traffic congestion due to automobile travel became a recurring ingredient of every day life, making millions of citizens suffer from significant time losses and environmental pollution, particularly those living in densely populated metropolitan areas.

Traffic congestion negatively affects inhabitants living in congested areas. It also has an impact on economic activities in these areas. Estimates of the costs of congestion vary depending on the estimation method applied. However, independent of the methodology, the size of these estimates makes clear that the costs of

---

[1]It is interesting to note that traffic congestion has been with us since cities began (Mogridge, 1990). Traffic congestion *due to automobile travel* has created a new type of congestion problem.

congestion are too large to be ignored. For example, Newbery (1990) suggests an annual figure for Great Britain of £12,750 million for 1989-1990 (see also Quinet (1990) and Verhoef (1994a)).

The traditional instrument to tackle the congestion problem is to build more road infrastructure. Nowadays, there seems to be academic consensus regarding the low feasibility of this solution, as the social and environmental consequences of building new roads could be far more severe than the beneficial effects to motorists (Mogridge, 1990). Moreover, in some urban areas it is physically impossible to enlarge the existing road infrastructure without undue expense. As a consequence, transportation research is placing more emphasis on using existing road networks more efficiently (Horn et al., 1995; Rienstra et al., 1996). In theory, this can be achieved using various government policy instruments ranging, for example, from subsidising public transport to fuel taxation and from stimulating tele-commuting to regulatory parking fees (Button, 1992b; Glaister, 1981; Small, 1992a).

In the present book, the economic impacts of two such instruments are studied: motorist information systems and road-pricing. Most of the emphasis will be placed on the economic effects of motorist information systems. However, the impact of road-pricing will also be considered as these instruments might have serious interaction effects.

1.2 MOTORIST INFORMATION SYSTEMS AND ROAD-PRICING

The potential of technologically advanced telecommunication and information systems - known as telematics - in general, and *motorist information systems*, in particular, to resolve part of the congestion problem, is currently attracting much attention. Witness, for example, the DRIVE I, II and III (Dedicated Road Infrastructure for Vehicle safety in Europe) programmes and PROMETHEUS (PROgraM for European Traffic with Highest Efficiency and Unprecedented Safety) in Europe (Nijkamp et al., 1996; PROMETHEUS, 1989; Stergiou and Stathopoulos, 1989); the ITS (Intelligent Transportation Systems) programme in the USA; the CACS (Comprehensive Automobile Control System) programme, and recently the VERTIS (VEhicle Road and Traffic Intelligence Society) programme in Japan; for an overview, see Westerman (1995). These programmes are concerned with a very broad range of techniques to improve traffic flows. To mention a few, one could, for instance, think of traffic management, demand management, driver assistance and cooperative driving, freight and fleet management etc. The topic of this book is also at the core of these mentioned programmes: systems that provide traffic and travel information to travellers.

Sophisticated systems that provide travellers with traffic and travel information are known under various names, such as, motorist information systems, Advanced Traveller Information Systems (ATIS) and Dynamic Route Guidance Systems (DRGS); in this book the first term will be used. The objective of these systems is to provide drivers with up-to-date traffic information that should enable them to improve individual trip decision-making. This, in turn, should lead to improved aggregate traffic flows, and hence, to a more efficient use of the transport system (Arnott et al., 1991).

The potential of motorist information systems to reduce travel times and congestion delays is one argument for adopting these technologies, and it is often regarded as the most important one (Ben-Akiva et al., 1991). However, it has also been claimed that motorist information systems have the potential to reduce driving stress and anxiety, increase safety, and diminish levels of pollution (Bonsall, 1992a; Bovy and Stern, 1990; Mahmassani and Herman, 1990; Rumar, 1990; Shladover, 1993). The present book focuses on the impact of motorist information systems on congestion solely. The other potential beneficial effects are be left aside.

Road-pricing is traditionally viewed as the so-called first-best instrument to deal with problems related to external effects in general, and traffic congestion in particular. In theory, a fluctuating road-price is able to provide the optimal incentive to travellers to achieve Pareto efficiency in the transport system. The concept of road-pricing, levying a tax equal to the difference between marginal social and marginal private costs - the so-called Pigouvian tax (Pigou, 1920) -, has been given much attention in the literature, but has been scarcely implemented due to public and political opposition.

A combination of road-pricing and motorist information systems might provide an attractive policy option from both a theoretical and a practical point of view. Seen from a theoretical perspective, there are two reasons for simultaneously analysing road-pricing and motorist information systems. On the one hand, the road-price has to be conveyed to the users of the road infrastructure by means of an information channel, for example, a motorist information system. On the other hand, to achieve optimal levels of road usage, the road-price should reflect the external congestion costs caused by individual road users. Therefore, the road-price should provide some information on the actual traffic situation (de Palma and Lindsey, 1994; El Sanhouri, 1994; Emmerink, Nijkamp and Rietveld, 1995a). Viewed from a practical point of view, implementing both systems simultaneously might lead to synergy effects on the cost side. A similar kind of technological road side infrastructure is needed for both. Furthermore, the information devices in vehicles needed for implementing a motorist information system might be easily transformed into multiple purpose systems: providing information to the road user and debiting the road-price (Brett and Estlea, 1989).

It has also been envisaged that governments are most likely to tackle the congestion problem with some kind of a package approach: a package containing several instruments that will be implemented simultaneously (Goodwin, 1989; Jones, 1991a). In order to analyse the overall effects of such a package it is essential to address the interactions of the different instruments. A partial analysis might provide a biased view, as due to the interaction of the instruments, the overall effect is not necessarily equal to the sum of the effects of the instruments in isolation.

1.3 OBJECTIVE OF THE BOOK

The international research programmes that address the impact of telematics technologies in road transport have been paying much attention to the technical feasibility of these technologies. However, analyses of the economic consequences of these sophisticated systems have been rather modest. In this book, an attempt is made to fill this gap by studying theoretical and applied models that focus on the economics of motorist information systems and road-pricing. The three main objectives of the book are:

1. To formulate theoretical models in order to analyse the economic impact of motorist information systems and road-pricing and the combination of both.
2. To build simulation models in order to investigate whether the theoretical results obtained under (1) also hold in a more realistic context.
3. To analyse empirically the impact of motorist information systems and road-pricing.

An attempt will be made to answer questions such as: "Who benefits most from these systems, and under which conditions do these benefits occur?". To facilitate the above mentioned objectives, five intermediate objectives have been formulated.

1. Reviewing the theory on the economic impacts of motorist information systems and road-pricing, Chapters 2 and 3.
2. Briefly reviewing the theory on traffic equilibria, and proposing an analytical framework which enables one to model the economic impacts of motorist information systems and road-pricing, Chapter 4.
3. Analysing the economic impacts of motorist information systems and road-pricing under various assumptions, using the framework proposed under (B), Chapters 5 to 10.
4. Building a simulation model with which realistic rules for driver behaviour can be implemented, based on so-called *satisficing* principles (Simon, 1955). This model is used in Chapters 11 and 12 to infer conclusions on the impact of motorist information systems.
5. An empirical analysis of the impact of two simple information systems (radio traffic information and variable message sign information) on driver's behaviour, Chapter 13. An empirical analysis of the work schedule flexibility of commuters, Chapter 14.

1.4 OUTLINE OF THE BOOK

In the light of the above objectives, the book is organised into 15 chapters, see Figure 1.1. Chapters 2 to 4 constitute the **introductory part** of the book. **Chapter 2** revisits the economics of motorist information systems. Special attention is paid to the concept of transport externalities and the role of motorist information systems therein. Then, **Chapter 3** critically reviews the literature on road-pricing. Although, in theory, road-pricing is the first-best instrument for efficiently dealing with traffic congestion, it has not yet received much public and political support. The reasons for this low level of social acceptance are addressed in this chapter. **Chapter 4** concludes the introductory part of the book by proposing an analytical framework which enables one to study the impact of motorist information systems and road-pricing from a theoretical economic perspective. This framework is compared with concepts prevailing in the literature.

The **theoretical models**, following from the framework proposed in Chapter 4, are presented in Chapters 5 to 10. **Chapter 5** discusses the basic case, where the transport network consists of one route only. The impacts of motorist information systems on Pareto efficiency are examined. In particular, the effects of information provision on uninformed travellers are investigated. In **Chapter 6**, the transport network is extended to two routes. This implies that, in addition to so-called *mode-*

split effects, *route-split* effects can be studied as well. Whereas the number of informed travellers is modelled as an exogenous input in Chapters 5 and 6, it is determined within the model (i.e., it is endogenous) in **Chapter 7**. Here, an individual will acquire information when the private benefits of doing so exceed the private costs of obtaining the information. Attention is paid to the issues of subsidising motorist information systems and the interaction of this subsidy with the optimal fluctuating road-price. In **Chapter 8** it is acknowledged that besides travel costs related to travel time, costs related to uncertainty are an important component as well. Information provision, then, reduces the costs of uncertainty for informed travellers. In **Chapter 9**, a network with two origin-destination pairs is studied in an attempt to generalise the results obtained in the previous chapters. A network is constructed that is probably among the least favourable (worst case) for uninformed travellers. The part constituting of the theoretical models is concluded with **Chapter 10**. In this chapter, the interaction between road-pricing and motorist information systems is rigorously discussed. The welfare implications of several policy options are investigated; among others, fluctuating road-pricing and non-fluctuating road-pricing in combination with the provision of information.

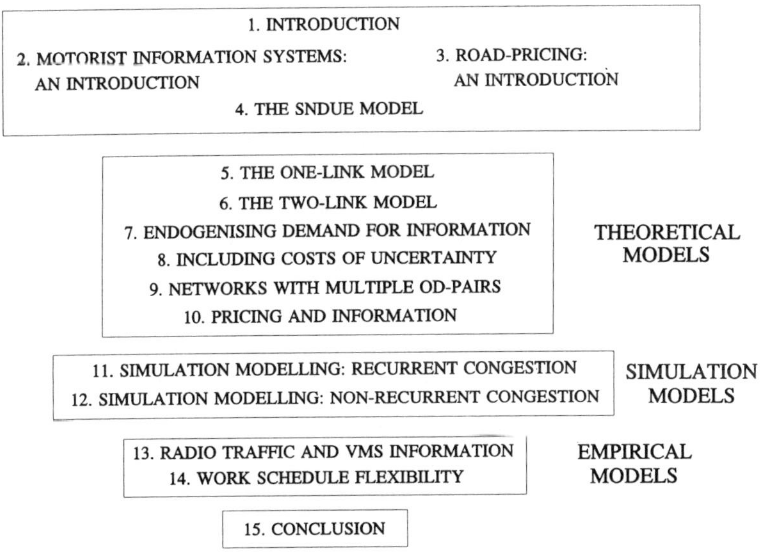

Figure 1.1 Structure of the book.

The **simulation models** of the book are provided in Chapters 11 and 12. Chapters 11 and 12 are based on results obtained with a simulation model, that has been developed for this book. In **Chapter 11**, the simulation model is presented and the rules used for driver behaviour are discussed. Four types of information provision will be considered. In **Chapter 12**, the simulation model is extended to deal with so-called *non-recurrent* congestion, that is, congestion caused by unpredictable incidents such as accidents, sudden lanc closures, weather conditions etc. The impact of motorist information systems on both the informed and uninformed drivers is addressed.

Empirical models are presented in Chapters 13 and 14. Chapter 13 contains an empirical study of route choice behaviour, while Chapter 14 addresses the issue of work schedule flexibility. In **Chapter 13** an empirical analysis is presented in which the impact of simple motorist information systems is assessed. The analysis is confined to radio traffic information and variable message sign information, as more advanced motorist information systems have not yet been implemented at a large scale. In **Chapter 14** the scope for adjustments in work start time of employees is empirically investigated. This is an important issue for the impact of both motorist information systems and road-pricing schemes, as both are partly intended to shift drivers' departure times.

The book concludes with **Chapter 15**. This chapter offers an overview of the research conducted in this book, draws some general conclusions and provides some directions for future research.

2 MOTORIST INFORMATION SYSTEMS: AN INTRODUCTION[1]

2.1 INTRODUCTION

Motorist information systems enhance drivers' knowledge of the situation in road networks, and thus improve driver's decision-making (Ben-Akiva et al., 1991; Bonsall et al., 1991; van Berkum and van der Mede, 1993). However, it is far less well understood whether and to what extent the interaction between the drivers themselves and the information and the drivers may reduce potential beneficial effects of these new technologies. A few models have been developed in order to assess the potential of motorist information systems. Whether these are able to capture the main characteristics of these systems is unclear (Watling, 1994). Using simulation experiments, Mahmassani and Chen (1991) and Mahmassani and Jayakrishnan (1991) found that if more than 20 per cent of the drivers are equipped with the motorist information system, the negative effects due to *concentration* and *overreaction* (Ben-Akiva et al., 1991), may begin to outweigh the beneficial effects. Concentration takes place if the information reduces the variations among drivers, increases uniformity of perceptions of network conditions and thus increases congestion. Overreaction occurs if drivers do not fully take into account the responses of other drivers provided with the same information, thereby shifting the congestion from one road to another.[2] In these circumstances Mahmassani and Jayakrishnan (1991) argued that provision of coordinated information is necessary. Other studies, for example Watling and van Vuren (1993), supported these findings, but stated that the level of market penetration[3] beyond which the system wide performance starts to deteriorate is dependent on the kind of information provided and the behavioural responses of the drivers.

The term *motorist information system* refers to an implemented information system that provides travellers with historical, current or predictive information on travel times of alternative routes, allowing travellers to choose their least costly alternative. Whether and to what extent this also leads to an increased efficiency of the transport network is the subject of the present book. Thus far, research has shown that the impacts of motorist information systems are rather uncertain, as the behavioural responses of the users towards these systems are difficult to trace. Some

---

[1]This chapter is based on Emmerink, Nijkamp, Rietveld and Axhausen (1994) and Emmerink, Axhausen, Nijkamp and Rietveld (1995c) published in *Transport Reviews* and *International Journal of Transport Economics*, respectively.

[2]The phenomena of overreaction and concentration are discussed in Section 2.3.

[3]The level of market penetration is the percentage of drivers equipped with a motorist information system.

studies addressing this issue have been conducted; see for instance Allen (1993), Bonsall (1992b), Caplice and Mahmassani (1992), Conquest et al. (1993) and Spyridakis et al. (1991). Although these studies give more insight into the behavioural issues involved, they do not indicate a clear pattern regarding users' responses. A large variance in the results is observed and strong conclusions can generally not be drawn.

A motorist information system affects both the equipped and the non-equipped drivers. In fact, one can argue that the following three performance measures are influenced:

1. the average road network travel time;[4]
2. the average road network travel time of the equipped drivers;
3. the average road network travel time of the non-equipped drivers.

The interactions between a motorist information system and the drivers - resulting in changes of the above listed travel times - affect the market potential and economic viability of these systems. With most scarce commodities, the benefits obtained from buying the commodity are independent of the level of market penetration. As mentioned above, this is not necessarily true for motorist information systems. The aim of this chapter is to discuss some of the economic implications of this dependency and to raise economic issues that are relevant when implementing motorist information systems. In particular, the focus is on the externalities (both positive and negative) caused by motorist information systems, and the implications of these externalities on the economic viability. In addition, the traffic generating potential of these new technologies is discussed.

In discussing these issues, other potential purposes of motorist information systems such as a decrease in stress or anxiety, an increase in safety, a decrease in pollution etc. will be ignored. In particular the potential of these technologies to reduce stress or anxiety may be heavily underestimated.

The discussion in this chapter initially focuses on the case of recurrent congestion, that is, congestion due to under-capacity of the road network. Later on, the more relevant case of non-recurrent congestion, congestion caused by incidents such as bad weather or traffic accidents, will be addressed. This is an important extension, since expectations of motorist information systems are particularly high for non-recurrent network conditions.

The chapter is organised as follows. First, Section 2.2 reviews the literature on potential benefits of motorist information systems, while Section 2.3 turns to potential adverse effects. Next, Section 2.4 and 2.5 discuss the use of the well-known Wardrop's principles in relation to motorist information systems; concepts that will prove to be useful throughout this book. Section 2.6 discusses the externalities involved in implementing a motorist information system, while Section 2.7 focuses on the traffic generating effects. Then, Section 2.8 touches the issue of market penetration of these systems, Section 2.9 extends the scope to the case of non-recurrent congestion, and Section 2.10 raises some issues of the government's role in dealing

---

[4]In a two player game in which the players represent the road users, de Palma (1992) showed that this is not necessarily true, even with the provision of *perfect* information. A similar result has been obtained by Arnott et al. (1991) under *imperfect* information.

with these new technologies. Finally, Section 2.11 contains some concluding comments.

2.2 BENEFITS FROM MOTORIST INFORMATION SYSTEMS

In the literature it has been suggested that road networks are not used as efficiently as possible. For instance, King (1986) and King and Mast (1987) found that about 6 per cent of all distance and 12 per cent of all travel time is wasted. In another study, Jeffery (1986) found an excess travel distance of 6 per cent. Hence, it seems sound to argue that there is some room for travel time and distance savings through better information. However, there is no consensus about the type and size of the gains that can be obtained from motorist information systems. In this section, relevant results from the literature will be reviewed.

Kanafani and Al-Deek (1991) presented estimated benefits of approximately four per cent in most cases in their theoretical model, without taking into account non-recurrent congestion. Their estimates were based on achieving the system optimum - Wardrop's second principle (Wardrop, 1952) - in the road network through the motorist information system at full market penetration. As discussed in Bonsall et al. (1991), this is in itself a highly debatable assumption and affects the plausibility of their figures.

Using a stochastic model in which a motorist information system is assumed to guide vehicles toward the expected shortest travel time routes, Tsuji et al. (1985) found benefits ranging from 9 to 14 per cent for equipped drivers, but add that even higher figures could be attained, when non-recurrent congestion would be considered. Tsuji et al. (1985) ignored an important aspect of these technologies; they assumed that the flow of guided vehicles did not affect the unguided vehicles, implying that unguided vehicles do not receive any direct or indirect benefit from the motorist information system. Building further on Tsuji et al.'s (1985) model, Jeffery et al. (1987) asserted that the benefits through route guidance could be as large as 10 per cent on average.

In other studies it has been suggested that most benefits of motorist information systems will be obtained in non-recurrent congested networks. See, for example, Dehoux and Toint (1991) and van Vuren and van Vliet (1992). Dehoux and Toint (1991) stated that one of the purposes of the application of systems that provide information to road users is to produce substantial changes in traffic patterns over short periods of time in order to dissolve traffic jams more quickly. Research efforts investigating the effectiveness of motorist information systems in such situations have been carried out by Hounsell et al. (1991), Mahmassani and Jayakrishnan (1988) and Rakha et al. (1989).

Mahmassani and Jayakrishnan (1988) analysed a transportation network during periods of perturbation and followed the system evolution to the *final equilibrium state*.[5] However, their simulation experiments only dealt with historical information

---

[5]The terminology *final equilibrium state* in Mahmassani and Jayakrishnan (1988) indicates a situation in which no driver has an incentive to switch alternative. The *final equilibrium state* is not necessarily unique as shown in Mahmassani and Chang (1987).

(the road users' own experience in previous periods) and thus did not involve the application of a motorist information system. Hounsell et al. (1991) and Rakha et al. (1989) found substantial travel time savings of up to 20 per cent in experiments conducted in a network with non-recurrent congestion.

To conclude, transportation researchers do not agree on the scale of the benefits obtainable from motorist information systems; estimates show large variations. One explanation for this divergence in results might be found in the actual definition of the motorist information system. Different researchers have used different methods / models / situations for assessing the benefits. Another explanation may be that the benefits depend on the shape of the network, the scale of the network and the current level of congestion. Thus, one of the properties of motorist information systems would be that the benefits are highly dependent on the setting of the situation. This has recently been suggested by Watling and van Vuren (1993) and will also be shown in this book. Seen from this perspective, the results obtained by different researchers are not likely contradictory, but rather complementary in nature.

Figure 2.1 depicts three possible effects of motorist information systems. On the y-axis the average network travel time is shown as a function of market penetration. The curves in Figure 2.1 are well-behaved (continuous and differentiable) and have only one local (and therefore global) minimum. Ill-behaved functions, for instance functions with multiple local minima, are not ruled out, but there is no clear rationale for such shapes.

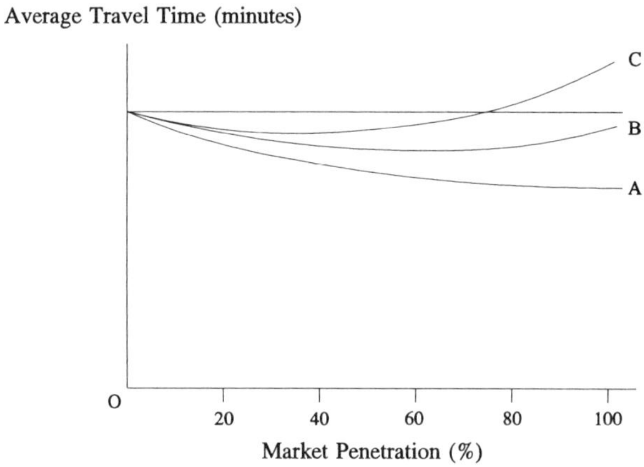

Figure 2.1 Network performance as a function of market penetration.

In Figure 2.1, case A depicts a situation in which the average network wide travel time is minimised at full market penetration, while in case B the optimal level of market penetration is below 100 per cent. In both case A and case B, at full market penetration the network is still better off with information. Case C depicts a situation in which this is no longer true due to overreaction and/or concentration. The next section is devoted to clarifying these adverse effects.

2.3 POSSIBLE ADVERSE EFFECTS OF MOTORIST INFORMATION SYSTEMS

In general, motorist information systems serve to increase the efficiency of transport networks and to alleviate negative externalities of transport behaviour. In the literature it has been recognised that motorist information systems could have adverse effects on the overall performance of a transport network (Arnott et al., 1991; Ben-Akiva et al., 1991; de Palma, 1992). Arguments supporting this possibility can, roughly speaking, be subdivided into two classes of conditions:

1. phenomena related to imperfect information;
2. phenomena regarding the behavioural responses of drivers towards the information.

The former case has been investigated by Arnott et al. (1991); their research confirmed that certain imperfect information could potentially lead to a worsening of actual network performance. Phenomena related to behavioural responses that may negate some of the beneficial effects of motorist information systems are mentioned in Ben-Akiva et al. (1991). They list three potential adverse effects of information provision:

1. oversaturation
2. overreaction
3. concentration.

Oversaturation occurs if drivers are unable to process the supplied information properly. For instance, information could overload the driver, thus distracting and impairing him from selecting the optimal route. As pointed out by Ben-Akiva et al. (1991), this is mainly a psychological-technical man-machine interaction problem which can in principle be overcome and will not be discussed further in this chapter. In Sections 2.3.1 and 2.3.2, the discussion is focused on the two remaining issues of overreaction and concentration, respectively.

2.3.1 Description of overreaction

A major potential flaw of motorist information systems is the possibility of overreaction. The phenomenon of *overreaction* has been mentioned repeatedly in the literature. Most rigorously it has been described by Ben-Akiva et al. (1991). In this section their description of overreaction forms the starting point of the analysis. Ben-Akiva et al. (1991) describe overreaction occurring in:

> "... a situation in which a substantial fraction of drivers receive information on traffic conditions. Overreaction occurs when drivers' reactions to traffic information cause congestion to transfer from one road to another. It may also generate oscillations in road usage. Overreaction may happen if too many drivers respond to information on current traffic conditions.
>
> When current traffic information is provided to drivers, overreaction may be avoided if drivers' decisions are based on correct expectations of the other drivers' reactions. For example, some drivers may not shift to a reportedly faster route because they anticipate a rush of the other drivers

to that route. Overreaction is likely to take place if drivers fail to consider or underestimate the potential responses of the other drivers. The task of anticipating the other drivers' responses is difficult to perform.

...

Overreaction may also occur when predictive information or route guidance is provided to drivers. However, in this case part of the blame for overreaction lies in the failure of the information provider to accurately predict driver behavior and reaction to information, including an estimate of the fraction of drivers who will follow the advice. This results in an inconsistency between predicted and realized traffic conditions." (p. 254)

They support their case with the example of holiday makers who overreact to supplied information on how to plan their journey optimally. Clearly, this kind of information is static in the sense that it is received some time before the journey is undertaken. In addition, exactly the same information is given to all holiday makers. As will be argued in Section 2.5.3, overreaction is not restricted to static information, but can occur with real-time - or even predicted - information as well.

From the description above, it follows that if overreaction takes place, some roads will be overused, while at the same time others will be underused. The travel time on the overused routes will obviously be larger than the travel time on the underused ones. By definition, this implies that the conditions for the *user equilibrium* (Wardrop's first principle) are not met, since at user equilibrium (Wardrop, 1952) for each origin and destination:

"The journey times on all the routes actually used are equal, and less than those which would be experienced by a single vehicle on any unused route." (p. 345)

In addition, it is clear that overreaction deals with a situation in which drivers are unable to predict the responses of those drivers provided with the same information correctly, and therefore causes congestion on some roads to be more severe than on others. In other terms, the information supplied to the drivers is imperfect in the sense that drivers still need to predict the responses of other drivers to the information.[6] In this respect it can be argued that overreaction is the consequence of providing drivers with imperfect information or of drivers reacting to actual information without including behavioural responses of other road users. Thus, overreaction is essentially the result of a partial equilibrium approach without the inclusion of sufficient positive and negative *feedback* and *feedforward* mechanisms.

Furthermore, it is fruitful to address the inter-relationship between the congestion externality and overreaction.

1. The congestion externality is caused by the fact that drivers do not take into account the impacts of their own decisions on the trips of other drivers during their decision-making process. This kind of behaviour leads to the user equilibrium in the network and *is* optimal from an individual perspective. However, from a collective perspective, this situation *is not* optimal if the user equilibrium does not coincide with the *system optimum* (a situation in which

[6] In this chapter, imperfect information refers to information that is not perfectly predictive.

"The average journey time is a minimum", Wardrop (1952, p. 345)); this is generally the case in congested networks (Sheffi, 1985).

2. The phenomenon of overreaction occurs if drivers fail to take into account the responses of the other road users to the information provided. This leads to a sub-optimal decision from an individual point of view.

Summarising, it is argued that for a driver to behave optimally from an individual point of view, his decision-making process *should not* take into account the implications of his behaviour for other drivers, but it *should* take into account the decisions of other drivers. If all drivers succeed in doing so, the situation in the network will be represented by the user equilibrium. Overreaction might take place, if travellers are not able to do so.

Whether the implications of overreaction are severe, mainly depends upon the degree of *irreversibility* of the decisions made by the drivers. If a *wrong* decision, due to overreaction, cannot be reversed, the full cost consequences of this decision have to be borne. However, if a decision can partly be altered after some of the consequences have been experienced, overreaction has evidently a less serious impact. An irreversible decision can be illustrated by a departure time decision. Once a driver has started his trip, the departure time cannot be changed. A reversible decision is, for instance, a route choice decision with options to switch to other available routes during the trip. It should be noted however, that in case of reversible decisions also all other road users may adjust their behaviour, so that at the end still severe adverse impacts might be expected.

2.3.2 Description of concentration

Concentration is a related concept frustrating the potential of motorist information systems. Ben-Akiva et al. (1991, p. 254) describe *concentration* as follows.

> "In general, a group of drivers travelling from an origin to a destination tend to use different routes and departure times because they have heterogeneous preferences and diverse perceptions of network conditions. Differences between perceived and actual network conditions imply that drivers who are not well informed may select alternatives which are not in their best interest. Information tends to reduce the variations among drivers because it increases uniformity of perceptions of network conditions around the true values. As a result a greater number of drivers may select the best alternatives (from their individual point of view) and consequently drivers with similar preferences will tend to concentrate on the same routes during the same departure times. Thus, more information could potentially generate higher levels of traffic congestion."

Ben-Akiva et al. (1991) illustrate the situation of concentration by means of an example in which the initial situation in the network (without information) leads to a better system performance than the situation with information in which they assumed that the user equilibrium would result. Thus, concentration occurs if differences in drivers' perceptions are reduced due to information provision. As a consequence, some drivers will make *better* decisions from an individual point of view, but the situation in the network as a whole will be *worse*.

The role that information might play in transport networks can be further illustrated by looking into the principles that were first defined by Wardrop, the user equilibrium and the system optimum (Wardrop, 1952).

2.4 WARDROP'S PRINCIPLES AND MOTORIST INFORMATION SYSTEMS

In 1952 Wardrop laid the foundation for equilibrium modelling in transportation networks (Wardrop, 1952). He discussed two principles, the user equilibrium and the system optimum, and examined their application to transportation networks. Later, these principles have often been referred to as Wardrop's first and second principle.

2.4.1 The role of Wardrop's first principle: The user equilibrium

It is well known from economic forecasting that each forecast may lead to behavioural responses that make this forecast futile. The example given in the previous section addresses in this context a crucial point. If *perfect predictive information* is supplied and it is assumed that drivers behave *rationally*, it is reasonable to assume that the *user equilibrium* according to Wardrop (1952) results. Here, perfect predictive information is defined as predictive information that will appear to be correct in the future. Provision of perfect predictive information is, however, highly hypothetical because it requires the provider of the information to forecast perfectly the drivers' responses towards the information. This is an unrealistic assumption since, to be able to provide perfect predictive information, drivers' responses towards the information need to be known in advance, while these responses, in turn, depend upon the information provided. Figure 2.1 shows the circular relationship between predictive information and drivers' responses, and illustrates that *as soon as the predictive information is given new predictions have to be made.*

Perfect predictive information possesses the property that the information remains correct after the drivers' responses. Figure 2.2 shows an iterative hypothetical method for obtaining perfect predictive information.[7] Here, it is assumed that the provider of the information knows the drivers' responses with certainty.

Information reflecting the traffic situation at user equilibrium is ceteris paribus perfectly predictive. This can easily be checked with Figure 2.2. At user equilibrium none of the drivers has an incentive to change routes since they will be worse off by doing so. In addition, this is the only kind of information having this property. This can be clarified as follows:

> Suppose that information is perfectly predictive, but is not consistent with traffic flows at user equilibrium. Then, some drivers will be better off by not complying with the information. Therefore, the traffic flows, after drivers have received the information, will not reflect the given information, and hence, the information is not perfectly predictive.

---

[7]Practical work, trying to model and predict traffic conditions on a highway network in real time, is currently being carried out in the DRIVE project DYNA; see Lindveld et al. (1992). Also, Koutsopoulos and Xu (1993) published an article addressing the prediction of travel times.

The argument above provides another justification for assuming Wardrop's user equilibrium if perfect predictive information is supplied and can be summarised as follows:

> *Perfect predictive information is consistent with the user equilibrium and leads to traffic flows implied by the user equilibrium.*

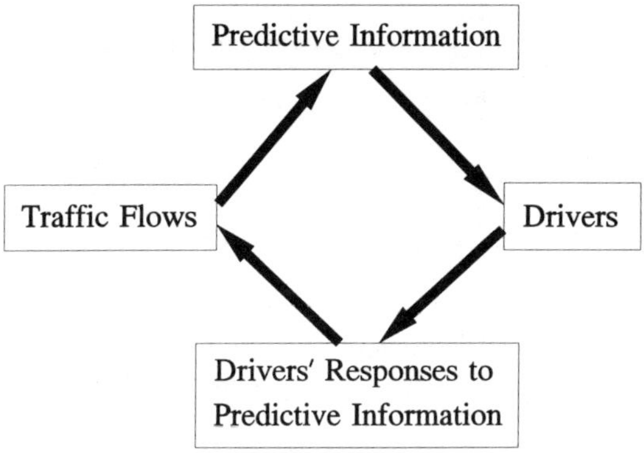

Figure 2.2 Predictive information and drivers' responses.

Wardrop's first principle has been widely accepted as giving a reasonable approximation of the prevailing situation in road transport networks without a motorist information system, and has been used repeatedly in modelling transport demand.[8,9] Recently, some researchers have questioned the appropriateness of the user equilibrium. Here, two arguments will concisely be discussed: one relates to information processing of drivers, the other one to behavioural rules of drivers.

First, Horowitz (1985) has argued that if link costs are flow dependent (as they are in real-life networks), travellers are unlikely to know the costs of a trip before it is actually taken. Rather, they must estimate the costs from past experience with the network. It can be shown (Horowitz, 1984) that depending on the drivers' information about the network's performance in the past, and the way they use the information to estimate the costs of a current trip, traffic flows may converge to their equilibrium values, oscillate around these values perpetually or converge to values that depend upon the initial conditions and may be considerably different from the user equilibrium. As pointed out by Dehoux and Toint (1991), the assumption needed for

[8]In particular, the user equilibrium is widely used in the assignment phase of the traditional four-step modelling process.

[9]Some peculiar properties of Wardrop's first principle have been mentioned in the literature; the most famous one is Braess' paradox (Braess, 1968; Frank, 1981). Braess showed that the addition of an extra link in a network could lead to a larger average (system wide) travel time under the user equilibrium principle.

the application of the user equilibrium (viz. everyone has complete knowledge on the situation in the network), is not met in real-life transport networks. On the contrary, this is exactly what motorist information systems would like to achieve.

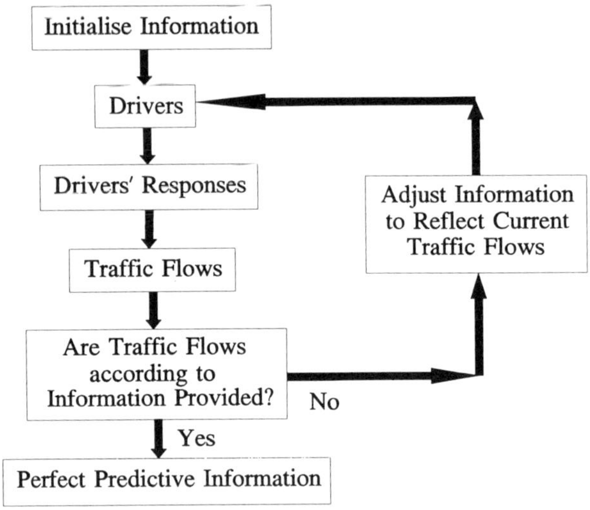

Figure 2.3 Iterative procedure for calculating perfect predictive information.

Second, Dehoux and Toint (1991) have claimed that one of the purposes of the application of systems that provide information to drivers is to produce substantial changes in traffic patterns over short periods of time in order to resolve traffic jams more quickly. Such situations, often referred to as non-recurrent congestion, do not fit into an equilibrium framework, since traffic flows will deviate from their equilibrium values during these periods. In view of these arguments, instead of an equilibrium principle, Dehoux and Toint (1991) proposed *behavioural rules* that could vary from one group of users to the next in order to capture some of the wide behavioural variations between drivers. The kernel of their approach thus consists of a *behavioural framework*. A similar modelling methodology has recently been proposed by Watling and van Vuren (1993). They argued that the component *driver behaviour* (Watling and van Vuren, 1993, Figure 1, p. 162) is one of the three key components of motorist information systems. Research according to this approach has been conducted by several authors. The reader is referred to the work of Iida et al. (1992), Mahmassani and co-authors (summarised in Mahmassani and Herman (1990)) and Stern et al. (1990).

2.4.2 The role of Wardrop's second principle: The system optimum

An important question is whether there are tools to achieve an optimal performance of a transport system. Wardrop's second principle, known in the literature as the system optimum, is the situation in which total travel time in the network is minimised. In congested networks, the system optimum does not necessarily coincide with the user equilibrium (cf. Sheffi (1985)). Thus only in uncongested situations it is guaranteed that the two principles lead to identical flows.[10] Furthermore, Mahmassani and Peeta (1993) found that the discrepancy between the system optimum and the user equilibrium can be large in heavily congested networks. If it is the government's aim to achieve the system optimum in road transport networks, an effective policy is required. Given the arguments in the previous section, it is unrealistic to believe that the implementation of motorist information systems is such a policy. It was seen that perfect predictive information leads to the user equilibrium, and therefore when congestion occurs a kind of imperfect information should be provided to the drivers when the regulator wants to achieve the system optimum. However, as Bonsall et al. (1991) point out, it is not very likely that drivers will comply with imperfect predictive information; they will tend to ignore it. In particular, given that drivers are able to assess the reliability of information (Vaughn et al., 1992), it is unlikely that the system optimum will be reached by providing imperfect information.

Hence, an alternative policy is required. The *best* policy (first-best), from a theoretical point of view, is the implementation of a congestion-pricing scheme. Such a policy charges the drivers for the difference between the marginal social and marginal private costs, and leads to the system optimum. However, implementing congestion-pricing is difficult in practice; see Chapter 3. Even if the technology required for electronic congestion-pricing were available, the implementation is not straightforward. The reason is that the optimal charge of a trip is dependent upon the level of congestion on the roads at the time of use. Therefore, the precise charge cannot be known prior to the trip. In such a situation, motorist information systems could provide drivers with information and, based upon this, a predicted charge for each route. The driver could then make a trade-off between the costs and travel time associated with the trip of the alternative routes. Therefore, it is plausible that motorist information systems could be an important complement to an advanced congestion-pricing scheme. More research in this direction is clearly needed. The contributions by de Palma and Lindsey (1994), El Sanhouri (1994) and in Chapters 7 and 10 of this book are the few attempts in this direction.

2.5 THE RELATIONSHIP AMONG CONCENTRATION, OVERREACTION, MARKET PENETRATION AND WARDROP'S PRINCIPLES

2.5.1 Wardrop's principles and concentration

Under the assumption, discussed in the previous section, that the user equilibrium corresponds to a situation in a network where perfect predictive information is provided, the phenomenon of concentration can be described in terms of both the system optimum and the user equilibrium (that is, both Wardrop's principles).

---

[10] Also, with inelastic demand, the system optimum and the user equilibrium might coincide, even with congestion.

Let *T(system optimum)* denote the average travel time in the system at system optimum, *T(user equilibrium)* the average travel time at user equilibrium, and *T(initial)* the average travel time in the system without information. Furthermore, *T(information)* is the average travel time in the network under the perfect predictive information.

Clearly, *T(initial)* is equal to the average travel time in the system under the initial conditions, or in other terms, the average travel time associated with the initial route and departure time decisions of the drivers in the network. Then it can easily be seen that concentration occurs if relation (2.1) holds.

$$T(\textit{system optimum}) \leq T(\textit{initial}) < T(\textit{user equilibrium}) \tag{2.1}$$

In this situation, supplying perfect predictive information, which by assumption makes *T(information)* equal to *T(user equilibrium)*, leads to a worsening in network performance. However, if relation (2.2) holds:

$$T(\textit{system optimum}) \leq T(\textit{user equilibrium}) < T(\textit{initial}) \tag{2.2}$$

then provision of perfect predictive information increases the network performance, since it brings the average travel time down to the average travel time at user equilibrium, which is in this case smaller than the average travel time under the initial conditions.

The example given in Ben-Akiva et al. (1991) - mentioned in this chapter in Section 2.3.2 - illustrates clearly these relationships. Another example taken from a game-theoretic context can be found in de Palma (1992). De Palma models the traffic flows in a transport network as a game between two drivers (two players). In this two-player game, he shows that under certain regimes (certain values of the entries in the pay-off matrix) information provision can shift the outcome of the play to a Nash equilibrium unequal to the system optimum, thereby decreasing social welfare.[11]

The analysis in this section, stressed one important point with respect to the effectiveness of perfect predictive information in a world with rational (utility maximising) drivers:

> Whether the information will be beneficial depends solely on the system performance under the initial conditions and at user equilibrium.

The importance of the initial situation in the network, has also been recognized by Mahmassani and Jayakrishnan (1991) and Mahmassani and Chen (1991). Their simulation experiments suggested that the usefulness of information provision was strongly dependent upon the initial conditions in the network, thereby underlining the arguments given in this section.

Given the arguments in Section 2.4.2, it is unlikely that drivers will follow system optimal routes if that will lead to an increase in their own travel time. This

---

[11] The Nash equilibrium may be regarded as being the game-theoretic equivalent of the user equilibrium. In game theory, a situation is a Nash equilibrium if the players respond optimally to each others' strategies.

means that a combination of congestion-pricing and motorist information systems will then be needed to reach the system optimum.

2.5.2 Market penetration and concentration

A major issue on the impact of information is the access to or user acceptance of motorist information systems. In the theoretical exposition given in Sections 2.3 and 2.5.1, the term market penetration or adoption of (telematics) information is not used. The discussion of which level of market penetration is best is irrelevant in a world with perfect predictive information and rationally behaving drivers. The situation with the smallest average travel time will always be reached at a market penetration level equal to 100 per cent if relation (2.2) in Section 2.5.1 holds; next, if relation (2.1) holds, the situation will be worst if the level of market penetration equals 100 per cent.

If the information provided is not perfectly predictive, the situation becomes more complicated. In general, given the arguments in Section 2.4.1, one could argue that the purpose of any kind of information provision system is to get closer to the user equilibrium. As a consequence, if relation (2.1) holds, the provision of information, even if the information is not perfectly predictive, will have a tendency to worsen the network performance.

Therefore, it is concluded that, if relation (2.1) holds, information provision will lead to concentration and hence has an adverse effect. From this perspective, it is important to analyse whether prevailing situations (initial conditions) in real networks are represented by relation (2.1) or (2.2), as this is decisive for the network performance of motorist information systems.

2.5.3 Overreaction and market penetration

In contrast to the previous section, the level of market penetration becomes crucial in situations with overreaction. The difference between overreaction and concentration is underlined by a quotation taken from Ben-Akiva et al. (1991):

> "Concentration is intrinsic to any system and holds even with perfectly rational drivers. Overreaction is a consequence of the fact that the drivers and/or the driver information system are unable to perfectly forecast the use of information by all drivers." (p. 254)

In light of these differences it is evident that concentration can take place even with perfect predictive information and rational drivers (as shown in Section 2.5.1), while overreaction is a consequence of either imperfect information or incompletely rational drivers. If perfect predictive information is given to rational drivers, overreaction cannot occur, since the perfect predictive information reflects, by definition, the situation in the network at any time in the future, and therefore takes into account the responses of all drivers towards the information provided. The driver simply has to choose his alternative according to the information and does not have to worry about the impacts of the decisions of other drivers on his trip, since these are incorporated in the supplied perfect predictive information.

However, as illustrated in Section 2.4.1, the provision of perfect predictive information is highly hypothetical. In any real-world situation there will be some uncertainty regarding drivers' responses to the information, hence making the information imperfect. As a consequence, one should be very careful in providing information (cf. Arnott et al., 1991). The full impacts of overreaction should be acknowledged and it is at this point that the issue of market penetration enters the discussion.

Overreaction taking place at a relatively low level of market penetration will not dramatically affect the situation in the network. Stronger even, as long as the benefits due to the information provision exceed the negative impacts of the overreaction (also due to the information provision), the network performance will improve. Therefore, the resulting situation in the network after the provision of information is a trade-off between the positive effects of the information (the fact that the drivers are told where the congestion takes place) and the negative effects (the fact that too many drivers change routes to the uncongested route, thereby shifting the congestion).

At a low level of market penetration the positive effects will outweigh the negative effects of overreaction. As the level of market penetration increases, the negative effects of overreaction will become more severe and will eventually outweigh the benefits generated by motorist information systems. In such a situation, according to economic principles, the optimal level of market penetration is the level at which the marginal benefits of the information are equal to the marginal costs due to overreaction. However, in practice this level of market penetration may be hard to find, and additionally, it may be dependent upon both the network and the kind of information provided, see also Section 2.8.

On theoretical grounds, one could argue that more reliable information allows a higher level of market penetration, a situation which will be underlined by the findings in Chapters 11 and 12. There it will be shown that the optimal level of market penetration is close to 20 per cent if after-trip information (information given to the drivers on the situation in the network during the last travel period) is supplied, while real-time en-route information permits a significantly higher level of market penetration. This stresses the importance of the quality of traveller's information. High quality information, being informative and accurate, seems to allow a relatively high level of market penetration; low quality information causes overreaction taking place already at low levels of market penetration. This trade-off between the level of market penetration and the quality of the information has as yet been given sparse attention in transport and telematics research. However, if the technological ability of motorist information systems does not allow continuous information updating, the question of the quality of the information, which in turn is directly related to the updating frequency of the information, becomes a crucial one. Further research should thus be directed to the sensitivity of overreaction with respect to the updating frequency of the information and the level of market penetration. This will be done in Chapter 12.

2.6 EXTERNAL EFFECTS AND MOTORIST INFORMATION SYSTEMS

2.6.1 Positive externality to non-equipped drivers

As discussed in Section 2.2, a motorist information system could produce significant benefits, the size depending on the characteristics of the information system and the road network under consideration. Assuming that equipped drivers will always be better off, these benefits are caused by one of the following two reasons:

1. Equipped road users are better off and non-equipped road users are worse off. The gains to the equipped road users more than offset the losses to the non-equipped ones.
2. All road users are better off.

The first option has been mentioned by Bonsall and Parry (1990). They argued that if equipped drivers change their behaviour from day to day in light of the information available, the non-equipped drivers could be faced with a less predictable system, thus reducing their ability to achieve user optimal routes. In this way, non-equipped drivers might be worse off due to the volatile behaviour of equipped drivers. This situation is however not regarded as very plausible; particularly from a theoretical perspective, this argument can be questioned. Information provided to drivers will spread the congestion more evenly throughout the network, or more precisely, information will direct the traffic flows towards the user equilibrium and therefore, travel times on alternative routes connecting an origin and a destination will slowly converge. Hence, non-equipped road users are faced with a network in which the traffic flows are more balanced and as a consequence, a positive externality arises. This externality implies that *the higher the level of market penetration of the motorist information system, the lower the travel times to the non-equipped road users.*[12] The exact size of this external effect cannot be inferred from the literature, and will, like the network wide benefits, be strongly dependent on the kind of information, the network under consideration etc. Besides the theoretical arguments favouring the existence of this externality, Chapters 4 to 10 will provide theoretical evidence, while Mahmassani and associates (summarised in Mahmassani and Herman (1990)) and Chapters 11 and 12 provide some (limited) evidence obtained by simulation experiments.

The existence of this externality raises an interesting question about the pricing policy of motorist information systems: Should these systems be partly subsidised because of the beneficial effects to non-equipped drivers, an issue that is also raised in Chapter 7. An answer to this question is not straightforward, since besides this positive externality, the introduction of motorist information systems induces another externality, which is addressed in the next section.

2.6.2 Negative and/or positive externality to equipped drivers

After having discussed the beneficial effects of motorist information systems on the network wide performance in Section 2.2 and the externalities to non-equipped road users in the previous section, it is now time to turn to the effects to the road users that are equipped with the system. In fact, they are causing the changes in the traffic flows. It is here assumed that - ceteris paribus - only the level of market penetration determines the travel times in the network. For the network under consideration it is assumed that the (hypothetical) relation between travel time and level of market penetration as depicted in Figure 2.4 holds.[13]

---

[12]For a rigorous definition of an external effect, we refer to the next chapter.

[13]Empirical evidence on the exact shape of these two travel time curves is not available, though many pilot studies are currently being carried out. The curves in Figure 2.4 are purely illustrative. Obviously, however, the non-equipped curve lies above the equipped curve.

With this figure, the curve for *information benefits to equipped drivers* can be derived. Given a certain level of market penetration *m*, the benefits to the equipped drivers are equal to the distance between the curve for non-equipped and equipped drivers, since these are the travel time savings obtained from buying the motorist information system. Dependent on the shape of the travel time curves for equipped and non-equipped drivers, the information benefits to equipped drivers curve will look like Figure 2.5 (having a peak at market penetration level A) or Figure 2.6 (monotonically decreasing). The theoretical and simulation experiments in this book in Chapters 11 and 12 will suggest Figure 2.6.

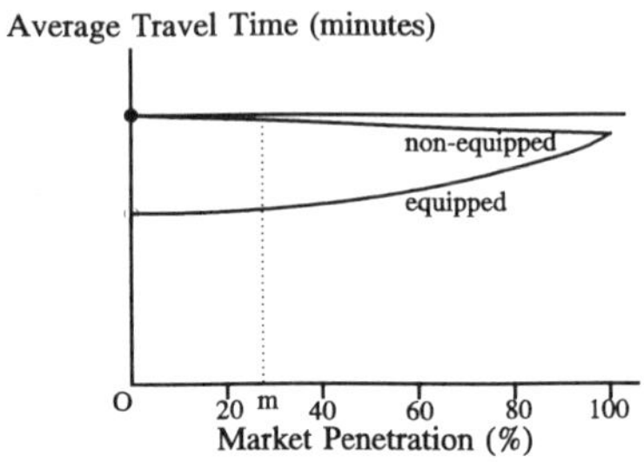

Figure 2.4 Travel time as a function of market penetration.

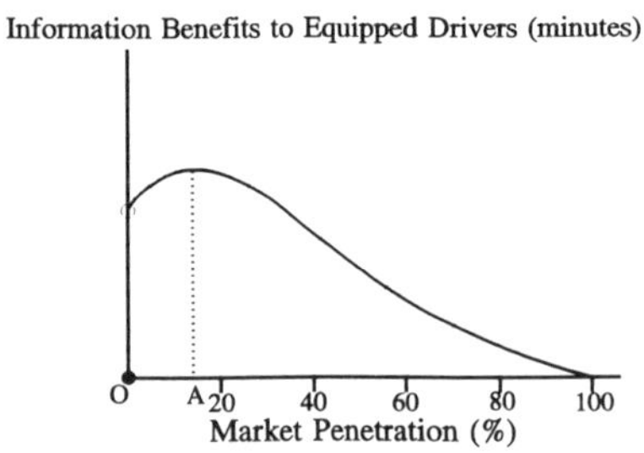

Figure 2.5 Information benefits to equipped drivers. Case 1.

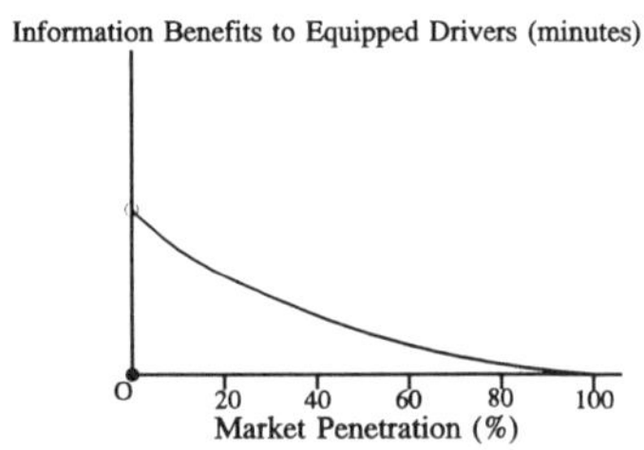

Figure 2.6 Information benefits to equipped drivers. Case 2.

The information benefits to equipped drivers curve is discontinuous at zero market penetration, since there will be clear travel time savings to the first driver to be equipped. Furthermore, it is assumed that the information benefits to equipped drivers is continuous at full market penetration, implying that the last road user to be equipped will not obtain any additional gains from the information system. Here, the underlying assumption is that at full market penetration the traffic flows in the road network will represent the user equilibrium. While justifiable on theoretical grounds for recurrent congestion, this assumption becomes doubtful when dealing with a network with non-recurrent congestion. Section 2.9 will address this aspect in greater detail.

In Figure 2.5, between O and A, the benefits to equipped drivers increase with the penetration level, implying that already equipped drivers benefit if an additional driver gets access to the system. In other terms, between O and A there is a positive externality for the already equipped drivers. There are not many economic goods possessing this property. One of the few goods one can think of are telecommunication networks or the currently quickly growing Internet-based information services. In these systems, the *willingness-to-pay* is strongly dependent on the number of subscribers (Allen, 1988), or in terms of this chapter the level of market penetration. From the government's point of view, the positive externality has the interesting implication that the buyer of the motorist information system should (on welfare economic grounds) be subsidised. According to the economic literature, the subsidy should be equal to the total external benefits generated for the already equipped drivers. Then road users are paying the *true* price for the information system; the externality is internalised. Moreover, recalling the arguments given in the previous section, the size of the benefits to the non-equipped drivers caused by the marginal equipped driver should also be added to this subsidy. In order to achieve maximum efficiency, the subsidy given to all road users should be equal to the external benefits minus the external costs generated by the last equipped road user.

In Figure 2.5 between A and 100 per cent and along the whole x-axis of Figure 2.6, the information benefits to equipped drivers curve is downward sloping. This implies that a marginal equipped driver adversely affects the already equipped drivers. But since the information benefits are still positive, it is beneficial for the marginal driver himself to buy the equipment. Hence, in the downward sloping part, there exists a negative externality to the already equipped drivers. This is an argument for levying a tax on these systems in this part of the curve. However, it should be realised that the positive effects to the non-equipped drivers might outweigh the negative effects to the already equipped road users, implying that the subsidising argument is a stronger one than the taxing one. Then the odd situation arises that a good that adversely affects the people that already possess the good is subsidised.

An additional complexity arises when considering the real-world implementation of a motorist information system. If a central computer collects the information via the computer unit in the vehicles of the equipped drivers (two-way communication links), then it is likely that the quality of the information shows a dependency on the level of market penetration as depicted in Figure 2.7.[14] In Figure 2.7, the quality of the information is low (the information will be strongly historically based) at small levels of market penetration, rendering the information system unreliable under these circumstances. Clearly, this is due to the fact that the network is not yet fully covered by equipped drivers. As the level of market penetration increases, the quality of the

[14]For instance, the primary source of real-time traffic information in the ADVANCE programme are the vehicles themselves (Boyce et al., 1991).

information improves quickly.[15] Therefore, it seems a prerequisite for motorist information systems implemented in this way, that the initial level of market penetration - that is, the level of market penetration directly after implementation - exceeds a certain (*critical*) threshold value in order to be able to provide a reliable information service. To achieve this critical level, the company/government selling the equipment might consider subsidising the system during the product's take-off phase.

Further, Figure 2.7 might have consequences for the shape of Figure 2.5 and Figure 2.6. The dependency between the quality of information and the level of market penetration might give rise to information benefits to equipped drivers shaped as Figure 2.3. This suggests that there is a rationale behind an information benefits to equipped drivers curve as in Figure 2.5, when information benefits strongly depend on the quality of the information.

However, if the information is collected via loop detectors in the road, the quality of the information is independent on the level of market penetration. But, due to the inaccuracy of these detectors, the information will be relatively unreliable compared to a system that collects information from equipped drivers; see Figure 2.8.

To summarise, at most levels of market penetration a motorist information system is most likely to generate a *positive* externality for non-equipped drivers and a *negative* externality for already equipped ones. However, in terms of travel time the equipped road users will still outperform the non-equipped ones. Hence, a motorist information system is likely to enhance the efficiency of the road network, but induces several externalities and possesses the awkward property that already equipped road users are adversely affected by marginal equipped drivers. The next section is devoted to the long run consequences of such an efficiency improvement.

---

[15]In Figure 2.7 we have assumed that eventually (as the level of market penetration increases), the information provided to the drivers is a perfect representation of the actual situation in the network. However, in real-world implementations it is questionable whether this quality level will be obtained. Sources for error in the traffic information collection and distribution process are the precision and reliability of the traffic measurement technique, the reliability of the broadcasting channel, the updating frequency of the information, the delay in the transmission of the information etc. (Watling and Van Vuren, 1993).

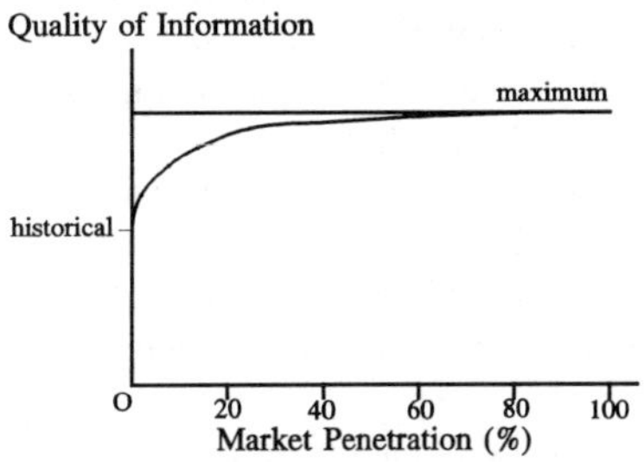

Figure 2.7 Relationship between quality of information and level of market penetration.
Information collected via equipped drivers.

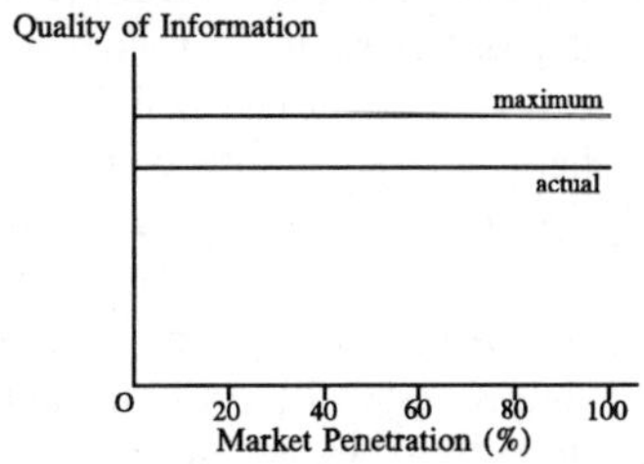

Figure 2.8 Relationship between quality of information and level of market penetration.
Information collected via loop detectors in the road.

2.7 TRAFFIC GENERATING PROPERTIES OF MOTORIST INFORMATION SYSTEMS

The introduction of a new technology affects the supply side of a production process, which in turn influences the economic equilibrium (if existing). From economic principles it is well known that an efficiency improvement on the supply side of a production process might result in more demand in the long run, due to a decrease in price. A similar kind of argument holds for motorist information systems, as these systems are designed to increase the efficiency of road transport networks, and therefore decrease the cost (price) of mobility. In the argument that follows the assumptions below are made:

1. there is full market penetration,
2. travel time at full market penetration is below travel time at no market penetration (case A or case B in Figure 2.1),
3. the analysis is static in nature: the adjustment process leading to the new economic equilibrium is not investigated.
4. An isolated road network is analysed.

Implementing a motorist information system leads to an improvement of the efficiency of the network which translates into a decrease in the road user's average travel time. Thus, an equilibrium situation - before adopting the motorist information system - is turned into a disequilibrium: given the decrease in travel time, more people would like to travel. Consequently, a larger number of motorists will use the road network as it has become more attractive owing to the implemented information system. This reasoning is underlined by Thomson (1977). He argued that the attractiveness of alternative travel modes should be equal in equilibrium:

> "... if the decision to use public or private transport is left to the free choice of the individual commuter, an equilibrium will be reached in which the attractiveness of the two systems is about equal, because if one

> is faster, cheaper, and more agreeable than the other, there will be a shift of passengers to it, rendering it more crowded while the other becomes less so, until a position is reached where no one on either system thinks there is any advantage in changing to the other." (p. 165)

Applying this theory to the present situation leads to the following. Due to the implementation of the motorist information system, the relative attractiveness of the private car is improved, and some people will shift from public transport to the private car. This process will continue until an equilibrium situation settles down in which both alternatives are again equally attractive.

It is likely that the network wide benefits under the implementation of a motorist information system are dependent on the level of road usage (and hence congestion) in the network. The benefits owing to motorist information systems are negligible at low levels of congestion; only beyond a certain level of congestion, network wide benefits start to accrue. With low levels of road usage - given that the drivers are familiar with the road network - the information provided to the drivers is of relatively little use. Consequently, the benefits of a motorist information system depend on the time of the day. On the one hand, during off-peak periods benefits might be small; on the other hand, benefits increase during peak-periods.

If it is assumed that some drivers possess little or no reliable information concerning travel and route alternatives and may be uninformed of road conditions on any specific day, then information might also be beneficial at low traffic densities. Such unawareness could lead to misperceptions on the part of the drivers as to the relative desirability of alternative travel decisions (Ben-Akiva et al., 1991). If this situation conforms more to reality, then there is clearly more scope for motorist information systems. Hence, in these circumstances there will be travel time savings and traffic generation at low traffic densities as well. Furthermore, if the motorist information system helps motorists to avoid searching for the location of the destination, there may also be a reduction in kilometres travelled. If the travel time savings are called *first order* benefits, then the reduction in travelled distance might be referred to as *second order* benefits. The total benefits accruing from introducing such a new technology then depends on the size of both types of benefits.

2.8 THE LEVEL OF MARKET PENETRATION

The previous section dealt with the traffic generating features of motorist information systems. It was argued that as these systems are designed to improve the efficiency of road networks, travelling by private car becomes more attractive and more people will use this mode. In this section, the market potential of these systems is addressed, given a fixed demand for car mobility. Hence, the analysis that follows is a short run one. Furthermore, only one travel mode - the private car - is considered. In addition, it is assumed that the market penetration of these systems can solely be determined by their potential to generate travel time savings. Other beneficial factors, such as a decrease in stress or anxiety, are not taken into account.

The benefits accruing to a marginal driver being equipped with a motorist information system - the information benefits to equipped drivers curve - were depicted in Figure 2.5 and Figure 2.6. The benefits in these figures were expressed in terms of time. In order to analyse the market potential of these systems, time has to be converted into monetary units. This can be done by using a *value-of-time* curve.

It is clear that road users with a high value-of-time are most likely to buy the motorist information system. If it is assumed that the rank order in which drivers decide to buy a motorist information system exactly corresponds with the rank order of drivers according to the value of time, the value-of-time curve can be *vertically multiplied* with the curve reflecting information benefits to equipped drivers to obtain the *willingness-to-pay* function for a motorist information system. Whether the willingness-to-pay curve touches the x-axis depends on whether recurrent or non-recurrent information is modelled.

Once the willingness-to-pay curve is known, the level of market penetration that will establish is completely determined by the cost structure of the motorist information system under consideration. Three different structures can be considered:

1. constant average costs to scale,
2. decreasing average costs to scale,
3. increasing average costs to scale.

Given the initially high infrastructure investment needed to implement motorist information systems, economies of scale are likely to take place. Then, a decreasing average cost curve will result. A decreasing average cost curve has severe welfare economic implications. Decreasing average costs imply that marginal costs are below average costs for each penetration level (production level). In such a situation, marginal cost pricing, which maximises social welfare (Pigou, 1920), leads to losses (deficits) for the company selling the system.

Therefore, it is highly unlikely that when a private company brings the system on the market, social welfare is maximised. A price will be charged that is higher than the marginal cost price. To achieve a social optimum, the company should be forced to adopt marginal cost pricing, but clearly the government then has to bridge the operational deficit.

In the analysis below, the *equilibrium* level of market penetration is derived under the assumption of marginal cost pricing. Figure 2.9 shows two hypothetical marginal cost curves (both derived from a decreasing average cost curve) combined with a willingness-to-pay curve. Marginal cost curve A has two intersection points with the willingness-to-pay curve, while marginal cost curve B crosses the willingness-to-pay curve only once. It can easily be seen that the point of intersection of curve B and the willingness-to-pay curve represents a stable equilibrium, in the sense that: (1) the system will converge to this level of market penetration, and (2) the system will return to this level for small deviations in market penetration.

The picture is different for marginal cost curve A. For low levels of market penetration, the marginal costs of the system exceed the benefits. Hence, the new technology will not *take-off*, but needs an exogenous impetus. A subsidy is needed to bring the level of market penetration beyond point 1. Intersection point 1 does not correspond to a stable equilibrium. For levels of market penetration exceeding point 1, the benefits to motorist outweigh the costs, and hence, more drivers will be attracted to the system. Road users will buy the system up to a penetration level corresponding with intersection point 2. This point reflects a stable equilibrium as discussed above.

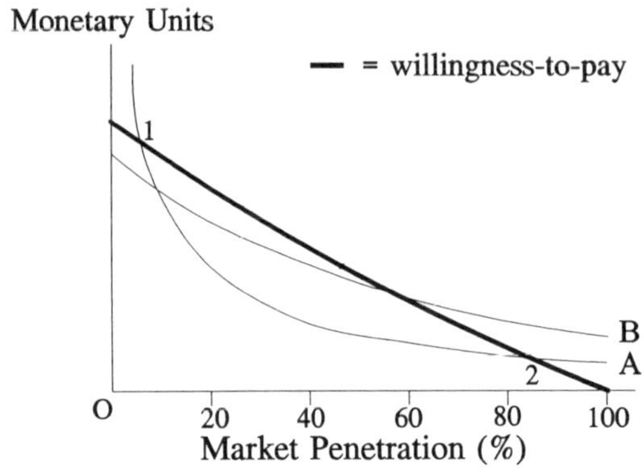

Figure 2.9 Market potential of motorist information system for two marginal cost curves (A and B).

Penetration level associated with point 1 can be referred to as a kind of *critical mass* value (Allen, 1988) for the motorist information system.[16] Here, the critical mass value might be defined as the minimum penetration level needed to render the motorist information system beneficial to motorists. A critical mass value is a well known phenomenon when dealing with telecommunication networks. These networks are from a consumer's point of view beneficial, only after a sufficient number of consumers (the critical mass) have subscribed to the system. Beyond this level benefits increase as penetration increases (Capello, 1993). In Figure 2.9, however, benefits do not increase for all levels beyond intersection point 1. Net marginal benefits remain positive only between intersection point 1 and 2.

An exact figure for the market potential of these technologies is clearly dependent on the precise shape of the curves. The ones shown in this section are purely illustrative, and point at different possible scenarios. In this section, it was the objective to provide a framework for analysing the market potential of these systems, which can be applied once the precise shape of these curves is known.

Two additional considerations are relevant with respect to the level of market penetration of motorist information systems. First, it is most likely that these systems will be introduced into new automobiles. For relatively old automobiles, it might not be profitable to install a new motorist information system. Second, if, for example, 50% of the cars is equipped with a motorist information system, these cars might account for more than 50% of the total amount of vehicle-kilometres driven. Hence, the level of market penetration in terms of share of total number of kilometres is not necessarily equal to the percentage of cars that is equipped with an information device.

[16]The term critical mass is generally used in a slightly different context referring to benefits rather than costs.

2.9 THE CASE OF NON-RECURRENT CONGESTION

The discussion in the previous sections was focused on the case of recurrent congestion. In this section the case of non-recurrent congestion is investigated. Non-recurrent congestion is congestion caused by incidents; one could, for instance, think of bad weather (fog, heavy rain, snow etc.) or traffic accidents. These incidents suddenly and unexpectedly decrease the capacity of a certain part of the road network by a significant amount, directly leading to congestion. This type of congestion will play a major role in Chapters 4 to 10 where a theoretical model is proposed. According to Lindley (1986, 1987, 1989), non-recurrent congestion accounts for 60 per cent of total congestion delay. However, this figure should not be misinterpreted, since non-recurrent congestion delays would not be nearly as large if road networks were not already overloaded due to the recurrent congestion.

Another argument for extending the scope of the analysis to the non-recurrent case is the observation that particularly in these circumstances the expectations of motorist information systems are high. These systems are said to be able to solve traffic jams in situations of non-recurrent congestion quicker (Dehoux and Toint, 1991).

Taking non-recurrent congestion into consideration, the assumption made in Figure 2.4 regarding equal travel times for equipped and non-equipped drivers at market penetration levels close to 100 per cent becomes doubtful. This is illustrated with the following hypothetical situation.

Suppose that origin O and destination D are connected by two different routes, $Route_1$ and $Route_2$. With recurrent congestion and 100 per cent of market penetration it is assumed that the user equilibrium condition holds, that is, the travel times on $Route_1$ and $Route_2$ are equal.[17] Now, suppose that at time t_1 an incident occurs on $Route_2$, thereby making $Route_1$ the more attractive one. Then, at full market penetration of the information technology, all drivers departing after t_1 will choose $Route_1$ to travel to destination D until the user equilibrium condition is restored, say at time t_2.

Then, two important observations can be made in case market penetration is below 100 per cent:

1. Non-equipped drivers departing between t_1 and t_2 and deciding to travel via $Route_2$ will clearly experience a longer travel time than if $Route_1$ had been chosen. This implies that the information benefits to equipped drivers curve is discontinuous at 100 per cent market penetration. As a consequence, the last road user buying the equipment will still obtain some benefits.
2. The higher the level of market penetration, the quicker the user equilibrium conditions will be restored in the road network.

In particular, the first observation has important consequences for the economic viability of motorist information systems. The fact that the information benefits to equipped drivers curve is discontinuous at full market penetration clearly increases the market potential of these systems; even at high levels of market penetration there are still substantial benefits to not yet equipped drivers. As a consequence, the

[17] This is obviously not true if one of the two routes is unused.

willingness-to-pay is not zero at full market penetration, leading to a larger market potential.

Recently, Al-Deek and Kanafani (1993) expanded the argument given above with a simple deterministic queuing model for a two-route network, and found similar results. The equipped drivers benefit during the time-span leading to the user equilibrium, while at user equilibrium there are no additional benefits to equipped drivers.

In the past, research assessing the impacts of motorist information systems was strongly focused on recurrent congestion, the reason being that modelling recurrent congestion is less complex. In future research, attention should shift to the non-recurrent case since:

1. The traffic situation in real road networks is non-recurrent in nature. The situation on the roads differs from day-to-day.
2. Motorist information systems are particularly designed for non-recurrent congestion. They are able to provide information on the day-to-day fluctuations on the roads.

In this book, non-recurrent congestion is considered in Chapters 4 to 10 and 12.

2.10 MOTORIST INFORMATION SYSTEMS AS A TOOL FOR ACHIEVING GOVERNMENT'S OBJECTIVES

If the purpose of any motorist information system is to minimise average network travel time, given a certain level of demand for transport, Figure 2.1 depicts which level of market penetration should be decided upon when the traffic generating abilities are not taken into account. However, besides minimising average travel time, there are other important issues that need to be considered and might affect the *optimum* level of market penetration:

1. The cost structure of a motorist information system might affect the level of market penetration. If the initial (fixed) costs of the system are high, it might - from a cost point of view - be viable only at relatively high levels of market penetration.
2. An important cost component of road traffic is pollution and noise. These costs are not necessarily minimised when the average network travel time is at a minimum.
3. An often underestimated cost component of road traffic are the safety costs. These in fact are a more than substantial part of the total costs of transport, the actual amount depending on the technique used to evaluate the value-of-life (Jones-Lee, 1990). These costs are generally not minimised when the average network travel time is at a minimum.
4. In addition, Rumar (1990) claimed that motorist information systems might interfere with the driver's primary task of driving and therefore have a potential of decreasing safety on the roads. This is obviously opposite to what these systems would like to accomplish (Stergiou and Stathopoulos, 1989).

Furthermore, the level of market penetration that will be realised might well be dependent on the policy of the government with respect to road-pricing. It is well known that these two technologies (road-pricing and motorist information systems) need a similar kind of road infrastructure and information technology. Then, if it is the government's policy to adopt road-pricing, there might be economies of scope, rendering the additional costs of implementing a motorist information system small. This could lead to relatively low marginal costs and, as a consequence, a relatively high level of market penetration.[18]

All the factors mentioned above add to the complexity of determining the *optimal* level of market penetration from a social welfare point of view. Recalling the other issues affecting the benefits of these systems, it is also clear that every implementation technique should be largely site and system dependent.

Although the outcomes of implementing these technologies are still highly uncertain, it is useful for the government to pursue the analysis of the potential of these systems. In particular, it is worthwhile paying more attention to the simultaneous implementation and application of congestion-pricing and motorist information systems. As argued in Bonsall et al. (1991), it is unlikely that road users can be diverted from user equilibrium routes in order to obtain a system optimum. To accomplish this, a motorist information system should be accompanied by some pricing mechanism, thereby making a strong case for the joint implementation of these two technologies.

Finally, it should be stressed that motorist information systems by themselves do not tackle the congestion externality. They provide a so-called *second-best* solution, that is, they do not cure the congestion problem at its source. However, when these systems are able to diminish the general level of congestion in the network, the monetary value of the externality decreases in size, and therefore, motorist information systems might implicitly help decreasing the impact of the congestion externality. But, as pointed out in Section 2.7, the impact is smaller than one might expect because lower travel times induce an increased demand for mobility.

2.11 CONCLUSION

In this chapter an introduction into the relevant economic issues of motorist information systems was given. Past research studies assessing the benefits of implementing motorist information systems show large variations. It seems that the benefits are strongly dependent on the kind of system, the behavioural responses of the road users, the particular network under consideration, the level of market penetration etc. Especially, the phenomena of overreaction and concentration might play an important role in adversely affecting the network wide performance.

The discussion in this chapter was based on the assumption that a properly working motorist information system is implemented, that is, a motorist information system that reduces the travel time for the equipped drivers, and in addition, makes them better off than the non-equipped ones. Under this assumption, and the assumption of Wardrop's user equilibrium at full market penetration, it can be derived

---

[18]Here, we are solely pointing at economies of scope regarding the cost structure of the simultaneous implementation of these technologies. de Palma and Lindsey (1994), El Sanhouri (1994) and Emmerink, Nijkamp and Rietveld (1995b) addressed the economic consequences of a combined implementation of these technologies. In the present study this is addressed in Chapters 7 and 10.

that the so-called information benefits to equipped drivers are a decreasing function of the level of market penetration. Therefore, a marginal equipped driver adversely affects the already equipped ones. However, this marginal driver will have a positive influence on the non-equipped drivers. Hence, it can be concluded that a motorist information system is an economic good that produces both positive and negative externalities.

Due to the efficiency improvement of the traffic flows in the road network, a motorist information system will generate more traffic. The market potential of these new technologies is strongly dependent on the value-of-time curve, the information benefits to equipped drivers curve and the cost structure of the motorist information system. The market potential increases when the road network is strongly volatile, that is, a network with non-recurrent congestion.

If a motorist information system can be characterised by decreasing average costs to scale - which seems a reasonable assumption given the large initial investment needed - a marginal cost pricing strategy, that maximises social welfare leads to a deficit for the company operating the system. As a consequence, the government has almost inevitably a role to play with these new technologies. Also, to prevent monopolistic behaviour in this market, the government might want to provide regulations.

A motorist information system is only one of the available tools to tackle the congestion problem. Applied on its own, the impact might be small in practice. However, combined with a congestion-pricing strategy, transport planners have a strong tool to influence traffic flows in road networks. Consequently, the next chapter is devoted to the so-called *first-best* instrument in road transport: road-pricing.

3 ROAD-PRICING: AN INTRODUCTION[1]

3.1 INTRODUCTION

Congestion on the roads is a pressing problem for most metropolitan areas around the world. It causes transportation problems in traffic networks, and, in addition, the consequential delays negatively affect economic activities as a whole. Furthermore, congestion has adverse effects on the environment; one could think of pollution, noise annoyance etc. Many measures or instruments have been suggested to resolve (part of) the congestion problem. In the previous chapter, the discussion was focused on one such instrument: the impact of implementing motorist information systems. In this chapter the feasibility of an economic instrument that has caused a great deal of controversy is analysed: road-pricing. Although on theoretical grounds it can be easily shown that road-pricing is the first-best solution for efficiently dealing with externalities in transport, this instrument cannot yet boost much public and political support. In this chapter, various aspects, problems and consequences associated with implementing road-pricing will be discussed by reviewing the relevant literature. Solutions provided by different researchers to enhance the acceptability of a road-pricing scheme will be compared, and conflicting viewpoints will be identified and brought together.

The chapter is organised as follows. In Section 3.2 the theoretical first-best concept of road-pricing and then the applications of this feature in practice are questioned. In Section 3.3 the recent literature on the positive and negative aspects of actually implementing a road-pricing scheme is summarised. In Section 3.4 attention is focused on the behavioural issues that are often overlooked, but yet of crucial importance. In Section 3.5 the methods of implementing road-pricing are addressed. A distinction will be made between the actual technology used to collect the money and the political way of introducing the scheme to the public. In Section 3.6 the Dutch attempts to introduce a kind of road-pricing scheme are discussed, and finally, in Section 3.7 the main arguments in the chapter are summarised and recommendations are made for further research.

3.2 THE THEORY: FIRST-BEST?

In this section, the theoretical first-best concept of road-pricing will first be analysed. Then the main focus is on the applicability of the first-best theoretical feature in real-world situations. In Section 3.2.1 the notion of transport externalities is introduced; in Section 3.2.2 the first-best concept of road-pricing is presented; and in Section 3.2.3 this concept, when applied in realistic situations, is analysed.

---

[1]This chapter is based on Emmerink, Nijkamp and Rietveld (1995a) published in *Environment and Planning B*.

3.2.1 Transport externalities

Mobility in general, and car mobility in particular, gives rise to a number of external effects, also called externalities. In general, an external effect can be defined as:[2]

> "An external effect exists when an actor's (the receptor's) utility function contains a real variable whose actual value depends on the behaviour of another actor (the supplier), who does not take these effects of his behaviour into account in his decision making process." (Verhoef, 1994a, p. 274)

Within the transport network, externalities follow from the nontraded mutual interactions between road users. Driving on a congested road does not only lead to a relatively long travel time for the driver involved, but also increases the travel times for all other drivers on the same road.[3] The latter (cost) effect is not taken into account in the driver's decision-making process, which causes the external effect. Besides the congestion externality, there are other externalities involved with road transport. One may in particular think of noise annoyance, visual intrusion, pollution and last but not least accidents (Verhoef, 1994a). Particularly, the safety externality is important, since it is (depending on the valuation of life; see Jones-Lee (1990)) in size often even larger than the congestion externality. Nevertheless, most road-pricing efforts have been directed towards the congestion externality, particularly because this is the most tangible one: a large share of the population faces congestion on a daily basis.

To complicate matters further, interactions between the different types of external effects have also been mentioned in the literature. For instance, recently Shefer and Rietveld (1994) reported positive effects of congestion on safety. They claimed that in congested situations, fewer fatal accidents take place than in uncongested situations. US, German and Japanese data on the distribution of fatal accidents over the hours of the day seemed to support their theoretical model. The data revealed that the number of fatal accidents in the morning peak hours (the most congested period of the day) is relatively low. On the other hand, environmental externalities (pollution) may generally be expected to increase with less smooth traffic flows. A final example is the interaction between noise and safety: a completely silent car may be expected to be very dangerous for other road users.

For the ease of the argumentation, the focus is mainly on the congestion externality; hence, in the remainder of the chapter the term congestion-pricing will be used to avoid confusion.[4]

3.2.2 The first-best concept of congestion-pricing

The theoretical concept of congestion-pricing, stemming from Pigou (1920) and Knight (1924), and further explored by Walters (1961), Smeed (1964) and Sharp (1966), is based on levying a tax equal to the difference between the marginal social

---

[2]For a literature review on externalities, see Baumol and Oates (1988) and Mishan (1971).

[3]Strictly speaking, the drivers behind the additional one are affected.

[4]The term congestion-pricing might not be the most appropriate one either. For instance, Sharp (1966) preferred congestion-tax.

costs and marginal private costs (Pigouvian tax). In this way, the congestion externality is *internalised*[5], implying that the negative effects on other road users, when joining a road, are accounted for in the user's decision-making process.[6] In economic terms, the congestion-price ensures that only economically efficient trips (in utility terms) are undertaken, and therefore, congestion-pricing is viewed as the *first-best* instrument for tackling congestion.[7]

The assumptions needed to arrive at the first-best property are listed below:

1. Individual behaviour is rational, based on utility maximisation (cost minimisation).
2. There is full information on all costs involved (including detours) for both the road manager (or government) and the road users.
3. Time is a normal economic good, i.e. it has a positive value.
4. Congestion-pricing is applied to all relevant road segments in the network.
5. Congestion-pricing is technically feasible and the transaction costs are reasonably low.

The rationality and full information assumptions are generally made in economic theories. The third assumption is needed to ensure that congestion is associated with disutility. The fourth assumption is straightforward, and the fifth ensures that the welfare gains of applying congestion-pricing are not outweighed by the costs of implementing the system.

In Figure 3.1 MPC denotes the marginal private costs and MSC the marginal social costs, i.e. the costs to other road users if an additional road user joins the route. By applying congestion-pricing, one can simply show that the net *social* benefits are maximised; the *dead-weight* welfare loss (given by the area of the triangle in the traditional diagrammatic analysis of congestion-pricing, see Figure 3.1) is avoided. In order to realise the efficiency maximising level of road usage N_{opt} (instead of the market outcome N_{eq}), a congestion-price equal to r_{opt} should be levied.

3.2.3 Objections to the first-best concept of congestion-pricing

---

[5]Here, the term internalisation refers to Pigouvian taxation. Strictly speaking, however, Pigouvian taxation does not satisfy the definition of internalisation, since the *correct* (welfare maximising) Pigouvian tax is not internal to the economic process (Mishan, 1971; Verhoef, 1994b).

[6]In the literature there has been some debate on whether the diagrammatical analysis of congestion-pricing should be centred on the relationship between the cost of using a road and the *flow of traffic*, or on the cost of using the road and the *number of vehicles on it*. Nowadays, the consensus seems to be that the latter is the better representation, as the decision to use a road is essentially a decision to add to the number of vehicles on it (Else, 1981; see also Nash and Else (1982) for some comments). Much of the debate between Andrew Evans and Hills (see Andrew Evans (1992) and Hills and Evans (1993)) concerns the same issue. Andrew Evans uses flow-based demand functions, whereas Hills rejects such an approach.

[7]For a recent, detailed and comprehensive analysis of congestion-pricing see Hau (1992) and Johansson and Mattsson (1995).

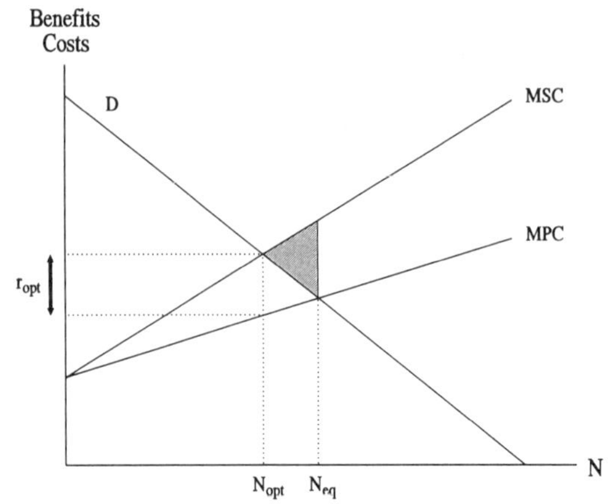

Figure 3.1 Traditional diagram of congestion-pricing with linear curves.

Several objections have been raised against the conventional (Pigouvian) concept of congestion-pricing; in this section, these will be listed.

First, the analysis is in nature a static one, whereas the real-world situation is obviously dynamic, especially with respect to congestion phenomena, even in the short run. A transportation system is highly dynamic, the traffic situation is changing continuously and affected by random disturbances. Ran et al. (1993) found that the static solution is not applicable (second-best) to dynamic congestion-pricing problems. Furthermore, when congested transport networks are modelled in a realistic manner, solving for the correct congestion-price leads to models that are mathematically intractable. Without simplifying assumptions, the models are impossible to solve for realistic networks (Boyce et al., 1990) and (if solvable) require much computational effort; simple analytical tools are not available (Ran et al., 1993). This is not a theoretical problem but a mathematical one. However, these practical problems become important the closer one gets to the phase of implementation.

Second, and in relation to the previous point, it is important to understand whether the congestion-price should be based on prevailing levels of congestion, or on predicted (future) levels of congestion. In the traditional, static theory this issue is irrelevant because, in such a framework, traffic flows do not evolve through time. However, in the real-world dynamic traffic situation this becomes an important issue. Below, it is argued that in a dynamic situation the theoretically correct congestion-price should be based on the predicted levels of congestion. The argument can be best illustrated by Figure 3.2 in which the driver's decision making process and the role of congestion-prices therein are depicted.

To induce the desired behavioural responses, the calculated congestion-prices in the upper box of Figure 3.2 should reflect the external costs caused by the driver in the lower box of Figure 3.2. Therefore, the theoretically correct congestion-price should not be based on prevailing levels of congestion, but on predicted levels of

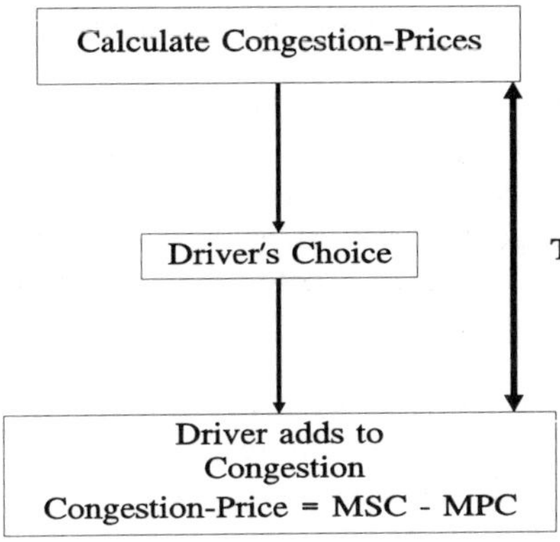

Figure 3.2 Congestion-prices: Based on prevailing or predicted levels of congestion?

congestion with a prediction horizon equal to the time span in Figure 3.2. Moreover, if the calculated levels (upper box) are not equal to the experienced levels of congestion (lower box), for example, owing to a traffic accident, then the driver should still be charged for the calculated (predicted) price. The reason is that the driver's behaviour was influenced by the predicted charge and hence by the predicted level of congestion, not by the experienced level of congestion.

Next, the question arises as to what extent one is able to predict future levels of congestion. Clearly, prediction of non-recurrent traffic incidents is by definition impossible. However, Ishtiaq and Hounsell (1993) suggest that short term forecasting of traffic congestion is feasible with a reasonable accuracy. Nevertheless, as in the previous chapter, the difficulty arises that the predicted congestion-price is dependent on drivers' behaviour, which, in turn, is dependent on the predicted congestion-price. Related to this point, the driver should be informed about the congestion-price. Without this information, the driver will not be able to adjust his or her travel decision properly according to the prevailing travel costs. Hence, a congestion-pricing scheme and a motorist information system cannot be viewed as independent instruments for tackling the congestion problem, but instead have to be implemented simultaneously. An equilibrium model for doing so is studied in Chapters 4 to 10.

Third, to arrive at the first-best conclusion, all the prevailing externalities have to be taken care of in the road-price. As stated in Section 3.2.1, this chapter will be mostly concerned with the congestion externality. Here, however, it is important to stress that also the safety and pollution aspects should be taken into account to tackle the problem in a first-best manner. Taking the congestion externality into account is a difficult task in itself, but it is even more complicated to account for pollution and safety as well. This is because: (1) not all pollutants correlate in a similar fashion with speed; (2) the economic value of pollution (that is, the many external environmental costs) cannot be calculated in any precise manner; and (3) the correlation between the different externalities is not very clear either (for safety and speed, see Shefer and

Rietveld (1994)). Hence, it might be questioned whether there is enough knowledge of the behaviour of the transport system and its users to implement road-pricing in a first-best manner.

Fourth, the conventional analysis assumes homogeneous traffic; an assumption that cannot be empirically validated. When dealing with heterogeneous drivers (in terms of vehicle type) the congestion-price should be discriminatory. In theory this can be easily accomplished; in practice, however, taking this forward may be much harder. In this context, Andrew Evans (1992) and FitzRoy and Smith (1993) preferred road rationing through priority lanes to congestion-pricing, in order to resolve (part of) the congestion problem.

Fifth, in the traditional analysis, an extremely simple network structure, without any traffic control components, is considered. In reality, one observes more complex networks and, as a consequence, the interaction between link flows and traffic control should be taken into account. This is (very) important, as traffic control plays a crucial role in directing traffic flows in large networks (Ghali and Smith, 1992; Smith and Ghali, 1992).

Sixth, Hills (see Hills and Evans (1993)) added that the complexity of travel behaviour (travel demand is a derived demand) is not accounted for in the conventional congestion-pricing analysis, and concluded that it is not feasible to model congestion and congestion-pricing in a simple way. In particular, the assumption of rational behaviour based on cost minimisation might be questioned in practice. This is so important that Section 3.4 will be devoted to clarifying these behavioural issues.

To conclude, although in a simplified theoretical framework congestion-pricing is the first-best instrument to tackle the congestion externality, in practice this first-best character may be difficult to achieve. A congestion-pricing scheme is *first-best* only if all the first-best assumptions are satisfied; otherwise the scheme is *second-best*. In a second-best congestion-pricing scheme, the optimal (welfare maximising) toll is not necessarily equal to the difference between marginal social and marginal private costs. Instead, the optimal second-best toll is generally some kind of weighted average of the costs caused by the relevant external effects (Verhoef, Nijkamp and Rietveld, 1995a). Viewed from this perspective, congestion-pricing might be seen as just another second-best tool for resolving (part of) the congestion problem, although it may often be expected to approach first-best standards more closely than other second-best tools.

3.3 CONGESTION-PRICING: PROS AND CONS

Having analysed the complexity of the theoretical concept of congestion-pricing in Section 3.2, this section brings some practical evidence into the discussion. In Section 3.3.1 the practical advantages of congestion-pricing are discussed, and in Section 3.3.2 potential practical problems are presented.

3.3.1 Advantages of congestion-pricing

Notwithstanding the objections to the simplified one-link homogeneous traffic context, congestion-pricing remains the first-best solution for tackling the congestion problem in this simple context. Other available instruments do not possess the potential of achieving the social welfare maximising solution in such a simplified context (Verhoef, Nijkamp and Rietveld, 1995a). These instruments, often indicated as second-best tools, implicitly attempt to internalise the congestion externality by affecting mobility related markets. A few examples of second-best instruments are: (1) parking charges, (2) public transport subsidies, (3) fuel taxation, (4) vehicle license fees (or vehicle quota systems[8]), (5) road rationing (special lanes for high occupancy vehicles), (6) the use of telematics to improve network efficiency. For a more exhaustive list of alternatives to congestion-pricing and useful references, see Button (1992b).

Besides being a first-best instrument in the simplified context, congestion-pricing has three additional attractive features:

1. Being a financial instrument, congestion-pricing addresses the problem of latent demand directly by using money rather than delays to ration road capacity. This in contrast to the more traditional solution of offering additional road capacity.
2. Congestion-pricing provides the government with an opportunity to privatise (part of) the road network operations.
3. Congestion-pricing raises government's revenues by taxing externalities rather than economic productivity.

Most second-best instruments do not possess the first characteristic. Frequently, measures taken by the government to curb traffic flows are frustrated by the unexpected appearance of latent demand. One of the most interesting results of a study addressing the effects of the completion of the Amsterdam ringroad is the *return to the peak* phenomenon (Dutch Ministry of Transport, 1991a, 1991b). The study showed that a relatively large number of drivers who previously (before the completion of the ringroad) used the motorway before 7 AM or after 9 AM moved to a more "convenient" departure time within the 7-9 AM time interval. In fact, these latent demand effects are consequences of Downs' (1962) *fundamental law of peak-hour expressway congestion*, implying that capacity expansion does not affect peak-period travel speeds, see also Emmerink and van Beek (1997).

The second point, on the one hand, widens the scope of government's policy options, whereas, on the other hand, the public may (and most likely will) be negatively affected by a privatised road network, particularly when the exploiting companies act as purely profit maximising entities.[9] In addition, owing to the market failures arising from the external effects in road transport, it is economically inefficient to rely on the *market* outcome. However, conversely one might expect that a private operator tends to be more efficient. Furthermore, private operators tend to be more willing to be innovative, to explore new technologies and techniques, which might be an advantage to the public (Gomez-Ibanez and Meyer, 1994). Therefore, privatisation of (part of) the network will create a trade-off between efficiency gains, on the one hand, and welfare losses on the other. In addition, the overcharging of the monopolist could partly coincide with the value associated with the remaining externalities. Verhoef, Nijkamp and Rietveld (1996) analysed the situation where road users can choose between a tolled and an untolled route. They showed, inter alia, under which

---

[8]See Koh and Lee (1994) and Phang (1993) for recent assessments of the vehicle quota system in Singapore.

[9]See Fielding and Klein (1993) for a paper on the franchising of highways.

circumstances it is more efficient to privatise both routes rather than to apply welfare maximising congestion tolling to one of the two routes.

The third point mentioned above stresses the intrinsic good feature of congestion-pricing. In contrast to most taxes, it does not penalise *economic productivity*; instead, it penalises *economic inefficient use of scarce resources*, that is, it penalises use of scarce resources that are external to the economic process. This is widely viewed as a good principle for taxation (Atkinson and Stiglitz, 1980).

3.3.2 Disadvantages of congestion-pricing

It is striking that most of the literature on congestion-pricing is concerned with the disadvantages associated with the policy. From the beginning, congestion-pricing has been perceived as a controversial method of curbing congestion, favoured by some experts but at the same time criticised by many opponents. While many economists tend to support congestion-pricing, the public and politicians are opposed to congestion-pricing schemes:

> "Although many transport and planning professionals are agreed that some form of road pricing is likely to be the most effective and flexible way of coping with the growth in urban traffic problems, it is clearly one of the least popular measures among the public at large." (Jones, 1991a, p. 194)

The papers by Borins (1988), Andrew Evans (1992), Goodwin (1989), Jones (1991a), Oldridge (1990) and Seale (1993) provide a flavour of the divergence of opinions on the desirability, practicability and usefulness of congestion-pricing. In the following sections, the major objections to congestion-pricing will be discussed, and methods to overcome these analysed. Altogether these arguments may be viewed as the main reasons for the public and political aversion towards congestion-pricing.

Potential Pareto improvement

Although it is a first-best solution in the simplified framework, the implications of congestion-pricing in practice are less clear. It is not without any reason that the political and public support to implement a congestion-pricing system is small (Jones, 1991a). It is true that congestion-pricing in the simplified context - as a first-best policy - maximises social welfare, but it is not true that this is accomplished by reaching a *strict* Pareto efficient allocation, that is, all parties being better off. Instead, it leads to a *potential* Pareto improvement, or in other words, congestion-pricing satisfies the Kaldor-Hicks welfare criterion (Hicks, 1939; Kaldor, 1939). In practice, however, the revenues obtained from congestion-pricing may not be properly redistributed to make everyone better off (Layard, 1977; Richardson, 1974; Segal and Steinmeier, 1980; Small, 1983). Without redistribution, most of the revenues obtained will accrue to the government. In this case both the *tolled* and *tolled off* road users are worse off; the government is the winning party (Hau, 1992).[10]

---

[10]To be precise, road users with a very high value-of-time will also be better off, and during periods of *hyper-congestion* (density is beyond the corresponding point of maximum flow) every one is better off (Hau, 1992).

Lave (1994) contradicted this general consensus by arguing that depending on the shape of the marginal net private benefits curve, the aggregate consumer surplus may increase independently of how the revenues obtained are spent. His argument hinges on shifts in the value-of-time curve for different speeds. In Lave's diagrammatical representation, different speeds are treated as different commodities. However, as pointed out by Verhoef (1995), Lave does not allow for shifts in the relative position of drivers in the diagram that reflects the marginal benefits associated with a trip.[11] Or, in other words, if driver A's marginal benefits at 40 km/h exceed driver B's marginal benefits at that particular speed, then this is also the case for any other speed level. Verhoef challenges Lave on this assumption by arguing that this is not necessarily so.

Redistributing the revenues

The issue of redistributing the revenues is crucial for making congestion-pricing a publicly acceptable instrument. Through a proper redistribution scheme, the potential Pareto improvement (Kaldor-Hicks welfare criterion) might be turned into a real Pareto improvement. The efforts made by both Goodwin (1989) and Jones (1991a) point into this direction. Goodwin (1989) argued that political and public support can be raised by dividing the revenues from congestion-pricing in such a manner that all parties involved have something to gain compared with the current situation. He proposed a simple rule (the *rule of three*) to do so. Jones (1991a) suggested that a package of measures be set up around the congestion-pricing instrument to gain public support. Evidence from surveys (Jones, 1991b) seemed to suggest that public support is sensitive to such approaches. Nevertheless, it is unlikely that a proper Pareto improvement (all parties being better off) will be accomplished.

However, Small (1992b) argued that the money raised through a congestion-pricing scheme is sufficient to implement practically a strict Pareto improvement. Following Burtraw's (1991) principle of *linked compensation*, Small attempted to offset the losses of motorists by taking measures that directly alleviate the harm done through the congestion-pricing scheme. However, Andrew Evans (1992) showed that in many cases the net benefits of a congestion-pricing scheme may be small in practice, compared with the redistributional impacts, so that the full compensatory features of Small's technique may be questioned. In addition, May (1992, p. 328) claimed that "it has to be accepted that any form of road pricing will introduce some inequities", when compared with the current situation (the *status quo*). Furthermore, Small did not address the implications of feedback mechanisms on motorists' behavioural responses. Section 3.4.2 of this chapter will be devoted to a further discussion of this phenomenon.

Attitudes towards congestion-pricing

In contradiction with the arguments above, Giuliano (1992) argued that equity and distributional considerations are *not* particularly important in regard to public acceptance. She argued that current investment policies often have a similar adverse

---

[11]For Lave's rejoinder, see Verhoef (1995).

effect on equity as congestion-pricing would have. Hence, she concluded that distributional considerations are not a primary concern of congestion-pricing opponents, but adds that "distributional equity may present an apparently legitimate basis for opposition that is actually motivated by other reasons." (p. 349)

The attitudes of politicians to congestion-pricing have recently been investigated by Seale (1993) for London. Seale's analysis led to some surprising results. For instance, there appeared to be a positive correlation between the knowledge on congestion-pricing and the support for congestion-pricing. More fundamentally, Seale discussed two kinds of problems: problems of *principle* and problems of *practicality*. It was found that the principle of giving up *freedom* by paying for something that was up to now free, will cost support and votes, and that the problem of *rat-running* appeared high on the list of practical problems. Rat-running is the term that refers to the problem of motorists trying to avoid the congestion-price by driving on unpriced roads. Generally, these roads are not built with the objective of carrying much traffic; too much traffic on these (often small) roads leads to problems regarding the environment, noise, pollution, and congestion. Politicians identified the relationship between the congestion-price and other taxes as an important issue. In order to broaden the support for congestion-pricing, the public should not be allowed to view congestion-pricing as *just another tax*.

Another problem associated with congestion-pricing refers to the way in which it is *perceived* by different groups in society. Owing to the wide impact of congestion-pricing, it is viewed by different interest groups as a means of achieving different ends (Giuliano, 1992). For instance, the government may perceive it as an instrument that increases social welfare, raises government revenues and provides a possibility of privatising roads. Environmentalists may view it as a means of stimulating carpooling and the use of public transport, and hence reducing pollution and the need for new roads. Business people may perceive the congestion-price as funding resources for new roads and continued economic growth. It is clear that the interests of the above groups do not necessarily converge. However, this problem is more of a practical (perceptual or psychological) than of a theoretical nature. In theory, a first-best congestion-pricing scheme would simultaneously (1) raise government's revenues, (2) decrease levels of pollution, and (3) increase the accessibility of congested areas, thereby rendering all the above mentioned interest groups better off.

Additional negative impacts

In addition to the previously mentioned problems, there are six additional negative aspects related to implementing congestion-pricing. First, the low-income groups will be hit hardest by implementing congestion-pricing; it will cause inequity, see Andrew Evans (1992). According to economic principles, their trips will be suppressed disproportionally, as a result of their normally lower marginal values-of-time and higher marginal value-of-money. However, a trip of a low value-of-time does not necessarily imply a trip of low social value. Andrew Evans (1992) also argued that the welfare gain (equal to the area of the dead-weight loss triangle) might be relatively small compared with the income shifts (from the road users to the government) for most levels of congestion. He questioned the desirability of gaining a relatively small welfare increase at the expense of large shifts in the distribution of income.

In the second instance, congestion-pricing will hit certain regions harder than others. Given the congestion-pricing concept, people living in congested areas will have to pay a high price for mobility, whereas no congestion-price is levied in low density areas, thus stimulating regional inequality and raising political problems. Imagine telling a regular road user living in a congested region that he or she has to pay for using the roads, whereas road users in rural areas have to pay nothing at all. The impact of congestion-pricing on economic activity is also not totally clear. On the one hand, congestion-pricing will improve the accessibility of congested areas and might therefore induce more economic growth in these regions. On the other hand, congestion-pricing might induce a shift from economic activities to relatively uncongested regions.

Third, and related to the previous point, congestion-pricing may (indirectly) alter land-use patterns. This is due to the fact that congestion-pricing will affect the productivity of certain sites. Therefore, the location behaviour of firms will be influenced: heavily congested areas may become less attractive compared with the less congested ones. In general, these indirect effects are not necessarily negative, but they are extremely difficult to predict. Therefore, the outcome of a congestion-pricing policy is uncertain and thus the parties involved (firms, urban authorities) may oppose congestion-pricing. A study in the Netherlands by Hols (1992) suggested urbanisation trends caused by employees moving to dwellings closer to their work location in order to avoid congestion charges.

Fourth, congestion-pricing might provide perverse incentives to governments to raise revenues at the expense of the road users (Andrew Evans, 1992). In the theory of congestion-pricing it is assumed that governments (as monopoly suppliers of roads) will not exploit their monopoly power. Unfortunately, there is no clear rationale underlying this assumption and it might be seriously questioned in practical applications. Without making this assumption, governments may overcharge motorists and undersupply infrastructure to increase the amount of revenues raised. Of course, this argument could also be put forward against the use of an emissions tax and in fact against any form of taxation.

Fifth, the argument that congestion-pricing may endanger the individual's privacy is often put forward. This argument is not always justifiable, particularly not when a kind of smart card system is proposed to implement the congestion-pricing scheme. Furthermore, evidence from various countries (for example, Italy) has shown that technologies such as automatic debiting systems have gradually been accepted at a wider scale. However, it is clear that any congestion-pricing system should be accompanied by privacy-protecting legislation.

Sixth, the costs of implementing and maintaining a congestion-pricing system should not be underestimated. However, recent evidence indicated that, for high density, highly congested regions, these may be relatively low. For instance, Small (1992b) estimated that for the Los Angeles region the collection costs would be about 4.4% of total revenues.

Summary

The problems associated with introducing a congestion-pricing scheme, as discussed in this section, may be best summarised with Borins' (1988) paper. Given the evidence of the electronic congestion-pricing experiment in Hong Kong, Borins concluded that overcoming the political implementation problems is extremely hard. In his view, congestion-pricing is an intrinsically unpopular policy; it will always be opposed from different angles. The source of the opposition towards the introduction of any congestion-pricing scheme may stem from the public misperceptions of who is causing the problem. Or as Sheldon et al. (1993) put it:

> "... no-one appears willing to accept that they contribute to the problem: it is typically something that is caused by *someone else*." (p. 141)

3.4 BEHAVIOURAL RESPONSES TOWARDS CONGESTION-PRICING

In this section light will be shed on the behavioural issues involved in modelling congestion-pricing. In Section 3.4.1 the appropriateness of utility based modelling principles is assessed. In Section 3.4.2 the indirect responses are dealt with, and in Section 3.4.3 the (limited) empirical evidence on behavioural responses is discussed.

3.4.1 Modelling behavioural responses

Modelling the behavioural responses of road users to congestion-pricing is a difficult task. Simple utility or cost/benefit based models can be developed to do so, but it is questionable whether they capture the real-world travel-choice situation. Two analytical-economic problems have to be faced here:

1. The assumption of some kind of *rationality*.
2. Travel demand is a *derived* demand; it usually does not have any value or utility by itself.

These two points are addressed below.

Assumption of rationality

The assumption of *rational* behaviour is often made in modelling human behaviour and was needed to arrive at the first-best feature of congestion-pricing in Section 3.2.2. This assumption, stemming from micro-economic theory, has been an issue for debate for over 30 years (Abelson and Levy, 1985). To provide a flavour of the debate, Simon (1990) recently argued that it might be impossible to develop quantitative laws that describe human behaviour. Gärling (1994) explored this argument for travel-choice modelling and suggested several psychological theories with more accurate assumptions. As travel decisions are made so frequently (repetitious choices), it seems particularly unlikely for an individual to make a rational decision each time. Habitual behaviour or inertia might play a more important role than utility maximisation.[12] Recently, van Berkum and van der Mede (1993) incorporated a habitual component in their model of drivers' behaviour under information provision. Other methodologies that leave some scope for adjusted behavioural notions are offered by the fuzzy logic and approximate reasoning approaches; for an application of these theories to route choice behaviour, see Lotan and Koutsopoulos (1993). Furthermore, one may use plausible heuristics to model

[12]In some circumstances, habitual behaviour may be captured within a utility-maximising framework.

drivers' behaviour. The boundedly rational model, used by Mahmassani and associates (see for example Mahmassani and Herman (1990)) and in Chapters 11 and 12, is an example of an intuitively appealing model.

Derived demand

Demand for travel is the result of the need to engage in certain activities, for instance school, work, shopping, leisure etc., viz. travel demand is a *derived* demand. Furthermore, travel demand is a derived demand with respect to the spatial organisation of economic activities (Verhoef and van den Bergh, 1994). The activity-based approach "emphasizes this need to consider travel within a broader context through the pattern or sequence of activities undertaken by individuals at various locations in space during a period of time" (Fischer, 1993, p. 25). An analysis of the effects of congestion-pricing on travel behaviour is a typical example of a situation for which the activity-based approach is most suited because congestion-pricing might have a great impact on individuals' daily activity patterns. Under congestion-pricing, individuals are stimulated to, for example, chain more activities together. These changes in activity patterns cannot be captured in the traditional four-step approach which ignores the derived demand characteristic of travel. Unfortunately, conducting an activity-based analysis in practice is a difficult task, because (1) operational tools are not (yet) available (opposed to, for instance, the case of the traditional four-step approach where such tools are available), and (2) the required high-quality detailed travel diary data are not available. For a discussion of the likely problems inherent in collecting this kind of high-quality data, see Axhausen and Alves (1994).

3.4.2 Indirect responses

The analysis of drivers' responses is further complicated by the indirect effects of introducing a congestion-pricing scheme. Most analyses of congestion-pricing only consider the direct impacts of the scheme; the indirect effects are ignored. In a theoretical context, this might be justified owing to the result derived in Mohring and Harwitz (1962) that if total user costs plus capacity construction costs exhibit constant returns to scale, then revenue from the optimal toll exactly covers the cost of constructing optimal capacity. However, when the revenues are not used for optimal infrastructure supply (as most likely will be the case in any practical context) it is crucial to consider the indirect effects of compensation schemes when assessing the usefulness of the redistribution scheme under consideration. This point will be clarified with a simple example.

Assume that there is a situation that can be described with the simple one-link, homogeneous traffic framework for congestion-pricing. Further, assume that the government raises the proper congestion-price (the difference between the marginal social and marginal private costs). Then, by measuring the changes in welfare, it can be easily derived that society as a whole will benefit from the scheme; there is an increase in social welfare. However, if the government decides to reimburse the road users that are worse off with a part of the obtained revenues, the road users might (partly) reverse their changed behaviour. The road users might again account for their marginal private costs only, and as a result the congestion-pricing scheme has no effect on traffic flows. Clearly, this is an extreme case and a typical example of the so-called *buy-back* effect: the government fully compensates *depending on* the individual's behaviour. However, the current debate on how to make congestion-pricing acceptable to the public and politicians is related to the use of certain reimbursing schemes, of which the indirect effects are not immediately clear but should be carefully investigated. Figure 3.3 shows a schematical representation of the above discussion.

In general, a change in the price of a commodity has two separate effects, the *substitution* effect and the *income* effect (see the Slutsky equation, Varian, 1992, p. 120). If compensation is carried out according to the theoretical concept of *lump sum* subsidies, then the buy-back effects as mentioned above will be very limited. By definition, the lump sum redistribution scheme will not induce a substitution effect; the income effect remains. In theory, the income effect will shift the (derived) demand curve for trips outwards, and would influence the optimal congestion-price. In practice, however, the income effect can be expected to be a very small percentage of total income, and hence might be negligible. The theoretical concept of lump sum compensation is difficult to implement in practice because any compensation scheme will (either in the short or long run) induce certain behavioural responses. Therefore, the lump sum concept should not be viewed as a realistic scheme, but rather as a theoretical construction that could serve as a benchmark for comparing schemes with the *ideal* situation. In practice, any scheme will induce behavioural changes, and, moreover, behaviourally oriented compensation schemes are more likely to be introduced to directly offset the *losers* of the congestion-pricing policy. Clearly, great care should be taken, as these schemes may suffer from the important drawbacks of the buy-back effects mentioned above.

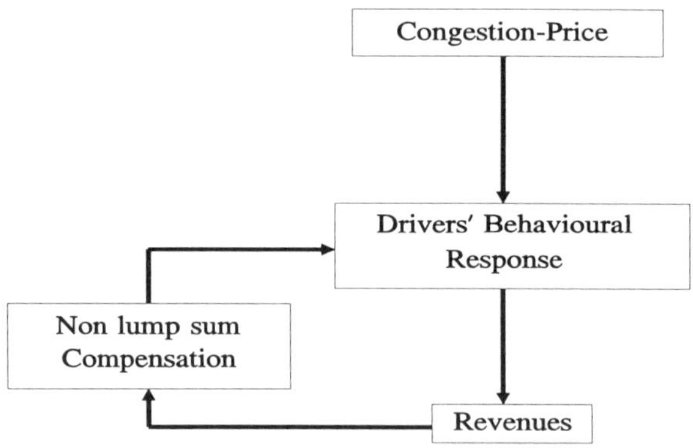

Figure 3.3 Compensation effects of congestion-pricing.

In addition to the indirect compensation effect mentioned above, Fearon et al. (1994) argue that congestion-pricing might also have indirect effects on the price level of some goods. In their survey results they found that:

> "... commercial-vehicle operators would continue to operate their vehicles during the priced periods and incur the road pricing costs. Any net costs,

after congestion benefits have been taken into account, would generally be passed on to the customer. If such net costs occur, this would lead to a small increase in the cost of goods and services associated with the GV (goods vehicle) operations." (p. 55)

Against the results of Fearon et al. (1994) one might argue that companies might try to diminish their transportation costs by increasing the transportation activities during off-peak periods. Furthermore, Button (1978) argued that the *inflationary impact* will be fairly small since transportation costs form only a small part of total production costs. Finally, one might argue that the behavioural changes induced by implementing congestion-pricing might (to some extent) be frustrated when the employer fully compensates the employee for the commuting costs.

To conclude, the analysis of these indirect effects is a difficult task, but as these effects may partly offset the direct ones, they should not be left aside in future work. As discussed before, the activity-based approach should play a more important role in dealing with these issues; the need for transportation is strongly linked with the engagement in and the chaining of activities.

3.4.3 Empirical evidence on behavioural adaptations to congestion-pricing

In contrast to the vast amount of literature on the problems linked to implementing congestion-pricing, the literature on the behavioural impacts of congestion-pricing is rather limited. This might be caused by the fact that worldwide only a few cities have adopted a certain kind of congestion-pricing; here, one could think of Singapore, Oslo, Bergen and Trondheim. Even for these sites, the charging structures are not as much dependent on the level of congestion as they should be according to the theory; simple cordon schemes are used. Hence, practical evidence of (revealed) behavioural responses to congestion-pricing is limited.[13]

Another methodology for assessing the effects of congestion-pricing on motorists behaviour is the use of stated preference (SP) surveys.[14] Although SP methods are getting more attention in the transportation literature, the technique has been sparsely used for investigating the impact of congestion-pricing. Therefore, the remainder of this section is based on some preliminary findings.

The results of Meland and Polak (1993) for the Trondheim toll ring[15] indicate a small shift from peak-period travel to off-peak periods. Furthermore, they obtained different responses from different groups of people; it seems important to distinguish

[13]In Hong Kong, the technical feasibility of a sophisticated electronic road pricing scheme has been shown (Catling and Harbord, 1985). However, owing to strong public opposition the government was unable to implement the scheme (Borins, 1988).

[14]See Polak and Jones (1993) for an overview of the issues involved in the use of SP methods in transport modelling.

[15]A fixed amount is charged when crossing the boundary of the Trondheim toll ring. This is mostly done by using automatic vehicle identification (AVI) and automatic debiting technology. Some 60% of the inhabitants of Trondheim live outside the ring. The ring operates Monday to Friday from 6 AM to 5 PM; after 10 AM a discount is given to account for peak or off-peak traffic. For more information on the Trondheim toll ring, see Meland and Polak (1993).

between commuters, business people, shoppers etc. This may be an indication of the importance of different activity patterns for determining travel behaviour. However, the Trondheim results may not be indicative for other congestion-pricing schemes, because the toll ring was particularly designed to raise money for new infrastructure: the main objective was not to price for externalities, and hence the charges were relatively low. Finally, Meland and Polak argued that the impacts on travel behaviour are also dependent on, for instance, flexible work arrangements, long shopping hours, quality of public transport etc. The impacts of congestion-pricing should be analysed within the whole context, not in isolation.

Polak et al. (1993) investigated how travellers might modify their time of travel under the influence of congestion-pricing charges which vary according to the time of day. They argued that, in order to tackle this issue, it is necessary to take into account the interrelationship between the timing of different journeys within an entire activity pattern. Polak et al. (1993) used the concept of *time of day utility profiles* (see also Supernak (1992)) to analyse the effects of charging on the travellers' choice of time to travel.

Brown et al. (1993) developed a set of demand elasticities for various congestion-pricing options in London. They argued that relatively little formal evidence exists of motorists' responses to an increase in the cost of travel as a result of the introduction of direct road user charges. The demand elasticities that are assumed to be known in the literature (for example, demand elasticities with respect to fuel prices or public transport fares) cannot simply be transferred to a congestion-pricing context.[16] In Tables 5, 6 and 7, Brown et al. (1993, p. 175) provide estimates of elasticities (ranging from -0.03 to -0.8) under different congestion-pricing regimes. Again, it is important to note that these values differ greatly for different groups of people.

3.5 IMPLEMENTING CONGESTION-PRICING

In the previous sections attention has been paid to the theoretical concept of congestion-pricing, the positive and negative aspects, and the behavioural responses that might be expected. In this section attention is turned to a more practical question: the actual implementation of a congestion-pricing scheme. A modest attempt is made to answer the question which congestion-pricing system is most likely to render the best results given the current situation. To answer this question, a distinction should be made between:

1. the choice of charging system; and
2. the political way of introducing it.

The first and second points are discussed in Sections 3.5.1 and 3.5.2, respectively.

[16]For evidence on elasticities, see Goodwin (1992) and Oum et al. (1992).

3.5.1 Which kind of implementation?

Milne (1993) discusses the four road user charging systems under consideration:

1. Toll cordons: a charge is levied at specific points in order to permit travel within a specified area, defined by a single boundary. The charge might be dependent on the time of day to account for congestion (for example, Trondheim).
2. Distance-based charging: the charge is based on the distance travelled in a certain area. In addition, the charge might be dependent on the time of day.
3. Time-based system: a charge is levied dependent on the time spent travelling in a specified area. Such a system has been proposed for Richmond in London. The charge may be varied depending on the time of day.
4. Congestion-metering system: charges are directly related to the prevailing level of congestion on the road network. The system proposed for Cambridge could be put into this category.

In Table 3.1, factors that have an impact on the attractiveness of different types of congestion-pricing schemes are depicted. It is important to stress that the results in Table 3.1 are exploratory in nature, and unfortunately not based on extensive scientific research. Given the lack of experience with real implementations of various types of congestion pricing, there is considerable uncertainty about their impacts. Therefore, only a qualitative expression of the level of these impacts is given, and it is inevitable that these qualitative statements are to a certain extent subjective.[17]

Important factors affecting the attractiveness of different types of implementations are the costs of implementing the system, the impact on congestion, the predictability of the congestion-price, the difficulties of implementing the technology etc. Given the tentative results in Table 3.1 and the risks associated with implementing the systems that have the greatest impact on congestion, the cordon system with time-dependent charges seems to be the most appealing policy option available.[18]

The importance of implementing a certain kind of congestion-pricing system is evident. However, it might be dangerous to proceed with technologically advanced systems. In a first phase, user awareness of the costs associated with mobility during congested periods should be realised. Then, in the next phase, there is a fair chance that more advanced systems can be implemented successfully with the backing of the politicians and the public. Having in mind the quote by Sheldon et al. (1993) (see Section 3.3.2) that people perceive congestion as a problem caused by *someone else*, it seems that user awareness of the costs of congestion is not yet present. One of the main objectives of a simple time-dependent cordon system is to give shape to this user awareness.

---

[17]The attractiveness of a particular scheme will also be dependent on other factors such as network structure, levels of congestion etc. These are not taken into account in this chapter.

[18]In fact, a formal multicriteria analysis should be carried out to arrive at this conclusion. If in such an analysis a substantial weight (relative to the other factors) is given to the costs, the impact on congestion, and the user friendliness, then the cordon system with time-dependent charges will turn out to be the most attractive system.

Type	Costs (one-off)	Impact on congestion	User friendliness[b]	Side effects[c]	Implementation[d]	Overall[e]
Cordon (F)	+	-	+	-	+	+
Cordon (T)	+	+	+	-	+	++
Distance-based (F)	-	-	+	-	-	-
Distance-based (T)	-	+	+	+	-	+
Time-based (F)	--	+	-	+	--	-
Time-based (T)	--	+	--	+	--	-
Congestion-metering	--	+[f]	--	+	--	-

Note: F fixed charges; T time-of-day dependent charges

[a]The impacts are measured at an ordinal scale ranging from most unfavourable (--) to most favourable (++) with - and + as intermediate outcomes.
[b]Reflecting the predictability of the congestion-price. This might have an important impact on the behavioural responses of the road users and the awareness of the costs of mobility during congested periods. In addition, it might positively affect public acceptance.
[c]Reflecting shifts in land use patterns. All these systems will induce certain shifts in land use patterns, though these might be most undesirable under a cordon system.
[d]Reflecting the difficulties associated with putting the theory into practice. For example, the problem of rat-running falls into this category.
[e]Overall assessment taking into account all relevant factors.
[f]From a theoretical perspective, the impact on congestion of this implementation scheme is optimal. However, owing to the low predictability of the congestion-price, the impact might be much lower in practice. (To properly inform the road users on the congestion-prices, a driver information system is required; for example, see Emmerink, Nijkamp and Rietveld (1995b) and Chapter 10.)

Table 3.1 Assessment of the attractiveness of different types of congestion-pricing.[a]

Practical and survey evidence (Seale, 1993; Sheldon et al., 1993) also seems to suggest that politicians prefer a kind of cordon toll system. Such systems have already been implemented in Singapore, some Norwegian cities, and were proposed for the Netherlands (Stoelhorst and Zandbergen, 1990).[19] Furthermore, Meland and Polak (1993) argued that a cordon system is relatively less risky compared with other proposed technologies.

The technology that will be chosen is also relevant for the applications that are available. For instance, Stoelhorst and Zandbergen (1990) mentioned Road Transport Informatics (RTI) systems as a possible application of the infrastructure provided by the congestion-pricing technology. The PAMELA-project within the EU DRIVE I programme showed the feasibility of the technology to implement congestion-pricing (Blythe and Hills, 1992) and these authors argued that it might also be suitable for RTI applications. At present, the PAMELA-project is being taken further and the

[19]See Section 3.6 for a discussion of the Dutch proposal and the public responses.

range of applications is being extended within the follow-up ADEPT-project which is part of the EU DRIVE II programme.

An important research question that needs to be considered is the percentage of welfare benefits that can be obtained with simple congestion-pricing systems. If, for instance, the time dependent cordon system is able to give 80% of the potential welfare improvement (the potential welfare improvement is given by the area of the dead weight loss triangle), then the marginal improvement of implementing a sophisticated system might be outweighed by the marginal costs of implementing such an expensive system. Hence, operationalisation of the theoretical concept of the *index of relative welfare improvement* as described in Verhoef et al. (1995a) is highly relevant as a tool for trading off the costs and benefits of different kinds of congestion-pricing systems. A modest attempt in that direction is undertaken in Chapter 10.

3.5.2 Method of introduction

Related to the political method of introducing, May (1992) discussed five models that can be used to introduce congestion-pricing:

- Policy-led approach: specify a broad policy of which congestion-pricing forms a part, seek public approval, and then implement it.
- Technology-led approach: develop complex technology to achieve congestion-pricing.
- Revenue generating approach: specify the objective of raising money to pay for infrastructure from the start.
- Make use of demonstration projects.
- Analysis-led approach: a careful and objective assessment is made of the issues to be resolved, and an analytical programme designed to answer these.

The limited evidence in the literature seems to suggest that the revenue generating approach is most suitable for gaining public support for congestion-pricing. Following this approach, the government should clearly state the manner in which the revenues will be spent. Furthermore, as regards the public and political opposition, a congestion-pricing system will be most successful if the redistribution scheme approaches a *strict* rather than a *potential* Pareto improvement (see Section 3.3.2).

The fourth model mentioned above raises another important issue: Which sites should be selected for introducing congestion-pricing? Poole (1992) argued that "the least desirable place to start would be existing freeways, no matter how congested. Putting a price on something that has traditionally been offered free at the point of use risks major public and political resistance" (p. 385). He claimed that the best sites for introducing the concept of congestion-pricing are (1) existing toll facilities and, (2) completely new facilities which give users a choice compared with existing, unpriced facilities. Giuliano (1992) holds the same opinion and moreover adds that it is unlikely that congestion-pricing will be implemented to any significant extent in the US.

Grieco and Jones (1994) are more optimistic for Europe, however. They indicate that public support for congestion-pricing will be strongest where current congestion is most intense. Given this localised character of congestion, they foresee an important role of municipalities - in addition to national governments - in introducing congestion pricing. A gradualist implementation strategy is called for in this case. Given the tendency that the public does not like high-tech charging and enforcement schemes, it may be better to start with low-tech solutions, according to Grieco and Jones. Furthermore, they observe that proposals to introduce congestion-pricing in particular countries have led to a reinforcement of support for this policy in other European countries, even when these proposals were finally not accepted in the countries where they were originally formulated.

3.6 THE DUTCH FAILURE TO IMPLEMENT CONGESTION-PRICING

In this section, the failure of the Dutch government to gain public support for the announced congestion-pricing scheme (the so-called "Rekening Rijden" scheme) will be discussed. Long before a congestion-pricing scheme should have been implemented, the scheme was removed from the political agenda.

The facts about the system that was planned can be found in the Second Structure Plan (1988) of the Dutch government. On page 116, the objective, mean, costs and time period of the plan were described. The objective of the congestion-pricing scheme, as stated in the Second Structure Plan, was to decrease road traffic during the peak-period. The means were by implementing a location-dependent and time-dependent electronic charging mechanism that influences the choices of departure time of car users and also the mode-split in favour of public transport and the bicycle. The costs, dependent on the type of system to be implemented, were estimated at $60 million. The system should have been operational in the years 1992-1995. The objective of the scheme (as state above) was not completely clear to everyone involved. Many people thought that the system was devised to be able to build an additional five tunnels in the western part of the country. This, to some extent, clouded the real arguments.

The failure of the system was announced in May 1990 by the Minister of Transport, Public Works and Water Management in a letter to Parliament, less than two years after it was first mentioned by the government.[20] In this letter the Minister wrote that she would refrain from any further steps leading to the introduction of this advanced congestion-pricing system. The reasons were rather vague: too much resistance from within society appeared to be the major one. In future, the Minister said, she would turn to more traditional tools to achieve her objectives, rather than to such a sophisticated instrument. It is interesting to note that these less sophisticated systems - such as yearly peak travel passes and toll booths - have never been implemented thus far. Apparently, the Rekening Rijden experiment has put all these instruments in a bad light.

In a note that has not been made public, the government was more precise about the reasons for abandoning the advanced congestion-pricing system (see Visser, 1992). It was argued that (1) the congestion-pricing scheme had to serve conflicting objectives simultaneously; (2) there was no political commitment to support the

---

[20]Letter dated 7 May 1990 from the Minister of Transport, Public Works and Water Management to the Chairman of the Second Chamber of the States General, Tweede Kamer, 1989-1990, 20922, number 9.

system; and (3) the great resistance from various parts within society was viewed to be a major obstacle. In Section 3.3, these three points were also identified in the list of *cons* to congestion-pricing. The public at large was not (yet) ready - or not (yet) prepared - to bear the consequences of such a radical, far-reaching change in the transport system. The introduction of a new market (a market for congestion) is clearly perceived as a large intrusion on the principle of freedom, paying for something that was up to now free of charge. The public awareness of the congestion problem was clearly not sufficient to generate enough support for this advanced policy.

In 't Veld (1991) adds another reason for the, in his view, "logical failure of road pricing" in the Netherlands. In 't Veld argues that the:

> "... elegance and potential effectiveness of congestion-pricing as an allocation device have not contributed to its popularity. ... the elegance of congestion-pricing is at the same time a major threat. ... The possibility that congestion-pricing could register each illegal passage and send a bill or a fine or a combination of both afterwards, is a menace to all; perfection in the repression of illegality is not at all desirable to most citizens." (p. 115)

In addition, In 't Veld (1991) argues that fragmentation within society induces differentiated behaviour from individuals. The introduction of this congestion-pricing scheme clearly suffered from this fragmentation:

> "Politics may for instance produce consensus or quasi-consensus into the direction of strong environmental policies on a very abstract level. As soon as the translation of these abstract policies into concrete measures is at stake, however, this consensus has disappeared altogether." (p. 119)

In short, it is concluded that the introduction of a congestion-pricing scheme in the Netherlands failed because of the population's general attitude of dislike, and as a consequence of the lack of political commitment and political will to explain the need for such a system. The positive aspects of congestion-pricing were given almost no attention in the debate on the desirability of the scheme.

It is interesting to note that the political climate in The Netherlands has suddenly changed in 1994/1995. Apparently, politicians recognised the increased need to do something about the congestion problem in the western part of the country, the so-called "Randstad".[21] In the last two years congestion has severely increased, and with the expected increase in the level of car ownership (from approximately 6 million cars now to approximately 8 million cars in the year 2010), politicians have realised the necessity to implement some kind of policy that restricts road usage.[22] If nothing is done, politicians have understood that future growth in GDP (Gross Domestic Product) might be hampered by congestion problems. A majority in parliament has agreed that

---

[21]This is in agreement with the predictions made by Grieco and Jones (1994) that eventually there will be support for congestion-pricing in those regions where congestion is most severe.

[22]See the recent letters - dated October 1994, 2-6-1995, and 23-6-1995 - from the Minister of Transport, Public Works and Water Management to the Chairman of the Second Chamber of the States General, Tweede Kamer, 1994-1995, and the amendment of a majority of Dutch parliament, dated 9-11-1994.

some kind of congestion-pricing will be implemented in the most congested part of the country, the Randstad, after the year 2001.

The observations made in this section stress the importance of the need to increase the individual's awareness of the costs of mobility, particularly during congested periods, and to change the congestion-pricing debate from the negative aspects to the intrinsic positive side. To start with, a simple, time-dependent cordon-based charging mechanisms might play an important role in changing the population's opinion towards greater understanding of these economical instruments.

3.7 CONCLUSION

In this chapter, the problems associated with the theoretical concept, the behavioural responses, and the implementation of congestion-pricing were analysed with the arguments provided in the literature. Conflicting viewpoints were identified and brought together.

Although congestion-pricing is a first-best instrument in theory, the assumptions needed to arrive at this conclusion oversimplify reality. In practicc, congestion-pricing will most likely not be a first-best instrument for tackling the congestion externality, although it may be expected to be able to approach first-best standards more closely than does any other instrument.

Congestion-pricing has some important advantages over other available measures for curbing levels of congestion. Nevertheless, implementing congestion-pricing will give rise to many problems: (1) it will lead only to a potential Pareto improvement; (2) it will cause horizontal inequity; (3) it might induce regional inequity; (4) it will provoke the question as to how to redistribute the revenues; (5) it may be perceived differently by different groups in society; (6) it might affect people's privacy; (7) it might be perceived as a loss of freedom; and (8) it may give rise to the problem of rat-running.

The behavioural consequences of introducing congestion-pricing have not been given much attention in the literature. Activity-based models and models that allow for a certain kind of irrational behaviour are most likely to be successful in predicting the responses of motorists under congestion-pricing. Future research should shift into this direction rather than using the traditional four-step approach. Furthermore, the impact of the compensation scheme - used to offset drivers that are worse-off under congestion-pricing - on the behavioural responses should be given more attention, because the compensation scheme might partly offset the behavioural changes induced (buy-back effects).

There are different technological ways to introduce congestion-pricing. If the potential opposition is taken into account, a cordon system - in which the price is dependent on the time of the day - seems currently to be the most attractive option. Such a simple scheme might increase drivers' awareness of the costs of mobility during congested periods. Moreover, a more sophisticated scheme can always be implemented once the simple system has proved successful and the awareness of the costs of mobility during congested periods have been established. The Dutch experience indicates that overcoming all the problems of introducing a relatively simple congestion-pricing scheme is already a cumbersome task.

Finally, future research should focus on the relative welfare improvement of simple congestion-pricing systems compared with more advanced ones. If simple systems are able to generate a large part of the welfare gains, then the need for implementing advanced systems declines.

In the next chapter, a theoretical route choice framework is presented that allows for the analysis of the impact of motorist information systems and congestion-pricing on network efficiency. This framework is used in the theoretical models presented in Chapters 5 to 10.

4 A STOCHASTIC ROUTE CHOICE FRAMEWORK[1]

4.1 INTRODUCTION

As discussed in Chapter 2, motorist information systems aim to improve the performance of transport networks by providing the users with information-oriented incentives for behavioural change. In the past, various theoretical modelling efforts have been conducted that attempted to identify the role played by new information technologies in transport (Bovy and Stern, 1990). Two different approaches have widely been used: simulation models and equilibrium models. In the present chapter, the focus is on the latter class.

The equilibrium concepts used in these models are the so-called deterministic and stochastic user equilibrium. The stochasticity in the stochastic user equilibrium reflects limited information on the traffic situation from the traveller's point of view. The deterministic user equilibrium a priori assumes that all travellers are perfectly informed on the traffic situation. Therefore, as it is argued, the discrepancy between the deterministic and the stochastic user equilibrium reflects, to some extent, the impact of motorist information systems in transport networks.

In the present chapter, a slightly different route choice model will be proposed; a concept that is applied throughout the theoretical body of the present book. This concept allows to analyse the impacts of traveller information in a natural way. The main difference between the proposed concept and most of the prevailing literature is that the travel costs are treated as flow-dependent *random variables*. Information, then, provides travellers with *realisations* of these random variables. Uninformed travellers, in contrast, will base their behaviour on *expected* rather than *actual* costs.

A route choice model, based on this concept, is then used to formalise the issues of user *perception* and user *uncertainty* in the context of travel behaviour and new information technologies. In the literature on user equilibrium models, the terms perception and uncertainty have often been used in a rather ambiguous and loose manner.

The chapter is organised as follows. Section 4.2 addresses the ambiguity in the use of perception and uncertainty in user equilibrium models. The route choice model to be used in the equilibrium concept in Chapters 5 to Chapter 10 is described in Section 4.3. Next, Section 4.4 adds an additional component to the route choice model,

---

[1]This chapter is partly based on Emmerink, Verhoef, Nijkamp and Rietveld (1996a) published in *Journal of Transport Economics and Policy*.

while Section 4.5 concludes. Finally, the appendix to this chapter pays attention to an alternative model that has been widely used in the literature: the bottleneck model.

4.2 PERCEPTION AND UNCERTAINTY IN USER EQUILIBRIUM MODELS: RELATION TO MOTORIST INFORMATION SYSTEMS

In the 1960s, the classical four-step (trip generation, trip distribution, mode-split and network assignment) transport model was developed. Despite major improvements in transport modelling techniques during the 1970s and 1980s, the basic structure of the model remained unaltered (Ortúzar and Willumsen, 1994). The fourth step of the model - assigning the origin-destination (OD) matrix to the transport network - has usually been based on Wardrop's (1952) first principle, the user equilibrium. According to this principle, each user of the network attempts to maximise his or her utility. In the literature, the standard user equilibrium principle is often referred to as the *deterministic user equilibrium*. A shortcoming of Wardrop's first principle is that it assumes that all users of the transport network have full information about the traffic situation; something that is unlikely to happen in practice due to all types of stochastic incidents in the transport network.

Daganzo and Sheffi (1977) recognized the above mentioned shortcoming of the deterministic user equilibrium concept, and introduced the notion of *stochastic user equilibrium*. This is defined as a situation in which "no traveler believes that his travel time can be improved by unilaterally changing routes" (Sheffi, 1985, p. 20). Sheffi and Powell (1981) were the first to apply the stochastic user equilibrium to a test network. The essence of the difference between the two equilibrium concepts concerns the difference between *beliefs* (user *perceptions of reality*) and *actual values*. The discrepancy between the drivers' perceptions and the actual traffic situation would however, disappear by the provision of perfect information. Therefore, it might be argued that the stochastic user equilibrium concept distinguishes itself from its deterministic counterpart by allowing for limited (incomplete) information from the traveller's point of view.

The above mentioned equilibrium concepts have been applied in analysing the impacts of driver information systems on network efficiency. First, the deterministic user equilibrium may be viewed as the *ideal* situation in which all users have perfect information on the traffic situation. Second, the stochastic user equilibrium characterizes a situation in which drivers are uncertain about route travel costs. The difference between the two concepts, then, is a measure for the (potential) impacts of driver information systems. In other words, driver information systems reduce driver uncertainty, and bring the stochastic user equilibrium closer to the deterministic user equilibrium. In the past, different researchers have been using stochastic network assignment models to assess the impact of driver information systems on network performance. See, for example, Hicks et al. (1992), Ran and Boyce (1994), Tsuji et al. (1985), and van Vuren and van Vliet (1992).

It is worthwhile to analyse the source of the uncertainty in the models that are based on the stochastic user equilibrium concept. In these models, the uncertainty stems from modelling the behavioural mechanism underlying the traffic assignment process as a discrete choice situation. The deterministic user equilibrium, then, follows under the assumption that travellers choose the least costly route and, in addition, have full information regarding the whole traffic situation. Conversely, stochastic models arise when it is assumed that, "due to variations in perception and exogenous factors (such as weather, lightning etc.) the path times are perceived differently by each driver" (Sheffi, 1985, p. 272). As Sheffi notes, these models assume that drivers perceive travel times differently. Put differently, the intrinsic path travel times may be constant in these models, but it is just the case that drivers have different perceptions of the actual costs.

		ASSUMED KNOWLEDGE OF TRAVELLER	
		Full (deterministic)	Limited (stochastic)
NETWORK CHARACTERISTICS	Deterministic	DUE	SUE
	Stochastic	SNDUE	SNSUE

Table 4.1 Combination of features of network and individual characteristics.

The discussion above is summarised by the first row in Table 4.1. The traditional deterministic and stochastic user equilibrium concepts are characterized as models in which the network is assumed to be deterministic, while the beliefs of the drivers (the assumed knowledge of the travellers) may either be according to reality (full information and therefore deterministic) or to subjective perceptions (limited information and therefore stochastic). Hence, in the traditional DUE (deterministic user equilibrium) and SUE (stochastic user equilibrium) concepts, the stochasticity stems from the travellers themselves, see the first row of Table 4.1.

However, as indicated by the second row of Table 4.1, one might also adopt an alternative perspective. Given the large variation in daily travel times that is witnessed on real road networks, it is reasonable to assume that the route travel costs are stochastic rather than deterministic, see also Mirchandani and Soroush (1987). In fact, this is for three reasons a natural approach for analysing the impact of driver information systems. First, this approach acknowledges that the daily fluctuations in travel costs for different road users are highly *correlated*, something that is not considered by the traditional class of models. Second, in this approach non-recurrent congestion is modelled explicitly by using stochastic route travel costs. Since motorist information systems are considered to be beneficial particularly in non-recurrent congested situations, this is an attractive feature. And third, this approach allows to provide informed travellers with *realisations* of the stochastic travel costs, which is a very natural way of modelling information. Consequently, the impact of information on network performance is modelled as an *endogenous* result of the drivers' responses to the information provided. Conversely, when the discrepancy between the DUE and SUE concepts is used as a measure of the impact of information, then it is assumed that information provision *exogenously* reduces the uncertainty on the network situation.

The models in the second row of Table 4.1 are referred to as stochastic network deterministic user equilibrium (SNDUE) and stochastic network stochastic user equilibrium (SNSUE). In this chapter, the underlying route choice model of SNDUE and SNSUE models is discussed. This route choice model will be used to analyse the impact of driver information on user uncertainty and perceptions.

4.3 A ROUTE CHOICE MODEL WITH STOCHASTIC TRAVEL COSTS AND DRIVER INFORMATION

In this section, the individual choice model in its simplest form is presented. In a slightly different context, van der Zijpp and Bovy (1994) have analysed a similar type of model. Their model is extended to the case of elastic demand, and, in addition, travel costs will be flow dependent in the next chapters. In the model, an individual would like to travel from a particular origin O to a particular destination D. There exists a - by the individual known - choice set of route alternatives, consisting of R different routes. Of course, it is allowed that certain routes have some overlap. It is assumed that the individual has a subjective probability distribution of travel costs for each route *r* denoted by C^s_r, where the superscript *s* refers to *subjective* probability and the subscript *r* to the *route*. If a road-price is levied on certain parts of the network, this is included in the travel costs. For the purpose of explaining the route choice model, it is assumed that travel costs are independent of the level of road usage. In the SNDUE models of the subsequent chapters this assumption is obviously relaxed.

In addition, it is assumed that there exists an objective stochastic route travel cost variable given by C^o_r, where the superscript *o* refers to the *objective* stochastic route travel cost variable. This stochastic variable reflects the intrinsic fluctuations in the transport network, for example, due to particular weather conditions, unpredictable lane closures, traffic accidents etc. In most of the equilibrium models that have been used so far, this function is assumed to be deterministic, as indicated by the first row of Table 4.1.

In the model considered here, it is not necessarily the case that the subjective and objective probability distributions are identical. They may differ, and when this is the case, the existence of perceptional errors is witnessed. It is further assumed that drivers are rational and are maximising some utility measure. For the time being, it will be assumed that utility is solely dependent on route travel costs, and in addition, that travellers are risk-neutral. Under these assumptions, the traveller's utility maximisation problem can be written in terms of a cost minimisation problem.

In the model the interesting case of elastic demand for travelling by car is addressed. This will be done, by including the possibility for a traveller to choose a so-called *alternative mode* labelled "0". This mode can either reflect a situation where the traveller decides to use public transport or a situation where the individual decides not to undertake the trip. Finally, as the model to be presented is static in nature, the *alternative mode* might reflect an individual using another departure time interval. But then, it is implicitly assumed that congestion during this interval is independent of the level of road usage in the model. Further, it is assumed that the subjective costs of choosing this alternative mode labelled "0" are given by the random variable C^s_0; the objective costs are given by the random variable C^o_0.

Without the provision of any additional kind of information, a driver will base his or her behaviour on the expected values of the subjective probability distributions. The traveller will decide to travel by car on the minimum expected cost route when these costs are below the expected costs of using the alternative mode labelled "0":

$$\min \left\{ E(C_0^s) \ , \min_{r \in \{1,\ldots,R\}} E(C_r^s) \right\} = \min_{r \in \{0,\ldots,R\}} E(C_r^s) \tag{4.1}$$

Assume that the above expression takes on its minimum for r is equal to r^*. Thus, r^* is the most preferred route according to perceived values. If $r^*=0$, then the most preferred "route" is the alternative mode.

Now, assume that an information device exists that provides drivers with perfect information on the route travel costs. Hence, it provides the *realisations* c_r^o of drawings from the stochastic route travel cost variables C_r^o, and in addition, it provides the *realisation* c_0^o of the costs of the alternative mode. As is customary in statistics, an upper case is used to denote a random variable; a lower case to denote the realisation (a drawing from a probability distribution).

Given the realisations c_r^o of the random variables C_r^o, an informed driver will choose the alternative that minimises costs:

$$\min_{r \in \{0,\ldots,R\}} c_r^o \tag{4.2}$$

It is straightforward to derive the expected decrease in costs owing to the perfect information provided for a traveller with objective perceptions. To do so, note that without information the expected costs are given by:

$$\min_{r \in \{0,\ldots,R\}} E(C_r^o) \tag{4.3}$$

On the other hand, when perfect information on the realisations of the stochastic route travel costs is available, expected costs of road users with objective perceptions are given by:

$$E(C_{\min}^o) \tag{4.4}$$

where $C_{\min}^o$ denotes the stochastic variable that is defined as the minimum of the sequence of stochastic route travel costs (including the alternative mode "0") C_0^o, C_1^o, ..., C_R^o: $C_{\min}^o=\min(C_0^o, C_1^o, \ldots, C_R^o)$. From probability theory it is well known that for random variables C_0^o, C_1^o, ..., C_R^o:

$$E(C_{\min}^o) \leq \min_{r \in \{0,\ldots,R\}} E(C_r^o) \tag{4.5}$$

The right-hand side denotes the expected route travel costs for the road user without information and objective perceptions; the left-hand side denotes the expected route travel costs for the road user with perfect information and objective perceptions. The difference between these values is a measure for the expected decrease in costs from perfect information for the road user with objective perceptions C_r^o. The *expected value of perfect information* (EVPI) can now be defined as:

$$EVPI = \min_{r \in \{0,\dots,R\}} E(C_r^o) - E(C_{\min}^o) \qquad (4.6)$$

The terminology EVPI stems from the theory of decision analysis and is defined as the maximum amount of monetary units that an individual is willing to pay for obtaining perfect information, see for example von Winterfeldt and Edwards (1986). As has been shown by van der Zijpp and Bovy (1994), the EVPI has the following three properties:

1. EVPI is non-negative: the value of information is always greater than or equal to zero.
2. EVPI is an increasing function of the variances of C_r^o: the value of information increases as the intrinsic stochasticity of the transportation system increases.
3. EVPI increases if the difference between the expected costs of the alternatives decreases: the value of information increases as the alternative routes are more competitive.

It is important to stress that the three results given above assume cost functions that are independent of the level of road usage.

Thus far, the analysis focused on the case where the stochastic subjective travel costs were equal to the stochastic objective travel costs. The benefits derived from information were called the expected value of perfect information (EVPI), and were caused by driver *uncertainty* regarding the traffic situation. Next, the situation in which the objective and subjective stochastic travel costs do not coincide is considered. Such a situation can be characterised by the existence of *perceptional errors*: Due to traveller-specific characteristics, the traveller under consideration is not perfectly aware of the prevailing objective stochastic travel costs. This, in fact, is the situation described by Noland (1995) when he studies so-called "biased perceptions" in the sample of road users.

The size of the perceptional error made by a driver with subjective perceptions C_r^s can be further formalised in the present model by:

$$\textit{Perceptional Error} = E(C_{r^*}^o) - \min_{r \in \{0,\dots,R\}} E(C_r^o) \qquad (4.7)$$

where, as previously, r^* is defined as the route that minimises expression (4.1). It is important to note that the perceptional error cannot be negative.

It is now also possible to decompose the total benefits available to a traveller owing to information provision into a *perceptional component* (as given by equation (4.7)) and an *information component* (as given by equation (4.6)). In the present route choice model, the provision of perfect information serves two purposes: (1) it reduces the perceptional error to zero; and (2) it provides additional benefits equal to the EVPI. Hence, the EVPI is the lower bound of the available information gains for an individual traveller.

4.4 AN EXTENSION OF THE ROUTE CHOICE MODEL: COSTS OF UNCERTAINTY

Thus far, it was assumed that the route choice decision is solely dependent on costs related to travel time. However, in the literature it has been argued that the variability in travel times is an important cost component of travel behaviour as well. For example, Hendrickson and Kocur (1981) identified the importance of schedule delay. More recently, Arnott et al. (1992), Jou and Mahmassani (1994) and Noland and Small (1995) returned to this issue and included schedule delay as one of the essential aspects in a departure time and route choice model. In the present section a term reflecting the variation in travel costs in the proposed route choice model is included, which can be interpreted as an element of stress.

For ease of presentation it is assumed that uninformed travellers are seeking to minimise expression (4.8):

$$E(travel\ costs) = \alpha\ E(travel\ time) + \beta\ Sd(travel\ time) \tag{4.8}$$

where *Sd* denotes the standard deviation. The parameter α can now be interpreted as the *value-of-time*, while the parameter ß measures the *monetary costs of uncertainty*. In fact, the second component in expression (4.8) is reflecting the risk-aversion of the traveller. The larger parameter ß, the higher the costs of uncertainty, and the more risk-averse the traveller.

Then, without information, the route choice problem faced by the traveller with objective perceptions is given by:

$$\min_{r \in \{0,\dots,R\}} \left(\alpha\ E(C_r^o) + \beta\ Sd(C_r^o)\right) \tag{4.9}$$

On the other hand, when information on the travel costs is provided to the traveller, then the within-day travel time variance disappears; the traveller is certain when he or she will arrive. Therefore, in this case, the traveller will choose the route that is the solution to:

$$\min_{r \in \{0,\dots,R\}} c_r^o \tag{4.10}$$

where the lower case indicates the realisations of the random variables C_r^o. Hence, in this model, information will provide the driver with certainty on the travel costs, and he or she will therefore not be faced with unpredictable variations in travel costs. The expected value of perfect information (EVPI) can be easily derived:

$$EVPI = \min_{r \in \{0,\dots,R\}} \left(\alpha\ E(C_r^o) + \beta\ Sd(C_r^o)\right) - \alpha\ E(C_{\min}^o) \tag{4.11}$$

Comparing the above expression with expression (4.6), the intuitively appealing result is found that information is of more value to risk-averse travellers.

Thus, three different types of gains from information provision were detected. Information provision reduces costs associated to incomplete information by:

1. reducing expected travel times (equation (4.6)),
2. removing the costs associated to risk-aversion (equation (4.11)).

In addition, in the case of perceptional errors, information provision reduces the perceptional error to zero (equation (4.7)).

In Chapter 8, a SNDUE model with costs of uncertainty is presented.

4.5 CONCLUSION

In the present chapter a stochastic route choice model was proposed. Within this framework, the issues of user perception and uncertainty in a motorist information systems context were formalised. First, it was concluded that the discrepancy between the traditional stochastic and deterministic user equilibrium stems from limited (or stochastic) information availability on the route choice context, rather than from so-called perceptional errors. Next, it was proposed to analyse the impact of motorist information systems on route choice behaviour by explicitly modelling the path travel times in the transport network as stochastic variables. This provides a natural means to supply informed drivers with realisations of these stochastic variables, while uninformed drivers will base their behaviour on expectations of the random travel costs. This in contrast with most of the prevailing literature in which it is assumed that information will exogenously decrease the travel time variability in the transport network. The model can further be used to decompose the total informational benefits available to a particular traveller into a perceptional and an informational component.

The route choice model discussed in this chapter is extended to a SNDUE model (with stochastic link travel costs that are a function of the level of road usage) in the subsequent chapters. In Chapter 5, the most basic case of a one-link network is studied. Chapter 6 extends the model to a two-link network, so that besides mode-split effects route-split effects can be studied as well. In Chapter 7, information is endogenised, that is, the number of informed and uninformed road users is determined within the model. In Chapter 8, costs of uncertainty are considered, while Chapter 9 deals with a network with multiple origin-destination pairs. Finally, Chapter 10 discusses the combination of congestion-pricing and information provision.

APPENDIX 4.A AN INTRODUCTION TO THE BOTTLENECK MODEL

In this appendix, a brief introduction into the bottleneck model is given. This model has been a popular tool to gain more insights into the effects of various policy instruments that aim at curbing levels of congestion. The seminal paper on the bottleneck model was written by Vickrey (1969). The term *bottleneck model* stems from the fact that Vickrey assumed that traffic congestion is modelled as a queue behind a bottleneck.

The main contribution of Vickrey's bottleneck model is that he endogenised travellers' departure times on the basis of a sound economic equilibrium principle. He assumed that in equilibrium none of the travellers should be able to decrease travel costs by changing departure time, implying that in equilibrium none of the travellers has an incentive to change. In recent years, this model has been extended into various directions by Arnott et al. (1990a, 1990b, 1991, 1992, 1993, 1994). This appendix draws heavily on their work.

In the basic version of the bottleneck model, all home to work commuters are assumed to *wish* to arrive at work at the same time, t^*. As this is physically impossible due to the fixed capacity of the bottleneck, some will arrive early, while others will arrive late. The costs associated with arriving early or late are called *schedule delay* costs. Travel costs are assumed to be a linear function of travel time and schedule delay costs:

$$C(t) = \alpha \cdot (\textit{travel time}) + \beta \cdot (\textit{time early}) + \gamma \cdot (\textit{time late}) \qquad (4A.1)$$

where C(t) denotes the travel costs of an individual departing at time t. It is realistic (Small, 1982) to assume that $\gamma>\alpha>\beta$, where α is the shadow cost of time spent travelling, β the shadow cost of time early, and γ the shadow cost of time late. A fixed number of N travellers is assumed to undertake a trip from a single origin to a single destination. In order to simplify the algebra, travellers are treated as a continuum. Travel between the origin and the destination is assumed to be uncongested except at a single bottleneck. The capacity of the bottleneck is deterministic and equal to s travellers per unit time. If the arrival rate at the bottleneck exceeds s, a queue will develop. For simplicity, it is assumed that travel time T(t) equals queuing time; hence, there is no fixed time component:

$$T(t) = T^w(t) \qquad (4A.2)$$

Here, $T^w(t)$ denotes the queueing time at the bottleneck. Given a queue length Q(t) at time t, the waiting time can be calculated using the bottleneck's fixed capacity s:

$$T^w(t) = \frac{Q(t)}{s} \qquad (4A.3)$$

If r(t) denotes the departure rate, then the queue length evolves according to:

$$\begin{aligned} \frac{dQ(t)}{dt} &= 0 \qquad \textit{for } Q(t)=0 \textit{ and } r(t)\le s \\ \frac{dQ(t)}{dt} &= r(t)-s \qquad \textit{otherwise} \end{aligned} \qquad (4A.4)$$

We define t_0 as the departure time of the first commuter, t_e as the departure time of the last commuter, and t_n as the departure time of the commuter who arrives exactly at time t^*, i.e., $t_n+T(t_n)=t^*$. Obviously, the first commuter who departs at t_0 does not face a queue, and arrives early. The last commuter departs at t_e and does not face a queue either. If he or she did, then departure time could be delayed without arriving later, thereby decreasing travel time, and hence, travel costs. In equilibrium the travel costs of all commuters should be identical. For the commuter departing at t_0 and the one departing at t_e this implies, using (4A.1), that:

$$\beta \cdot (t^* - t_0) = \gamma \cdot (t_e - t^*) \tag{4A.5}$$

In equilibrium, a queue should exist in the departure time interval $[t_0, t_e]$. If this were not the case, then a commuter facing no queue incurs lower schedule delay costs than the first or last commuter, and would have equal zero travel time. Consequently, during the departure time interval, the bottleneck works at capacity s, and therefore:

$$t_e - t_0 = \frac{N}{s} \tag{4A.6}$$

From (4A.5) and (4A.6), t_0 and t_e can be calculated:

$$\begin{aligned} t_0 &= t^* - \frac{\gamma}{\beta + \gamma} \cdot \frac{N}{s} \\ t_e &= t^* + \frac{\beta}{\beta + \gamma} \cdot \frac{N}{s} \end{aligned} \tag{4A.7}$$

Also for the commuter departing at t_n travel costs should be equal to the travel costs of the commuters departing at t_0 and t_e:

$$\textit{equilibrium trip costs} = \frac{\beta\gamma}{\beta + \gamma} \cdot \frac{N}{s} \tag{4A.8}$$

Therefore the following equality should hold:

$$\alpha \cdot (t^* - t_n) = \frac{\beta\gamma}{\beta + \gamma} \cdot \frac{N}{s} \tag{4A.9}$$

Hence, t_n is equal to:

$$t_n = t^* - \frac{\beta\gamma}{\alpha(\beta + \gamma)} \cdot \frac{N}{s} \tag{4A.10}$$

The evolution of travel time T(t) over the morning peak follows from the linear travel cost function and equilibrium trip price:

$$\begin{aligned} \alpha \cdot T(t) + \beta \cdot \big(t^* - t - T(t)\big) &= \frac{\beta\gamma}{\beta + \gamma} \cdot \frac{N}{s}, \quad t \in [t_0, t_n] \\ &\Leftrightarrow \\ T(t) &= \frac{\beta}{\alpha - \beta} \cdot (t - t_0), \quad t \in [t_0, t_n] \end{aligned} \tag{4A.11}$$

In a similar manner, it can be shown that:

$$T(t) = \frac{\gamma}{\alpha + \gamma} \cdot (t_e - t), \quad t \in [t_n, t_e] \tag{4A.12}$$

Consecutively, using the definition of queue length Q(t) - see equation (4A.3) - and the definition of the departure rate r(t) - see equation (4A.4) - the explicit expression for the departure rate over the morning peak is given by:

$$r(t) = \frac{\alpha}{\alpha - \beta} \cdot s \ , \quad t \in \langle t_0, t_n \rangle$$
$$r(t) = \frac{\alpha}{\alpha + \gamma} \cdot s \ , \quad t \in \langle t_n, t_e \rangle \tag{4A.13}$$

From (4A.13) it can be seen that for early arrival the departure rate is constant and above the capacity of the bottleneck s; for late arrival the departure rate is constant and smaller than capacity, so that the queue dissipates linearly. The standard picture of the bottleneck model, showing time on the x-axis and cumulative departures and arrivals on the y-axis, is depicted in Figure 4A.1.

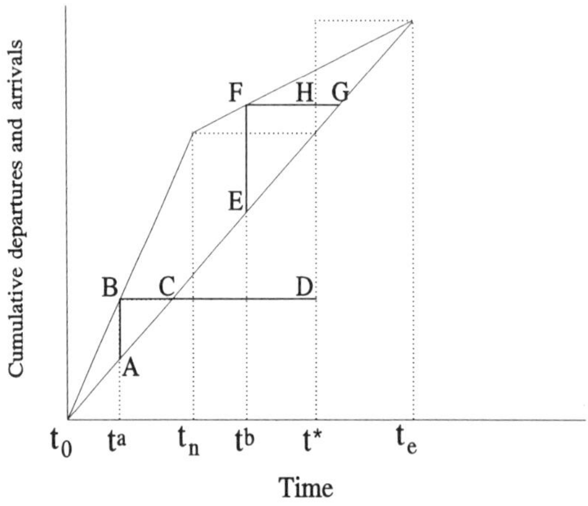

Figure 4A.1 Standard diagram of bottleneck model.

Note that in Figure 4A.1 the traveller departing at t_n arrives exactly at t^*. For a traveller arriving early - for example the traveller departing at t^a - travel time is equal to BC and schedule delay early equal to CD. The queue length experienced by this traveller is AB. For a traveller arriving late - for example the one departing at t^b - travel time is equal to FG, while schedule delay late equals HG. The queue length encountered by this traveller is given by the length EF.

Some characteristics of the bottleneck model

Five characteristics of the bottleneck model are worth mentioning:

- It can be shown that total schedule delay costs are equal to total travel time costs. In the standard textbook analysis of congestion one cannot explicitly distinguish between travel time costs and schedule delay costs. Both are contained in the generalised travel costs.
- In the bottleneck model queueing is purely a dead-weight loss. Consequently, in the system optimum there exists no queue.[2]
- The pattern of travel time departures from day to day is the same. The bottleneck model is not dynamic in the sense that it evolves over time. It is dynamic in the sense that departure time choices are endogenised.

[2]However, de Palma and Jehiel (1994) have shown that in some networks it is possible to have queues in the system optimum.

- Even if the number of travellers N is very small and the capacity of the bottleneck s very large, the user equilibrium departure time pattern will always result in a queue. On the one hand, this is a consequence of the remark that a queue should exist between t_0 and t_e. On the other hand, this is due to all travellers having the same desired arrival time t^* and all having the same free flow travel costs equal to zero. Hence, one can conclude that at user equilibrium *there is always a bottleneck in the bottleneck model.*
- The social feasibility of congestion-pricing is no problem in the bottleneck model as even *without* a redistribution of the revenues all commuters are equally well off in both the system optimum and the user equilibrium.[3] The costs of waiting in the user equilibrium are exactly equal to the optimal congestion-price that directs traffic flows to the system optimum.

Some differences between the bottleneck model and SNDUE

The model proposed in this Chapter lacks the novelty of the bottleneck model, that is, it lacks the endogenisation of departure time choices. In that respect, the model presented in this chapter is static, while the bottleneck model can be labelled as dynamic. However, the framework presented in this chapter, and taken forward in Chapters 5 to 10, saves algebraic complexity compared with the bottleneck model. Consequently, this model leaves room for analytical extensions into various directions; directions in which it is very difficult to analytically extend the bottleneck model. Realism will be added in the following four directions:

- two groups of travellers, informed and uninformed ones;
- stochasticity in link travel costs;
- endogenising demand for information;
- elastic demand.

Some of these extensions have also been dealt with in the bottleneck model, but not in a simultaneous fashion.

---

[3]This is clearly in conflict with the analysis in Chapter 3.

PART II

THEORETICAL MODELS

5 WELFARE ECONOMIC ANALYSIS OF MOTORIST INFORMATION IN A ONE-ROUTE NETWORK[1]

5.1 INTRODUCTION

In this chapter, a stochastic network deterministic user equilibrium model (SNDUE), as discussed in Chapter 4, is analysed. In this model the reason of the stochasticity relates to the capacity of the network, not to the perceptions of the individual drivers. The objective of this model is to enhance the insights into the welfare economic effects of information provision to a group of drivers, and to gain insight into the mechanisms affecting the impact of providing information. This is done by theoretically analysing the impact of information provision on network efficiency. In addition, the equity aspects of information provision are considered by answering questions such as "who benefits (or disbenefits) most from information?" and "do uninformed drivers also benefit?". In order to study these questions, the analysis is confined to the economic fundamentals of information provision and therefore limited to using simple economic equilibrium models rather than complex equilibrium assignment models. Although the latter ones can be extremely useful, particularly in the well-known four step transport model (Ortúzar and Willumsen, 1994), they are less appropriate for the objective in this chapter, as the complexity of these models may frustrate the objective of obtaining clear insight into the fundamentals of information provision.

The model presented here is a static SNDUE model, while allowing for elastic demand. An increasing cost curve represents the costs for travel (including congestion costs). The elasticity of demand and the stochastic cost functions distinguish the approach taken here from the existing literature on information provision. The equilibrium models reported on in the literature generally assume that demand is fixed or inelastic (Al-Deek and Kanafani, 1993; Arnott et al., 1990a, 1990b, 1991, 1992, 1994; Lindsey, 1995; Tsuji et al., 1985). The assumption of inelastic demand is a rather severe limitation of the analysis of information provision, because changes in usage due to changes in costs are then completely ignored. In the past, Ben-Akiva et al. (1986) studied an elastic demand version of Vickrey's (1969) bottleneck model, and so did Arnott et al. (1993). The present chapter is complementary to their work since it adds (1) information provision to drivers, and (2) stochasticity in terms of link travel costs.

In Section 5.2, the SNDUE one-link model is presented; in Section 5.3, the impact of information provision on equity and efficiency is studied; Section 5.4 concludes the chapter.

---

[1]This chapter is partly based on Emmerink, Verhoef, Nijkamp and Rietveld (1996a) published in *Journal of Transport Economics and Policy*.

5.2 ONE-LINK STOCHASTIC NETWORK DETERMINISTIC USER EQUILIBRIUM MODEL

In this section, the model is presented in its simplest form. In this form, the network is limited to one link only. Clearly, in such a simple network, potential beneficial route-split effects of information provision cannot be analysed, that is, beneficial effects of information provision owing to the fact that informed drivers change route in circumstances of incidents. In Chapters 6, 9, and 10 attention is focused on a simultaneous consideration of route-split and mode-split effects in relation to information. Mode-split effects however, can be addressed in the present context and these are precisely the ones that have thus far been widely ignored in the literature, due to the underlying assumption in most models that demand is inelastic.

The current section is structured in the following manner. In Section 5.2.1, the assumptions in the model are presented. Next, in Sections 5.2.2, 5.2.3, and 5.2.4, model P, model N, and model IMP are studied, respectively. In model P it is assumed that a certain (exogenously determined) group of travellers is provided with perfect information, while in model N no information is available. In model IMP, on the other hand, a particular group of travellers is provided with imperfect information. The properties of the three models are explored in Section 5.2.5.

5.2.1 Assumptions of the model

The demand-side is modelled using two groups of drivers, labelled group x and group y. The inverse demand function for x-travellers is denoted by D_x, and for y-travellers by D_y. These functions relate the number of road users in the network to total travel costs. Hence, for all levels of travel costs κ the inverse total demand function is, by definition, given by $D^{-1}(\kappa)=D_x^{-1}(\kappa)+D_y^{-1}(\kappa)$.

The supply-side of the system is modelled using link travel cost functions. It is assumed that the link travel cost function has either the functional form C^0 or C^1, depending on the state of the system. State 1 reflects *low capacity*, occurring with a probability p, while state 0 denotes *high capacity*, which occurs with a probability 1-p. The distinction between these states lies in the higher travel costs in state 1, that is, $C^1(N)\geq C^0(N)$ for all N, where N denotes the number of drivers using the one-link network. In addition, it is assumed that $dC^1(N)/dN\geq dC^0(N)/dN$. Consequently, both the travel costs itself and the rate at which travel costs increase are higher under state 1. This increase in travel costs is caused by random (unpredictable) incidents such as traffic accidents, sudden lane closures etc. It might be worth noting that according to measurements made in The Netherlands approximately 44% of congestion is related to unpredicted incidents (AVV, 1995).

The above introduced random cost component in combination with the elastic demand function render an analysis of the impacts of information provision relevant. To visualise the kind of information that is provided, one could best think of *pre-trip* information as the model is static. Then, without the provision of pre-trip information, it is assumed that uninformed drivers use the probability p to determine the *expected* cost function in the transport system, while informed drivers base their behaviour on the *actual* cost function as these are provided with the pre-trip information. *Consequently, an informed road user will use the network if private benefits are at least equal to actual private costs for the prevailing state of the transport system. An uninformed driver uses the network if private benefits are at least equal to expected private costs.* The SNDUE that is reached in this way conforms to Wardrop's first principle, the user equilibrium (Wardrop, 1952), as both may be characterized by individual maximizing behaviour.

It is important to stress that uninformed travellers are in fact assumed to be very familiar with average traffic conditions. They base their decision-making on *historic average traffic conditions*, whereas informed travellers make trip-decisions using *actual traffic conditions.*

In order to study the impact of information provision three models (model P, model N, and model IMP) are compared. In model P information is provided to the group of x-travellers, while no information is provided to y-travellers. In model N neither x nor y-travellers are supplied with information. In model IMP, x-travellers are provided with imperfect information, while no information is provided to y-travellers. Then, a comparison of these models allows to isolate the impact of information provision on both the informed (x) and uninformed (y) travellers. Furthermore, this gives the opportunity to study the effects of information on network performance.

5.2.2 Model P

In model P, x-travellers are provided with information, whereas y-travellers are not. Therefore, x-travellers base their behaviour on *actual* travel costs, whereas y-travellers use *expected* travel costs instead.

Transferring the verbally explained equilibrium conditions into mathematical expressions yields model P in expressions (5.1) to (5.3), where $N^0_{p,x}$ and $N^1_{p,x}$ denote the number of informed (x) drivers using the one-link network in state 0 and state 1, respectively, and $N_{p,y}$ the number of uninformed (y) drivers. Subscript p (referring to model P) denotes the equilibrium road usage values of model P. Expressions (5.1) and (5.2) ensure that the marginal informed (x) driver, that is, the informed driver who is indifferent between using the one-link network and an alternative (implying zero marginal net private benefits), equates marginal private costs and marginal private benefits for both state 0 and state 1. In a similar fashion expression (5.3) guarantees that the marginal uninformed (y) driver experiences zero expected marginal net private benefits. For the non-marginal drivers (expected) net private benefits are larger than zero, due to the downward sloping demand function. Finally, the additional condition that road usage is non-negative has to be imposed, that is, $N^0_{p,x}$, $N^1_{p,x}$ and $N_{p,y}$ have to be greater than or equal to zero.

$$D_x(N^0_{p,x}) \leq C^0(N^0_{p,x}+N_{p,y}) \ , \quad N^0_{p,x} \geq 0 \quad \text{and} \quad N^0_{p,x}\left(D_x(N^0_{p,x})-C^0(N^0_{p,x}+N_{p,y})\right)=0 \tag{5.1}$$

$$D_x(N^1_{p,x}) \leq C^1(N^1_{p,x}+N_{p,y}) \ , \quad N^1_{p,x} \geq 0 \quad \text{and} \quad N^1_{p,x}\left(D_x(N^1_{p,x})-C^1(N^1_{p,x}+N_{p,y})\right)=0 \tag{5.2}$$

$$\begin{aligned} &D_y(N_{p,y}) \leq (1-p)C^0(N^0_{p,x}+N_{p,y})+pC^1(N^1_{p,x}+N_{p,y}) \ , \quad N_{p,y} \geq 0 \quad \text{and} \\ &N_{p,y}\left(D_y(N_{p,y})-\left((1-p)C^0(N^0_{p,x}+N_{p,y})+pC^1(N^1_{p,x}+N_{p,y})\right)\right)=0 \end{aligned} \tag{5.3}$$

In the analysis below, it is assumed that the *group-regularity* condition applies for each group, that is, *for each state and each group of drivers the network will at least be marginally used.* For a one-link network this is a plausible assumption, since it seems likely that for each state at least some uninformed and informed drivers will use the network. Imposing this restriction implies that expressions (5.1) to (5.3) can be rewritten to equations (5.4) to (5.6).

$$D_x(N^0_{p,x})=C^0(N^0_{p,x}+N_{p,y}) \tag{5.4}$$

$$D_x(N^1_{p,x})=C^1(N^1_{p,x}+N_{p,y}) \tag{5.5}$$

$$D_y(N_{p,y})=(1-p)C^0(N^0_{p,x}+N_{p,y})+pC^1(N^1_{p,x}+N_{p,y}) \tag{5.6}$$

Figure 5.1 provides a diagrammatic representation for the situation where $C^j(N)$ (j=0,1) is linear and, in addition, the slope of the cost functions are identical.[2] In the left-hand panel of Figure 5.1, uninformed drivers (y-travellers) equate expected marginal link travel costs to their marginal benefits. In doing so, they take account of the effect that the expected number of informed drivers (x-travellers) will have on their costs. Given the assumptions on the cost functions, the expected number of uninformed drivers can be found by equating their demand to expected user cost. This leads to a total number of $N_{p,y}$ uninformed individuals using the network.

Next, informed drivers shift the prevailing cost curve C^j (j=0,1) with an amount $N_{p,y}$ to the left (see the dashed cost curves in Figure 5.1) to account for travel demand of uninformed road users. Then, for each state, the number of informed road users is found by equating demand with prevailing costs (as given by the dashed cost curve), leading to $N^0_{p,x}$ informed drivers using the network in state 0, and $N^1_{p,x}$ informed drivers in state 1. Under the assumptions on the cost functions C^0 and C^1, it follows that $N^1_{p,x}$

---

[2]A meaningful graphical illustration is impossible when these conditions (linear C^j (j=0,1) and the same slope) are not imposed on the cost functions. In such a situation the expected cost function (as shown in the left-hand panel of Figure 5.1) does not solely depend on $E(N_{p,x})$ but also on the exact values of $N^0_{p,x}$ and $N^1_{p,x}$.

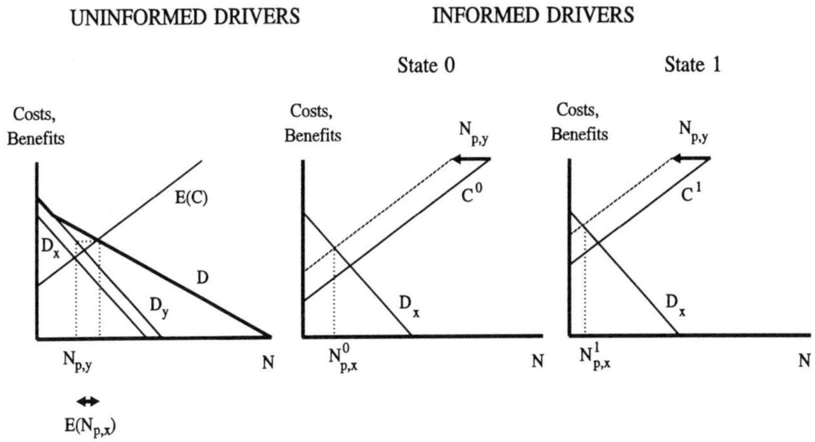

Figure 5.1 Graphical illustration of equilibrium model with informed and uninformed individuals.

is smaller than $N^0_{p,x}$: when low capacity prevails, some informed drivers will not use the network, but rather turn to the alternative mode (or stay at home).

It can be demonstrated that $N_{p,y}$, $N^0_{p,x}$, and $N^1_{p,x}$ are the only equilibrium values. In general, in user equilibrium models it can be shown that the equilibrium route travel flows are unique (Sheffi, 1985). Conversely, route flows are not unique. In the one-link network, however, a route is obviously equivalent to a link.

5.2.3 Model N

Next, the model in which no information is available to both x and y-travellers is investigated. This model is referred to as model N (no information). The equilibrium values of road usage of model N are studied in order to assess the welfare economic impacts of information provision. In model N, both x and y-travellers base their behaviour on *expected* travel costs. The equilibrium conditions of this model are given in (5.7) and (5.8).

$$D_x(N_{n,x}) \le (1-p)\cdot C^0(N_{n,x}+N_{n,y})+p\cdot C^1(N_{n,x}+N_{n,y}) \ , \quad N_{n,x}\ge 0 \ \text{ and} \\ N_{n,x}\left(D_x(N_{n,x})-\left((1-p)\cdot C^0(N_{n,x}+N_{n,y})+p\cdot C^1(N_{n,x}+N_{n,y})\right)\right)=0 \tag{5.7}$$

$$D_y(N_{n,y}) \le (1-p)\cdot C^0(N_{n,x}+N_{n,y})+p\cdot C^1(N_{n,x}+N_{n,y}) \ , \quad N_{n,y}\ge 0 \ \text{ and} \\ N_{n,y}\left(D_y(N_{n,y})-\left((1-p)\cdot C^0(N_{n,x}+N_{n,y})+p\cdot C^1(N_{n,x}+N_{n,y})\right)\right)=0 \tag{5.8}$$

The subscript n (referring to model N) is used to distinguish the equilibrium values of this model with the equilibrium values of the other models. Expression (5.7) shows that the x-travellers now also base their behaviour on the *expected* link travel costs, rather than on the *actual* link travel costs as in model P. By imposing the group-regularity condition, as defined in the previous section, these expressions can be simplified to two equalities.

5.2.4 Model IMP

In the current section a SNDUE model is presented in which part of the travellers is uninformed and the other part *imperfectly* informed. By considering the case of imperfect information it is acknowledged that any real-world motorist information system is likely unable to supply perfect information. Some sources affecting the quality of the information are: (1) measurement errors; (2) delays in transmitting the information; (3) a discrete (rather than continuous) updating frequency; (4) the format in which the information is presented etc. For more details, see Watling and van Vuren (1993). It is essential to understand in which direction and to what extent these kinds of (random) errors have an impact on the performance of the transport network.

In order to present the model in which drivers who have an interest in up-to-date information are supplied with imperfect information, the concept of *substate* is introduced. A substate can be seen as an "information state", involving the provision of imperfect information, which enables the informed drivers to adopt the initial probabilities p (of having low capacity) towards more likely levels p_j in both substates. In the model below it is assumed that there are two possible substates, denoted with 0 and 1, respectively. Substate 0 occurs with probability 1-q, while substate 1 takes place with probability q. In substate j (j=0,1), the link travel cost function C^1 prevails with probability p_j, while the link travel cost function C^0 is the relevant one with probability $1-p_j$. It will be assumed that the information system does not directly provide information on the prevailing state, but only on the substate. This allows informed users to make a better prediction of the actual state compared with the uninformed drivers.

Clearly, the following relationship between p (the probability that state 1 occurs), p_j (the probability that state 1 occurs conditioned on the occurrence of substate j) and q (the probability that substate 1 occurs) should hold:

$$P(state\,1) = p = p_0(1-q)+p_1 q = \sum_{j=0}^{1} P(state\,1 \mid substate\,j)P(substate\,j) \qquad (5.9)$$

The left-hand side of the above expression gives the probability that link travel cost function C^1 prevails, while the right-hand side denotes the same probability, but now calculated by conditioning on the occurrence of the substate.

If the above expression, representing unbiasedness of information provision, does not hold, then either informed or uninformed road users are basing their trip-making decisions on biased estimates of the traffic conditions. On economic grounds, an equilibrium founded on these assumptions cannot be viewed to represent a long-run stable situation. Moreover, in the literature on information provision in transport networks, it is regarded infeasible to deliberately supply faulty information, since it is impossible to consistently cheat (potential) road users. The empirical evidence in Bonsall and Joint (1991) and Bonsall and Parry (1991) suggests that drivers will not always follow the information that is provided by a motorist information system. Particularly, drivers that are familiar with the network are likely to ignore guidance that they do not perceive as being best for themselves, or which lacks credibility. As a consequence, the deliberate manipulation of drivers' decision-making to maximise system performance rather than individual benefits is unlikely to succeed in the long-run.

In the model it is assumed that informed drivers know with certainty which substate is prevailing (they are provided with imperfect day-specific traffic conditions), and in addition, know the relevant probabilities p_j (j=0,1), that is, the probability that a certain state occurs conditioned on the occurrence of the substate j. As before, uninformed drivers are familiar with average traffic conditions, and therefore know the probabilities p, p_j (j=0,1), and q, but are not aware of any day-specific traffic situation. Without loss of generality, it is assumed that p_0 is smaller than or equal to p_1, indicating that there is a greater probability of low capacity when substate 1 occurs.

Since informed drivers are informed on the relevant substate, they will base their behaviour on the expected link travel costs given the occurrence of a substate, while uninformed drivers will simply base their behaviour on expected link travel costs unconditioned on the occurrence of a substate. Hence, informed drivers need not use the probabilities q, but do use p_1 and p_2, while uninformed drivers merely use p, which will be split out into q, p_1, and p_2. The model that follows these principles is given in expressions (5.10) to (5.12).

$$D_x(N^0_{imp,x}) \le (1-p_0)C^0(N_{imp,y}+N^0_{imp,x})+p_0C^1(N_{imp,y}+N^0_{imp,x}) \;,\quad N^0_{imp,x}\ge 0$$
$$\text{and} \tag{5.10}$$
$$N^0_{imp,x}\Big(D_x(N^0_{imp,x})-\big((1-p_0)C^0(N_{imp,y}+N^0_{imp,x})+p_0C^1(N_{imp,y}+N^0_{imp,x})\big)\Big)=0$$

$$D_x(N^1_{imp,x}) \le (1-p_1)C^0(N_{imp,y}+N^1_{imp,x})+p_1C^1(N_{imp,y}+N^1_{imp,x}) \;,\quad N^1_{imp,x}\ge 0$$
$$\text{and} \tag{5.11}$$
$$N^1_{imp,x}\Big(D_x(N^1_{imp,x})-\big((1-p_1)C^0(N_{imp,y}+N^1_{imp,x})+p_1C^1(N_{imp,y}+N^1_{imp,x})\big)\Big)=0$$

$$D_y(N_{imp,y}) \le (1-q)\big((1-p_0)C^0(N_{imp,y}+N^0_{imp,x})+p_0C^1(N_{imp,y}+N^0_{imp,x})\big)+$$
$$q\big((1-p_1)C^0(N_{imp,y}+N^1_{imp,x})+p_1C^1(N_{imp,y}+N^1_{imp,x})\big) \;,\quad N_{imp,y}\ge 0$$
$$\text{and} \tag{5.12}$$
$$N_{imp,y}(D_y(N_{imp,y})-((1-q)((1-p_0)C^0(N_{imp,y}+N^0_{imp,x})+p_0C^1(N_{imp,y}+N^0_{imp,x}))+$$
$$q((1-p_1)C^0(N_{imp,y}+N^1_{imp,x})+p_1C^1(N_{imp,y}+N^1_{imp,x}))))=0$$

It can easily be seen that the above model collapses to the one with perfect information (model P) when p_0 is equal to 0 and p_1 equal to 1. In this situation a substate is identical to the overall state, and hence, the probability p is equal to the probability q. Conversely, when p_0 is equal to p_1, then the above model collapses to the one without information (and $p=p_0=p_1$). Knowing the prevailing substate is irrelevant in this situation.

5.2.5 Properties of models P, N, and IMP

In the present section, the properties of the models P, N, and IMP specified in Sections 5.2.2, 5.2.3, and 5.2.4 are explored. In order to keep the analysis manageable and the outcomes tractable linear demand and cost functions are assumed over the relevant ranges considered (that is, the ranges containing the levels of usage in each of the possible states and in each of the possible regulatory regimes). Although the use of linear functions may be criticized, they are in any case sufficient to serve the general goal of the current chapter, which is to enhance the insights into the welfare economic effects of information provision to a group of drivers. Furthermore, it might be interesting to note that for inelastic demand Arnott et al. (1992) have proven that the equilibrium travel cost functions in Vickrey's dynamic congestion model (Vickrey, 1969) of the morning rush hour with two groups and two parallel routes are special cases of the linear cost functions chosen here.

As model IMP turns out to be an intermediate case of models P and model N, attention is focused on models P and N. An analytical comparison of the model P (in which information is available to x-travellers, and no information is available to y-travellers) and the model N (in which no information is available to both x and y-travellers) leads to the following proposition for a system with linear demand and cost functions.

Proposition 5.1: Assuming linear demand (D_x, D_y) and cost (C^0, C^1) functions, group-regularity, and $C^1(N)\geq C^0(N)$ and $dC^1(N)/dN\geq dC^0(N)/dN$ for all relevant levels of road usage N, then the following relationships hold:

1. *expected road usage increases due to information*: $N_{n,x} + N_{n,y} \leq N_{p,y} + (1-p)N^0_{p,x} + pN^1_{p,x}$;
2. *road usage in state 0 increases due to information*: $N_{n,x} + N_{n,y} \leq N_{p,y} + N^0_{p,x}$, hence $C^0(N_{n,x}+N_{n,y}) \leq C^0(N_{p,y}+N^0_{p,x})$;
3. *road usage in state 1 decreases due to information*: $N_{n,x} + N_{n,y} \geq N_{p,y} + N^1_{p,x}$, hence $C^1(N_{n,x}+N_{n,y}) \geq C^1(N_{p,y}+N^1_{p,x})$;
4. *expected link travel costs decrease due to information*: $(1-p)C^0(N_{n,x}+N_{n,y}) + pC^1(N_{n,x}+N_{n,y}) \geq (1-p)C^0(N_{p,y}+N^0_{p,x}) + pC^1(N_{p,y}+N^1_{p,x})$;
5. *number of y-travellers increases due to information*: $N_{n,y} \leq N_{p,y}$;
6. *expected number of x-travellers increases due to information*: $N_{n,x} \leq (1-p)N^0_{p,x} + pN^1_{p,x}$;
7. *System welfare in model P $\geq$ system welfare in model N.*

Proof: Appendix 5.A

The Proposition has an interesting interpretation. It shows that information increases the expected road usage for both the drivers with information (x-travellers) and without information (y-travellers). However, at the same time the expected travel costs in the network will decrease. Hence, an increase in expected road usage is achieved while expected network travel costs havc decreased. Therefore, system welfare, measured as the sum of the individual benefits minus the sum of the individual costs, increases due to information. This result stems from the fact that - when provided with information - more road users will use the network when it is relatively cheap (state 0 occurs), while less informed drivers will use it when it is relatively expensive (state 1 occurs). Moreover, knowing that informed drivers behave in this fashion, more uninformed drivers will also find it profitable to use the network, because the informed drivers will relieve part of the congestion under high cost circumstances (state 1 occurs). However, it is important to note that equilibrium link travel costs in state 0 are higher under model P than under model N, which is a direct consequence of the second point in the Proposition. Nevertheless, the expected equilibrium link travel costs are smaller when information is provided.

The Proposition also shows the relevance of using elastic rather than fixed demand patterns. As shown in the Proposition, information does in fact alter the system performance even in a one-link network. Under fixed (inelastic) demand however, it is clear that information does not affect the performance of the system, since, independent of the prevailing link travel cost function, the same number of drivers will always use the network.

Even though the above results are appealing, the merits of information provision for government's policy purposes should be based on changes in social welfare rather than on some (derived) performance indicator as expected road usage or expected network travel costs. Social welfare, measured by the total system benefits minus the total (expected) system costs, is the most apt criterion on which to judge thé network's performance in terms of efficiency. Even though the Proposition shows that information provision increases system welfare in the model presented here, it is worthwhile to further analyse: (1) the relative size of this welfare improvement; (2) which travellers benefit most due to the information. The last issue, also referred to as an equity issue, is highly relevant when analysing the political feasibility of policy measures. Policies that have a strong effect on the current equity situation are likely to provoke resistance. In Section 5.3 these issues are addressed.

5.3 EFFICIENCY AND EQUITY ISSUES OF INFORMATION

The welfare economic aspects of information provision are now studied using the previously specified models P, N, and IMP. As before, most attention is paid to models P and N, as model IMP is the intermediate case. In the following, a link travel cost function C without superscript denotes the expected link travel costs, that is, $C=(1-p)C^0+pC^1$. By substituting the results obtained from Proposition 5.1 in the respective link travel costs functions, the following relationship for the equilibrium link travel costs is derived:

$$C_n^0 \leq C_{imp}^0 \leq C_p^0 \leq C_p \leq C_{imp} \leq C_n \leq C_p^1 \leq C_{imp}^1 \leq C_n^1 \qquad (5.13)$$

where these link travel cost functions obviously have to be evaluated at their relevant equilibrium levels of trip demand, for example, C_n^0 denotes $C_n^0(N_{n,y}+N_{n,x})$.

In Section 5.3.1, it is analysed which travellers benefit most from the information (the *equity* issue). In Section 5.3.2, the relative size of the welfare improvement due to information provision is studied (the *efficiency* issue).

5.3.1 Information and equity

Informed drivers (x-travellers)
The situation for the informed drivers is schematically depicted in Figure 5.2. In this figure, D_x gives the demand curve, while the horizontal lines denoted C give equilibrium values of costs, and hence should not mistakenly be seen as cost curves.

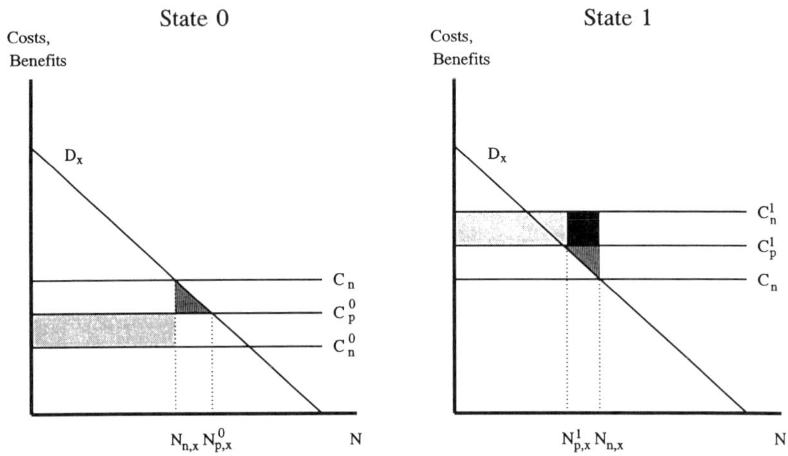

Figure 5.2 Welfare effects for x-travellers.

When state 0 occurs, then $N_{n,x}$ is less than or equal to $N_{p,x}^0$ and C_n^0 is less than or equal to C_p^0. This situation is depicted in the left-hand panel of Figure 5.2. The drivers on the left-hand side of $N_{n,x}$ will always use the network under state 0. With information provision, their link travel costs will be larger than in the absence of information. Hence, in state 0, these drivers suffer a cost disadvantage that is equal to the size of C_p^0 minus C_n^0, and is given by the shaded rectangle. It is interesting to point out that this cost disadvantage is an *increasing congestion externality*, since it is caused by the fact that *other* road users are informed. The size of this negative external effect decreases as less drivers have access to the information, since the difference between C_p^0 and C_n^0 will then decrease.

For the drivers between $N_{n,x}$ and $N_{p,x}^0$, information on the actual occurrence of state 0 induces them to change their behaviour. Without information they will not use the network because expected costs exceed their benefits, whereas they will use the network when they are provided with the information that low costs prevail. The size of the total welfare improvement for drivers between $N_{n,x}$ and $N^0_{p,x}$ is equal to $\frac{1}{2}(N^0_{p,x}-N_{n,x})(C_n-C^0_p)$ and is given by the shaded area in the left-hand panel of Figure 5.2. It is important to note that these welfare gains are *internal* in nature, since these arise from better decision-making by the informed drivers themselves. Therefore, these information benefits are called *internal decision-making benefits*. The size of the internal decision-making benefits decreases as more drivers are informed; with more informed drivers, the difference between C^0_p and C^0_n will increase, thereby (ceteris paribus) decreasing the difference between C_n and C^0_p. This negative effect for already informed drivers of equipping an additional driver is clearly external in nature. In this state, the marginally equipped driver will gain benefits from the information, while the information benefits for the already equipped drivers will dwindle. See also Chapter 2, where the same phenomenon was addressed.

If state 1 occurs, then $N_{n,x}$ is greater than or equal to $N^1_{p,x}$ and $C^1(N_{n,x}+N_{n,y})$ is greater than or equal to $C^1(N_{p,y}+N^1_{p,x})$. The situation is depicted in the right-hand panel of Figure 5.2. First, drivers on the left-hand side of $N^1_{p,x}$ will always use the network. Owing to the information provision, these will incur benefits equal to the difference in link travel costs C^1_n minus C^1_p. This cost advantage is a *decreasing congestion externality*, since it arises from the fact that *other* road users are provided with information and these are lowering link travel demand when state 1 occurs. In the right-hand panel of Figure 5.2 this external beneficial effect is shown by the large shaded rectangular area. Second, drivers between $N^1_{p,x}$ and $N_{n,x}$ will, knowing that state 1 occurs, change their travel decision and refrain from using the network. As a consequence, these drivers will benefit from a cost advantage equal to C^1_n minus C^1_p, and, in addition, from a decision-making advantage equal to the size of $\frac{1}{2}(N_{n,x}-N^1_{p,x})(C^1_p-C_n)$. Notice that the decision-making advantage is an *internal* effect, while the cost advantage is *external* in nature. The former arises from the fact that the driver himself is informed on the prevailing traffic condition, not from the fact that other drivers are informed. These two beneficial effects are illustrated in the right-hand panel of Figure 5.2 by the black (coloured) rectangular area (decreasing congestion externality) and the shaded triangular area (decision-making benefits).

To summarise, drivers on the left-hand side of $N^1_{p,x}$ (that is, drivers who always use the network independent of the occurring state) will suffer from an external cost disadvantage if state 0 occurs and an external cost advantage if state 1 occurs. Drivers between $N_{n,x}$ and $N^0_{p,x}$ benefit from an internal decision advantage if state 0 occurs. Drivers between $N^1_{p,x}$ and $N_{n,x}$ incur an external cost increase if state 0 occurs, and an external cost and internal decision-making advantage if state 1 prevails. Finally, drivers to the right of $N^0_{p,x}$ never use the network and are therefore indifferent between obtaining information or not.

When the equity aspects of information provision are considered in the model, one can derive that no informed individual is worse off due to information provision. Above, it was noticed that informed individuals on the left-hand side of $N^1_{p,x}$ are worse off in state 0 and better off in state 1. In terms of expected individual welfare (net private benefits), however, these drivers are at least as well off as without information, since C_n is larger than or equal to C_p as stated in Proposition 5.1. Therefore, $p(C^1_n-C^1_p)\geq(1-p)(C^0_p-C^0_n)$. Using the same argument, it follows that individuals between $N^1_{p,x}$

and $N_{n,x}$ are also better off, as they incur the same external cost advantage as drivers to the left-hand side of $N^1_{p,x}$ and in addition benefit from an internal decision-making advantage when state 1 occurs. Informed individuals between $N_{n,x}$ and $N^0_{p,x}$ are also individually better off as they gain when state 0 occurs and are indifferent when state 1 occurs. Finally, individuals to the right-hand side of $N^0_{p,x}$ never use the network and are therefore indifferent between obtaining information or not. *Therefore, in the model the provision of information will always lead to a welfare improvement for the group of informed drivers.* A typical individual (expected) welfare pattern (net private benefits) as generated by the model is shown in the left-hand panel of Figure 5.3. The shaded area under the bold curve shows the expected welfare gains due to perfect and imperfect information.

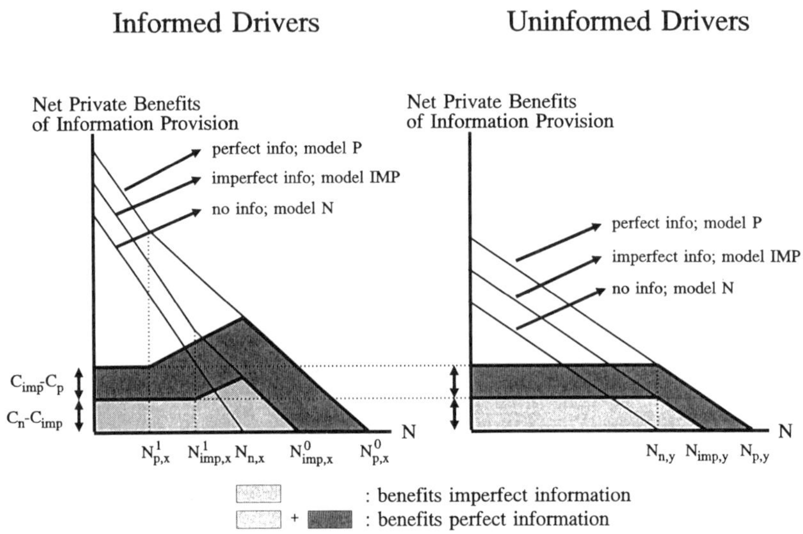

Figure 5.3 Expected net private benefits for x (informed) and y (uninformed) travellers.

The left-hand panel of Figure 5.3 indicates that individuals close to $N_{n,x}$ are best off due to information provision. This is an intuitively appealing result, since individuals close to $N_{n,x}$ are exactly those who doubt most whether or not to use the network. For these individuals, information provision will enhance their knowledge and will affect their travel decisions. On the other hand, individuals on the left-hand side of $N^1_{p,x}$ will never change their travel decisions regardless the kind of information provided. Thus it is clear that the information benefits for these drivers are external in nature, that is, owing to an improved network efficiency due to information provision to other individuals. Finally, individuals who never use the network have obviously nothing to gain (or lose) from information provision.

In practice, it is most likely that individuals with a low marginal utility of money are the first adopters of these types of technologies since in-vehicle navigation systems tend to be standard vehicle equipment in expensive vehicles. In that case, motorist information systems are viewed as luxury goods.

Uninformed drivers (y-travellers)

The situation for the y-travellers is schematically depicted in Figure 5.4. As before, the horizontal lines denote equilibrium values for costs. First of all, it is important to note that (following Proposition 5.1) $N_{n,y}$ is smaller than or equal to $N_{p,y}$, that is, the number of uninformed drivers will increase when information is provided to others. Then in state 0, the total benefits minus the total costs for uninformed individuals when no information is provided is shown by the polygon ABCD in the left-hand panel of Figure 5.4. When information is provided to informed individuals, then the total net benefits for uninformed drivers are given by the polygon AEFG. Hence, in state 0 the change in total welfare for uninformed drivers due to information provision to informed individuals is equal to the surface of the shaded polygon minus the surface of the shaded rectangle in the left-hand panel of Figure 5.4.

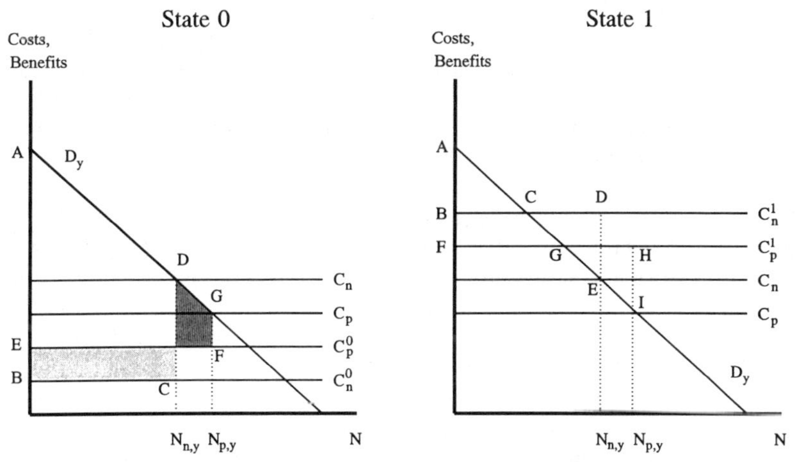

Figure 5.4 Welfare effects for y-travellers.

When the prevailing network condition is state 1, then changes in total welfare are as depicted in the right-hand panel of Figure 5.4. Total welfare of the uninformed drivers when no information is available is given by the area ABC minus the area CDE. When information is provided, total welfare changes to the area AFG minus area GHI.

As uninformed individuals between $N_{n,y}$ and $N_{p,y}$ decide to use the network when information is provided to informed individuals, they will experience individual expected benefits; if this were not the case, they would not decide to use the network in the first place. Also, the uninformed drivers on the left-hand side of $N_{n,y}$ benefit from the information provided to informed drivers. This is due to the fact that expected link travel costs decrease when information is provided (C_p is smaller than C_n; see Proposition 5.1). Therefore, *information provision to informed drivers will also lead to a welfare improvement for uninformed drivers*, see also Chapter 2. Clearly, these beneficial effects to uninformed drivers are external in nature; they are induced by behavioural responses from other (the informed) road users. A typical expected welfare pattern for uninformed drivers is shown in the right-hand panel of Figure 5.3.

Summary
In the model presented so far, information provision will lead to a *strict* Pareto improvement: both the informed and uninformed drivers are at least as well off. Furthermore, due to the provided information the expected level of road usage will increase, while the expected link travel costs will decrease. Finally, it is worth noting that these beneficial effects can in theory be reached by providing a very limited amount of drivers with information: only drivers between $N^1_{p,x}$ and $N^0_{p,x}$ have to receive information to obtain the results discussed, because these are the informed drivers who might change their travel behaviour due to the information on the prevailing link travel cost function. In practice, it is of course hard to identify this group. Although in a free market system, with perfect information on the costs and benefits of being provided with traffic information, this group of drivers would identify themselves. An attempt to endogenise information is undertaken in Chapter 7.

5.3.2 Information and efficiency

In the previous section the important result was obtained that, under certain conditions, information provision in a one-link network leads to a strict Pareto improvement. In the current section, the size of this efficiency improvement on the basis of some experiments is addressed. This will be done by comparing the effects on total (expected) welfare of the following three regimes:

1. information provision as studied in model P;
2. no information provision as studied in model N;
3. system optimal behaviour.

Under system optimal behaviour, the number of individuals using the network is derived in such a manner that total expected welfare, as measured by total system benefits minus total expected system costs, is maximised. It is well known that this can be implemented (in theory) by means of a fluctuating congestion-pricing scheme, see Chapter 3.

The effects of these three regimes on expected welfare are captured in the performance indicator ω (see Arnott et al., 1991; Verhoef, Nijkamp and Rietveld, 1995a), which in the present chapter indicates the relative welfare improvement of providing information to a group of drivers. The index ω is defined as:
Hence, ω gives the achievable welfare gains as a proportion of the theoretically possible welfare gains. Clearly, ω cannot exceed the value one. In addition, ω cannot

$$\omega = \frac{Welfare(model\ P) - Welfare(model\ N)}{Welfare(System\ Optimum) - Welfare(model\ N)} \tag{5.14}$$

be smaller than zero, since it was shown in the previous section that information provision leads to a strict Pareto improvement, implying that the numerator of expression (5.14) cannot take on negative values.

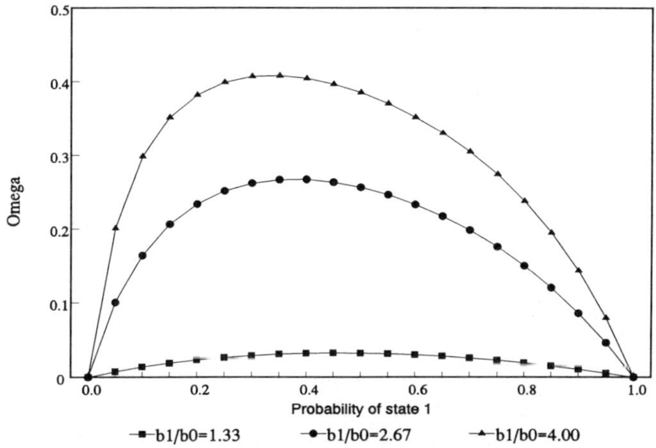

Figure 5.5 Impacts of probability of low capacity on ω.

The experiments have been conducted with linear inverse demand $D_j(N)=d_j-a_jN$ (j=x,y) and cost $C^m(N)=k^m+b^mN$ (m=0,1) functions. The base case parameters were set equal to $d_j=50$, $a_j=0.03$ (j=x,y), $k^m=20$ (m=0,1), $b^0=0.015$, $b^1=0.04$, and p=0.25.
Figure 5.5 shows the impact of changes in the probability of having low capacity on ω. Clearly, if there is complete certainty on the link travel cost function, then ω falls to zero (p=0 and p=1). For values in between, ω reaches a maximum of 0.4, depending on the size of the low capacity congestion cost parameter b^1. It should be noted that the probability for which ω takes on a maximum is dependent on b^1. In addition, the value of ω is surprisingly stable for large ranges of p. For example, when b^1 is equal to 0.04, then ω falls in between 0.2 and 0.3, for p-values in the interval ranging from 0.15 to 0.70.

Next, the impact of the elasticity of the demand functions on ω is analysed. To do so, the two demand curves were simultaneously *tilted* around the original intersection of model N, varying from high elasticities on the left-hand side to almost perfect inelasticity on the right-hand side. The reason for changing both a and d parameters (of the demand function) simultaneously is to avoid very (small) large levels of road usage when demand approaches complete (in)elasticity.

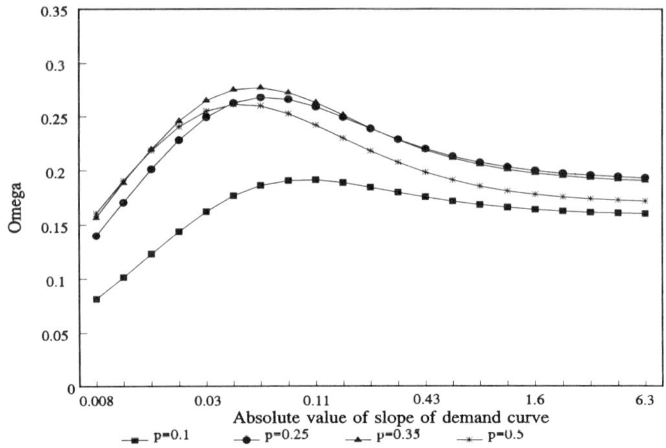

Figure 5.6 Impact of demand elasticity on ω.

In Figure 5.6, the x-axis follows a logarithmic scaling with base number 1.4. The results in Figure 5.6 indicate that as demand becomes less elastic, the welfare improving properties of information provision first increase and then decrease. Apparently, there exists some level of elasticity for which ω is maximised. Although with almost inelastic demand ω is still significantly different from zero, the available welfare improvement relative to the total welfare is in these circumstances very small; the equilibrium values for the three different models contained in the ω index are practically the same. It can be noted however, that with more routes available (and under certain conditions, see, for example, Chapter 10), ω may approach unity at inelastic demand due to the beneficial impact of information on route choice.

Finally, some experiments that focus on the impact of the group size on the relative efficiency indicator ω were conducted. In order to do so, the total demand curve was kept at the same position, while the respective demand curves of the informed and uninformed drivers were varied from a few to many informed drivers. For the base case parameters, and the total inverse demand curve equal to D(N)=50-0.015N, the results are depicted in Figure 5.7. The x-axis of this figure presents the percentage of informed drivers using the network; in the literature this is known as the level of market penetration.[3] Figure 5.7 indicates that the relative efficiency indicator ω increases as the expected number of informed drivers increases (hence, as the demand curve of the informed drivers rotates outwards). However, the concavity of the curve demonstrates that this increase takes place at a decreasing rate. Therefore,

[3]Observations below 10 per cent market penetration are not available, because these would violate the group-regularity condition; that is the network is not marginally used by both x and y-travellers.

from an efficiency point of view (taking the costs of implementing the technology into account), it might not be optimal to supply all drivers with the information.

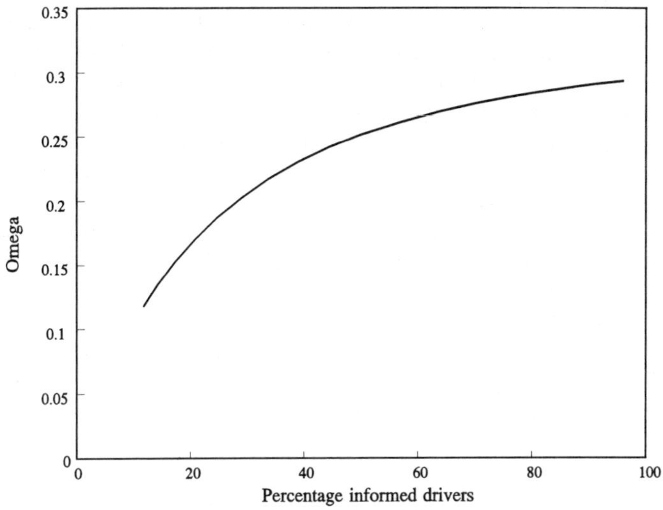

Figure 5.7 Impact of market penetration on ω.

The above observations have shown that the size of the welfare improving properties of information provision depend on a number of complex interactions between the probability of having low capacity (p), the impact of such an incident (b^1-b^0), the elasticity of demand (a_j, j=x,y), and the respective group sizes. The experiments suggest that for a linear system the maximum achievable welfare gains, expressed as a proportion of the theoretically possible efficiency gains, will most likely not exceed 0.4. For achieving larger values of ω, some form of congestion-pricing is inevitable. The combination of congestion-pricing and information is analysed in El Sanhouri (1994), de Palma and Lindsey (1994), and Chapters 7 and 10.

5.4 CONCLUSION

This chapter studied the welfare economic effects of information provision to a group of drivers. For that purpose an equilibrium model with elastic demand for road usage was used, the so-called SNDUE model. The model was introduced in a one-link network with stochastic capacity and two groups of (potential) users, informed and uninformed ones. Informed users base their decisions on *actual* prevailing traffic conditions, while uninformed drivers use *expected* traffic conditions.

It was found that information provision is welfare improving for both the informed and uninformed drivers. Hence, information leads in the model to a strict Pareto improvement. With information provision, the user equilibrium is nearer to the system optimum. Information provision will however, not close the gap between the two concepts. Even when all road users are well informed, the user equilibrium will still be different from the system optimum. Another interesting result of the analysis is that information increases expected road usage (generating more traffic), while at the same time decreasing expected link travel costs.

Furthermore, the analysis showed that many of the beneficial effects (and some of the adverse effects) of information provision are external in nature; they arise from changes in trip-making decisions by others. For example, the beneficial effects of information provision to uninformed drivers are clearly external in nature. They arise from behavioural adaptations by the informed drivers, rather than from changes in travel decisions by the uninformed road users. The existence of these external effects raises an interesting question about the government's role in introducing these technologies. It is well known that without proper government intervention, external effects distort the market mechanism, and result in an inefficient allocation of scarce resources. The issue of government intervention will be dealt with in Chapter 7.

In contrast to most of the literature on the impacts of information provision, the equilibrium model allows for elastic demand. In doing so, the important economic relationship between demand and supply is acknowledged. The results in this chapter indicate that the elasticity of demand is an important factor in determining the welfare improving properties of information provision. Information was found to be less useful at both low and high levels of demand elasticity.

The analysis in this chapter was confined to a one-link network with the provision of perfect and imperfect information to informed drivers. In the next chapter, the analysis is extended to a two-link network, which will provide the opportunity to simultaneously study mode and route-split effects of information provision.

APPENDIX 5.A PROOF OF PROPOSITION 5.1

Proposition 5.1: Assuming linear demand (D_x, D_y) and cost (C^0, C^1) functions, group-regularity, and $C^1(N) \geq C^0(N)$ and $dC^1(N)/dN \geq dC^0(N)/dN$ for all relevant levels of road usage N, then the following relationships hold:

- (1) *expected road usage increases due to information*:
  $N_{n,x} + N_{n,y} \leq N_{p,y} + (1\text{-}p)N^0_{p,x} + pN^1_{p,x}$;
- (2) *road usage in state 0 increases due to information*:
  $N_{n,x} + N_{n,y} \leq N_{p,y} + N^0_{p,x}$, hence $C^0(N_{n,x}+N_{n,y}) \leq C^0(N_{p,y}+N^0_{p,x})$;
- (3) *road usage in state 1 decreases due to information*:
  $N_{n,x} + N_{n,y} \geq N_{p,y} + N^1_{p,x}$, hence $C^1(N_{n,x}+N_{n,y}) \geq C^1(N_{p,y}+N^1_{p,x})$;
- (4) *expected link travel costs decrease due to information*:
  $(1\text{-}p)C^0(N_{n,x}+N_{n,y}) + pC^1(N_{n,x}+N_{n,y}) \geq (1\text{-}p)C^0(N_{p,y}+N^0_{p,x}) + pC^1(N_{p,y}+N^1_{p,x})$;
- (5) *number of y-travellers increases due to information*:
  $N_{n,y} \leq N_{p,y}$;
- (6) *expected number of x-travellers increases due to information*:
  $N_{n,x} \leq (1\text{-}p)N^0_{p,x} + pN^1_{p,x}$;
- (7) *System welfare in model P ≥ system welfare in model N.*

Proof:

First, we will prove that $N^0_{p,x} \geq N^1_{p,x}$. Since D_x is a decreasing function, and C^j (j=0,1) an increasing function, and since $C^0(n) \leq C^1(n)$ for all n the equilibrium conditions imply that $N^0_{p,x} \geq N^1_{p,x}$.

Next, we will show that $N_{n,x}+N_{n,y} \leq N_{p,y}+(1\text{-}p)N^0_{p,x}+pN^1_{p,x}=N_{p,y}+E(N_{p,x})$ [Proposition 1 (1)]. Using the equilibrium conditions and the linearity of D_x^{-1} it follows that:

$$\begin{gathered} N_{p,y} + (1-p)\cdot N^0_{p,x} + p\cdot N^1_{p,x} = \\ D_y^{-1}\big((1-p)\cdot C^0(N^0_{p,x}+N_{p,y}) + p\cdot C^1(N^1_{p,x}+N_{p,y})\big) + D_x^{-1}\big((1-p)\cdot C^0(N^0_{p,x}+N_{p,y}) + p\cdot C^1(N^1_{p,x}+N_{p,y})\big) \end{gathered} \tag{5A.1}$$

Next, the linearity of C^0 and C^1 is applied, thereby assuming that $C^i(n)$ has the following functional form: $C^i(n)=k^i+b^in$, i=0,1. Then, expression (5A.1) equals:

$$\begin{gathered} D_y^{-1}\big((1-p)\cdot C^0(E(N_{p,x})+N_{p,y}) + p\cdot C^1(E(N_{p,x})+N_{p,y}) + (1-p)b^0p(N^0_{p,x}-N^1_{p,x}) + pb^1(1-p)(N^1_{p,x}-N^0_{p,x})\big) + \\ D_x^{-1}\big((1-p)\cdot C^0(E(N_{p,x})+N_{p,y}) + p\cdot C^1(E(N_{p,x})+N_{p,y}) + (1-p)b^0p(N^0_{p,x}-N^1_{p,x}) + pb_1(1-p)(N^1_{p,x}-N^0_{p,x})\big) \end{gathered} \tag{5A.2}$$

We distinguish between two cases. First, consider $b^0=b^1$, then expression (5A.2) collapses to:

$$\begin{gathered} D_y^{-1}\big((1-p)\cdot C^0(E(N_{p,x})+N_{p,y}) + p\cdot C^1(E(N_{p,x})+N_{p,y})\big) + \\ D_x^{-1}\big((1-p)\cdot C^0(E(N_{p,x})+N_{p,y}) + p\cdot C^1(E(N_{p,x})+N_{p,y})\big) \end{gathered} \tag{5A.3}$$

Further, using the equilibrium conditions it follows that:

$$\begin{gathered} N_{n,y} + N_{n,x} = \\ D_y^{-1}\big((1-p)\cdot C^0(N_{n,x}+N_{n,y}) + p\cdot C^1(N_{n,x}+N_{n,y})\big) + D_x^{-1}\big((1-p)\cdot C^0(N_{n,x}+N_{n,y}) + p\cdot C^1(N_{n,x}+N_{n,y})\big) \end{gathered} \tag{5A.4}$$

Writing $N_{n,x}+N_{n,y}$ as s and $N_{p,y}+E(N_{p,x})$ as t, then the above equations state that:

$$s = D_y^{-1}\big((1-p)\cdot C^0(s) + p\cdot C^1(s)\big) + D_x^{-1}\big((1-p)\cdot C^0(s) + p\cdot C^1(s)\big) \tag{5A.5}$$

and

$$t = D_y^{-1}\big((1-p)\cdot C^0(t) + p\cdot C^1(t)\big) + D_x^{-1}\big((1-p)\cdot C^0(t) + p\cdot C^1(t)\big) \tag{5A.6}$$

Therefore, when b^0 is equal to b^1, then $N_{n,x}+N_{n,y}$ is equal to $N_{p,y}+(1-p)N^0_{p,x}+pN^1_{p,x}$.

Second, consider the situation that $b^0<b^1$, then:

$$(\Delta =)\ \ (1-p)b^0p(N^0_{p,x}-N^1_{p,x}) + pb^1(1-p)(N^1_{p,x}-N^0_{p,x}) < 0 \tag{5A.7}$$

since $N^0_{p,x}\geq N^1_{p,x}$. Again, using the previous notation of s and t it follows that:

$$s = D_y^{-1}\big((1-p)\cdot C^0(s) + p\cdot C^1(s)\big) + D_x^{-1}\big((1-p)\cdot C^0(s) + p\cdot C^1(s)\big) \tag{5A.8}$$

and

$$\begin{aligned} t = {} & D_y^{-1}\big((1-p)\cdot C^0(t) + p\cdot C^1(t) + \Delta\big) + D_x^{-1}\big((1-p)\cdot C^0(t) + p\cdot C^1(t) + \Delta\big) = \\ & D_y^{-1}\big((1-p)\cdot C^0(t) + p\cdot C^1(t)\big) + D_x^{-1}\big((1-p)\cdot C^0(t) + p\cdot C^1(t)\big) + \Delta_1 \end{aligned} \tag{5A.9}$$

for some positive Δ_1 (due to the fact that both D_y^{-1} and D_x^{-1} are decreasing functions). Again, using the decreasing property of both functions D_y^{-1} and D_x^{-1} the result that s is smaller than t follows.

The fact that inequality (5A.7) holds also implies that $(1-p)C^0(N_{n,x}+N_{n,y}) + pC^1(N_{n,x}+N_{n,y}) \geq (1-p)C^0(N_{p,y}+N^0_{p,x}) + pC^1(N_{p,y}+N^1_{p,x})$ [Proposition 1 (4)].

Next, we will show that $N_{n,y}\leq N_{p,y}$ [Proposition 1 (5)]. Because

$$N_{n,y} = D_y^{-1}\big((1-p)\cdot C^0(N_{n,x}+N_{n,y}) + p\cdot C^1(N_{n,x}+N_{n,y})\big) \tag{5A.10}$$

and

$$N_{p,y} = D_y^{-1}\big((1-p)\cdot C^0(N^0_{p,x}+N_{p,y}) + p\cdot C^1(N^1_{p,x}+N_{p,y})\big) \tag{5A.11}$$

and since $(1-p)C^0(N_{n,x}+N_{n,y}) + pC^1(N_{n,x}+N_{n,y}) \geq (1-p)C^0(N_{p,y}+N^0_{p,x}) + pC^1(N_{p,y}+N^1_{p,x})$, and D_y^{-1} is a decreasing function, the result follows.

Next, we will show that $N_{n,x}\leq E(N_{p,x})$ [Proposition 1 (6)]. First, notice that:

$$N_{n,x} = D_x^{-1}\big((1-p)\cdot C^0(N_{n,x}+N_{n,y}) + p\cdot C^1(N_{n,x}+N_{n,y})\big) \tag{5A.12}$$

Second, note that:

$$\begin{aligned} & (1-p)\cdot C^0(N^0_{p,x}+N_{p,y}) + p\cdot C^1(N^1_{p,x}+N_{p,y}) = \\ & (1-p)\cdot D_x(N^0_{p,x}) + p\cdot D_x(N^1_{p,x}) = D_x\big((1-p)N^0_{p,x} + pN^1_{p,x}\big) = \\ & D_x\big(E(N_{p,x})\big) \end{aligned} \tag{5A.13}$$

where the first equality follows from the equilibrium conditions, and the second from the linearity of D_x. Therefore,

$$E(N_{p,x}) = D_x^{-1}\left((1-p)\cdot C^0(N_{p,x}^0 + N_{p,y}) + p\cdot C^1(N_{p,x}^1 + N_{p,y})\right) \tag{5A.14}$$

The result follows from the fact that $(1\text{-}p)C^0(N_{n,x}+N_{n,y}) + pC^1(N_{n,x}+N_{n,y}) \geq (1\text{-}p)C^0(N_{p,y}+N_{p,x}^0) + pC^1(N_{p,y}+N_{p,x}^1)$, and that D_x^{-1} is a decreasing function.

Now, we will show that $N_{n,x} + N_{n,y} \leq N_{p,y} + N_{p,x}^0$ and $N_{n,x} + N_{n,y} \geq N_{p,y} + N_{p,x}^1$ [Proposition 1 (2) and (3)].

Since $N_{p,x}^0 \geq N_{p,x}^1$ it follows (using the equilibrium conditions, the fact that C^j (j=0,1) is an increasing function, and D_x^{-1} is a decreasing function) that:

$$\begin{gathered} N_{p,x}^0 + N_{p,y} = \\ D_x^{-1}\left(C^0(N_{p,x}^0 + N_{p,y})\right) + D_y^{-1}\left((1-p)\cdot C^0(N_{p,x}^0 + N_{p,y}) + p\cdot C^1(N_{p,x}^1 + N_{p,y})\right) \geq \\ D_x^{-1}\left((1-p)\cdot C^0(N_{p,x}^0 + N_{p,y}) + p\cdot C^1(N_{p,x}^1 + N_{p,y})\right) + D_y^{-1}\left((1-p)\cdot C^0(N_{p,x}^0 + N_{p,y}) + p\cdot C^1(N_{p,x}^1 + N_{p,y})\right) \end{gathered} \tag{5A.15}$$

The result $N_{n,x} + N_{n,y} \leq N_{p,y} + N_{p,x}^0$ follows by applying the equilibrium conditions for $N_{n,x}$ and $N_{n,y}$, and noting that $(1\text{-}p)C^0(N_{n,x}+N_{n,y}) + pC^1(N_{n,x}+N_{n,y}) \geq (1\text{-}p)C^0(N_{p,y}+N_{p,x}^0) + pC^1(N_{p,y}+N_{p,x}^1)$. In a similar fashion we can derive that $N_{n,x} + N_{n,y} \geq N_{p,y} + N_{p,x}^1$.

Using the linear forms of the cost and demand functions and the equilibrium conditions of model P and model N, respectively, it follows that:

$$E(\textit{Welfare Model P}) = \frac{1}{2}\cdot a_x \cdot E((N_{p,x})^2) + \frac{1}{2}\cdot a_y \cdot (N_{p,y})^2 \tag{5A.16}$$

$$E(\textit{Welfare Model N}) = \frac{1}{2}\cdot a_x \cdot (N_{n,x})^2 + \frac{1}{2}\cdot a_y \cdot (N_{n,y})^2 \tag{5A.17}$$

Then Proposition 5.1 (7) follows by applying Proposition 5.1 (5) and Proposition 5.1 (6).

6 WELFARE ECONOMIC ANALYSIS OF MOTORIST INFORMATION IN A TWO-ROUTE NETWORK[1]

6.1 INTRODUCTION

In this chapter, the analysis conducted in the previous chapter is advanced by extending the model to a two-link network. In this way, the analysis can be extended to study the impacts of mode-split and route-split effects simultaneously. The models presented are based on the SNDUE framework which was discussed in Chapter 4. Within this framework, the interactions between, on the one hand, the informed and uninformed drivers, and on the other, the drivers and the stochastic network conditions can be properly analysed.

The chapter is organised as follows. In Section 6.2, some definitions are provided that play an important role in the remainder of the chapter. Next, in Section 6.3, the model for two groups of drivers (informed and uninformed ones) and two parallel routes is studied. In Section 6.4 experiments with the two-link model are described and the implications of information provision in terms of expected travel costs and welfare are highlighted. Finally, Section 6.5 contains concluding remarks.

6.2 SOME DEFINITIONS

In this section, the terminologies of *network-regularity*, *route-regularity*, *group-network-regularity*, and *group-route-regularity* are introduced. These new concepts play an important role in explaining the impacts of information provision in a two-link network.

Network-regularity:	For each possible state, at least one route of the transport network is used.
Route-regularity:	For each possible state, at least one individual uses the specified route.
Group-network-regularity:	For each possible state, at least one individual from the specified group uses at least one route of the transport network.
Group-route-regularity:	For each possible state, at least one individual from the specified group uses the specified route.

It should be noted that the last two definitions combine group and network specific characteristics, while the first two definitions address the network situation. If the

[1]This chapter is based on Emmerink, Verhoef, Nijkamp and Rietveld (1996d) forthcoming in *European Economic Review*, and Emmerink (1995) and the comments by Sloof (1995) both published in *Tinbergen Institute Research Bulletin*.

network-regularity condition does not hold true, that is, if there exists a state in which no driver is inclined to use the network, then that particular state is unrealistic. In real-world situations it is unlikely that no individual is inclined to use the network. In fact, the same holds true if the route-regularity condition does not apply for a specified route. Then, that route is so unattractive that an unrealistic route has been included into the transport network.

It is straightforward to see that group-network-regularity does not necessarily imply group-route-regularity for all routes; however, group-route-regularity for at least one route, of course, implies group-network-regularity. Likewise, network-regularity does not imply route-regularity for all routes, although, route-regularity for at least one route implies network-regularity.

6.3 TWO-ROUTE STOCHASTIC NETWORK DETERMINISTIC USER EQUILIBRIUM MODEL

Two groups of drivers are considered, informed (x) and uninformed (y) ones, with respective inverse demand functions D_x and D_y, and a network consisting of two parallel routes connecting a single origin with a single destination. Furthermore, there are two by two average link travel cost functions giving the private costs of road use; for both routes one representing (so-called) high capacity, denoted by superscript 0; and another one representing the occurrence of low capacity, denoted by superscript 1: C_1^0, C_1^1, C_2^0, and C_2^1, where the subscript refers to the route.[2] State 1 implies relatively low capacity; hence the relationships:

$$C_r^0(N) \leq C_r^1(N) \quad \text{and} \quad \frac{\partial C_r^0(N)}{\partial N} \leq \frac{\partial C_r^1(N)}{\partial N} \quad \text{for } r=1,2 \tag{6.1}$$

hold for all feasible levels of road usage N. The probability that state 1 occurs on route r (reflecting low capacity) is denoted by p_r (r=1,2).

For notational ease the following probabilities for the four possible states s^{jk} (j=0,1; k=0,1) are introduced: $\rho^{00}=(1-p_1)(1-p_2)$ to denote the probability of state s^{00} with high capacity on both routes; $\rho^{10}=p_1(1-p_2)$ and $\rho^{01}=(1-p_1)p_2$ to denote the probability of low capacity on only route 1 (s^{10}) and route 2 (s^{01}), respectively; and $\rho^{11}=p_1p_2$ to denote the probability of simultaneous low capacity on both routes (s^{11}). The probabilities of high and low capacity on both routes are therefore assumed to be independent. Furthermore, attention is confined to the case that the information is perfect and concerns the prevailing state. This implies that informed drivers know which state s^{jk} is relevant, and can adapt their travel choice to this knowledge, while uninformed drivers only know the probabilities with which the four possible states take place, i.e they know ρ^{jk} (j=0,1; k=0,1). In addition, all individuals are assumed to know the functional forms of D_m (m=x,y) and C_r^j (r=1,2; j=0,1), that is, the common knowledge condition applies to the users of the system. The underlying rationale behind the above specified assumptions is that uninformed (potential) road users are familiar with historic average traffic conditions, while informed (potential) road users

[2]The terms link and route are interchangeable in the present context.

in addition obtain perfect information on the day-specific conditions. Finally, to avoid irrelevant links, the assumption of route-regularity is made for all routes to ensure that in each state at least a few individuals are using every route of the transport network.

The resulting equilibrium model is given in expressions (6.2) to (6.5), where $N^{jk}_{r,x}$ and $N_{r,y}$ denote the number of informed (x) and uninformed (y) drivers using route r (r=1,2) in state s^{jk} (j=0,1; k=0,1). Owing to information provision, the informed drivers will, unlike the uninformed ones, adapt their trip-making decision to the actual traffic situation, so that - depending on the prevailing state - a different number of informed individuals is using the network.

$$\begin{aligned} &D_x(N^{jk}_{1,x}+N^{jk}_{2,x}) \le C^j_1(N_{1,y}+N^{jk}_{1,x}) \;,\quad N^{jk}_{1,x}\ge 0 \;\text{ and} \\ &N^{jk}_{1,x}\left(D_x(N^{jk}_{1,x}+N^{jk}_{2,x})-C^j_1(N_{1,y}+N^{jk}_{1,x})\right)=0 \quad \text{for all } j=0,1 \text{ and } k=0,1 \end{aligned} \tag{6.2}$$

$$\begin{aligned} &D_y(N_{1,y}+N_{2,y}) \le \sum_{j=0}^{1}\sum_{k=0}^{1} \rho^{jk} C^j_1(N_{1,y}+N^{jk}_{1,x}) \;,\quad N_{1,y}\ge 0 \;\text{ and} \\ &N_{1,y}\left(D_y(N_{1,y}+N_{2,y})-\left(\sum_{j=0}^{1}\sum_{k=0}^{1} \rho^{jk} C^j_1(N_{1,y}+N^{jk}_{1,x})\right)\right)=0 \end{aligned} \tag{6.3}$$

$$\begin{aligned} &D_x(N^{jk}_{1,x}+N^{jk}_{2,x}) \le C^k_2(N_{2,y}+N^{jk}_{2,x}) \;,\quad N^{jk}_{2,x}\ge 0 \;\text{ and} \\ &N^{jk}_{2,x}\left(D_x(N^{jk}_{1,x}+N^{jk}_{2,x})-C^k_2(N_{2,y}+N^{jk}_{2,x})\right)=0 \quad \text{for all } j=0,1 \text{ and } k=0,1 \end{aligned} \tag{6.4}$$

$$\begin{aligned} &D_y(N_{1,y}+N_{2,y}) \le \sum_{j=0}^{1}\sum_{k=0}^{1} \rho^{jk} C^k_2(N_{2,y}+N^{jk}_{2,x}) \;,\quad N_{2,y}\ge 0 \;\text{ and} \\ &N_{2,y}\left(D_y(N_{1,y}+N_{2,y})-\left(\sum_{j=0}^{1}\sum_{k=0}^{1} \rho^{jk} C^k_2(N_{2,y}+N^{jk}_{2,x})\right)\right)=0 \end{aligned} \tag{6.5}$$

Since both expression (6.2) and (6.4) yield, in fact, four expressions, the present model consists of ten expressions and ten variables. Expressions (6.2) and (6.3) ensure that on route 1 private benefits do not fall short of (expected) private costs for each road user, while expressions (6.4) and (6.5) guarantee the same for route 2. Furthermore, when $N^{00}_{1,x}$ and $N^{00}_{2,x}$ are both positive, then expressions (6.2) and (6.4) for j=0 and k=0 guarantee that the link travel costs on route 1 and 2 are identical in state s^{00}, thereby satisfying Wardrop's first principle (the user equilibrium (Wardrop, 1952)) for informed drivers, stating that travel costs on the two competing routes should be equal when they are both in use. Moreover, when $N^{10}_{1,x}$ is zero while $N^{10}_{2,x}$ is positive, then expressions (6.2) and (6.4), for j=1 and k=0, show that $C^0_2(N_{2,y}+N^{10}_{2,x}) \le C^1_1(N_{1,y})$, thereby also satisfying the second condition of Wardrop's first principle for informed drivers that link travel costs can be different if one of the links is not used by informed drivers. Finally, expressions (6.3) and (6.5) guarantee Wardrop's first principle for the uninformed drivers, that is, the expected link travel costs are identical for both routes if (and only if) $N_{1,y}$ and $N_{2,y}$ are both positive; if expected link travel costs are not identical, then one of the two routes is not used by the uninformed drivers. To conclude, the model given in expressions (6.2) to (6.5) satisfies Wardrop's first principle (the user equilibrium) for both the informed and uninformed drivers.

6.3.1 Two-link model with group-route-regularity condition satisfied

To further analyse the model presented in the previous section, first assume that the group-route-regularity condition applies for both informed and uninformed drivers and for both routes, that is, $N_{r,y}>0$ and $N^{jk}_{r,x}>0$ (j=0,1; k=0,1; r=1,2). In this case Wardrop's first principle asserts that for each state s^{jk} (j=0,1; k=0,1), travel costs are identical on both routes. This, in turn, implies that the model presented above does not yield a unique solution. In order to verify this, notice that under these circumstances expression (6.3) directly implies that expression (6.5) is satisfied. Hence, the model consists of only nine independent expressions to solve for ten variables, namely $N^{jk}_{r,x}$ and $N_{r,y}$ (j=0,1; k=0,1; r=1,2). As a consequence, under the condition of group-route-regularity for both groups and both routes, an additional (independent) expression is needed to obtain a unique solution for the model.

Note that in static user equilibrium assignment models with fixed travel demand, depicted by the origin-destination matrix, the equilibrium link flows are uniquely determined (Sheffi, 1985). Hence, the introduction of two rather than one group of (potential) drivers in combination with the allowance for elastic demand complicates the model significantly.

The intuitive explanation for the occurrence of this phenomenon is that - under the assumption of group-route-regularity for both groups and both routes - uninformed drivers are indifferent between choosing route 1 or route 2. In fact, these two routes are perfect substitutes for the uninformed drivers, since for each possible state they yield identical link travel costs. Under the group-route-regularity assumption the informed road users ensure this equality of route travel costs. If route travel costs were not the same, while informed drivers are still using both routes, then the transportation system is not in equilibrium. In such a situation, some informed road users can reduce their travel costs by switching to the other (low travel cost) route.

In order to solve the model uniquely, the most logical additional independent expression is to require that half of the uninformed drivers choose route 1, while the other half chooses route 2. This can be justified by pointing out that for an uninformed driver there is no difference in terms of travel costs between the two competing routes, and therefore, he or she might as well choose route 1 with probability ½ and route 2 with probability ½. This would imply that on average half of the uninformed drivers chooses either route. In fact, there is no clear rationale why an uninformed driver should prefer one of the routes to the other when in each state the travel costs on both routes are exactly the same.[3]

---

[3]Sloof (1995) argues that another probability might be logical as well. Furthermore, he showed that Proposition 6.1 still holds with another probability, as long as the group-route-regularity condition is satisfied for both informed and uninformed drivers and for both routes,

An additional consequence of the group-route-regularity assumption is that informed individuals who will always use the transport network (that is, informed individuals on the left-hand side of the demand curve) will not be better off in terms of expected link travel costs than the uninformed drivers who use the network.[4] Moreover, informed drivers who do not always use the transport network (for example, when state s^{11} is prevailing) will only benefit from the information provided due to mode-split effects, since route-split effects do not exist when link travel costs are identical on both links. *Hence, under the condition of group-route-regularity for both groups and both routes, the welfare economic impacts of information provision in a two-link network can be analysed on a one-link network as well.* This case was discussed in depth in the previous chapter.

Using the results in Chapter 5, it should not be surprising that under the assumptions of group-route-regularity, linear demand, and linear link travel cost functions, information provision is welfare improving for both informed and uninformed drivers, and thus leads to a strict Pareto improvement. In addition, it can be shown that under these conditions information provision will lead to an increase in expected road usage by both informed and uninformed individuals, and a decrease in expected link travel costs.

Proposition 6.1: In a two-link network assuming linear inverse demand (D_x, D_y) and cost (C^0_r, C^1_r) functions, $C^0_r(N) \leq C^1_r(N)$, and $dC^0_r(N)/dN \leq dC^1_r(N)/dN$ ($r=1,2$), and assuming that the group-route-regularity condition holds for both groups and both routes, then the following results can be derived:

1. *expected road usage increases due to information;*
2. *expected link travel costs decrease due to information;*
3. *the number of uninformed road users in the network increases due to information;*
4. *the expected number of informed road users in the network increases due to information;*
5. *expected welfare increases due to information.*

Proof: Appendix 6.A1

Thus far, it was argued that the analysis of the welfare economic impacts of information provision in a two-link network can be conducted in a one-link network when the condition of group-route-regularity holds for both groups. It is therefore important to find out when the group-route-regularity condition holds, or in other words, which factors make it more likely for the group-route-regularity condition to hold.

---

and the equilibrium conditions (6.2) to (6.5) are met.

[4]In a dynamic model, informed drivers would only be better off during the phase leading to an equilibrium, see Al-Deek and Kanafani (1993).

Below, it will be shown that the number of informed drivers using the network is the crucial factor. The more informed drivers use the network, the more likely it is that the group-route-regularity condition holds. The argument leading to this conclusion is as follows. The first informed driver who uses the transport network will choose the route with minimum travel costs (say route 2). The second informed driver will choose the same route as travel costs on route 2 are still below those on route 1. By choosing route 2, these two informed drivers will increase the travel costs on route 2, and therefore, the cost difference between route 1 and route 2 will dwindle as more informed drivers are making use of the transport network. This process of informed drivers choosing route 2 will proceed until the number of informed drivers using the transport network is large enough to ensure that travel costs on route 2 are exactly the same as those on route 1. From then onwards, informed road users will ensure the equality between the link travel costs on both routes, and the group-route-regularity condition will be satisfied for both groups of users and both routes.

Another factor influencing the group-route-regularity condition is the size of the stochastic shock in the link travel cost function. The number of informed drivers that is necessary for travel costs on both routes to converge clearly depends on the initial cost discrepancy between route 1 and 2. The smaller the stochastic shock in the link travel cost curve, the less informed drivers are needed to ensure equality of travel costs on both routes, and the sooner the group-route-regularity condition will hold.

6.3.2 Two-link model without group-route-regularity condition satisfied

If the group-route-regularity condition does not hold true for the group of informed drivers, then a sufficient condition for information provision to be welfare improving is that $k_1^0=k_1^1=k_2^0=k_2^1$, where k_r^m denotes the fixed cost component of the travel cost function of link r under cost condition m (m=0: high capacity; m=1: low capacity). Hence, the additional assumption that free flow travel costs are state independent and equal for both routes is needed. A formal proof of this statement, summarised in Proposition 6.2, is given in Appendix 6.A2.

Proposition 6.2: In a two-link network assuming linear demand (D_x, D_y) and cost (C_r^0, C_r^1) functions, $C_r^0(N) \leq C_r^1(N)$, and $dC_r^0(N)/dN \leq dC_r^1(N)/dN$ (r=1,2), and assuming that $k_1^0=k_1^1=k_2^0=k_2^1$, then the following result can be derived:
expected welfare increases due to information.

Proof: Appendix 6.A2

The intuitive rationale behind this result is that an informed driver who switches to the (by informed drivers) unused route will increase expected travel costs in the network, and hence decrease welfare. Or in other words, with identical k-parameters for all states and all routes, private costs on a particular route A will exceed private costs on route B, if and only if, marginal social costs on route A exceed marginal social costs on route B. Since information provision will allow drivers to choose the individually optimal route in all prevailing states, it is welfare improving to provide information to as many as possible individuals. However, it is worth stressing that even with all drivers equipped, the user equilibrium (second-best solution) will not coincide with the system optimum (first-best solution) due to the existence of (external) congestion costs.

When the condition of identical k-parameters is not met, then it might be optimal (in terms of a second-best solution; the first-best solution cannot be reached without congestion-pricing) to equip a limited number of drivers with the information device. To see this, consider the example given in Figure 6.1 in which the marginal private and marginal social cost curve *for state* s^{01} are given. Here, MPC_r^m and MSC_r^m represent the marginal private and marginal social costs on route r with congestion level from state s^{01} (r=1,2).[5] Further, assume that the two competing routes are *identical in all respects*. When the inverse demand curve D_x holds for informed drivers, then these will only use route 1, since $MPC_2^1(0)$ is larger than $MPC_1^0(N_{1,x}^{0\,1})$. However, the marginal social cost curves show that $MSC_2^1(0)$ is smaller than $MSC_1^0(N_{1,x}^{0\,1})$. Hence, for the transport system as a whole it would be beneficial to switch an informed road user from route 1 to route 2. Clearly, this would increase the travel costs for the informed road user himself. In terms of informed and uninformed drivers the above argument implies that it would be welfare increasing to take the information device away from two informed drivers. Then, (on average) one of these would use route 2 rather than route 1 (since expected travel costs on identical routes are the same), which would increase system welfare.

Also, consider the inverse demand curve D_x^* for informed drivers. Up to this level of demand, information provision will be welfare increasing since $MSC_1^0(N)$ is smaller than $MSC_2^1(N)$ for N smaller than $N_{1,x}^*$. Finally, consider demand curve D_x^{**}. From this level of demand onwards, the group-route-regularity condition will be satisfied, since for road usage level $N_{1,x}^{**}$ the relation $MPC_1^0(N_{1,x}^{**})=MPC_2^1(0)$ holds true. Beyond $N_{1,x}^{**}$, it might still be beneficial to switch some informed drivers from the higher marginal social cost route to the lower one. This clearly stems from the fact that in a congested facility the second-best solution (user equilibrium, Wardrop's first principle) does not coincide with the first-best one (system optimum, Wardrop's second principle) with elastic demand. However, beyond $N_{1,x}^{**}$, it is welfare reducing to remove the information from all informed road users, as was formally proven in Proposition 6.1.

---

[5]In the graphical illustration of Figure 6.1 it is assumed that there are no uninformed drivers. A similar reasoning holds when there are uninformed drivers, however, a meaningful graphical illustration is rather cumbersome in this case.

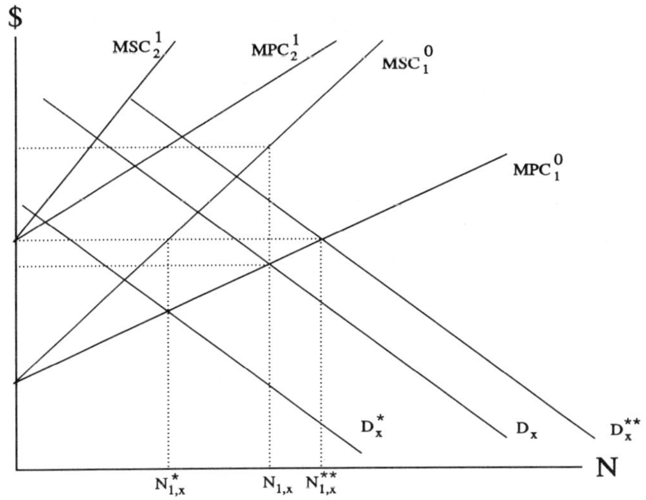

Figure 6.1 Example of an increase in welfare due to route switching by an informed road user.

6.4 MODEL EXPERIMENTS

In this section some experiments with the model presented in Section 6.3 are discussed. The experiments are conducted with linear inverse demand $D_j(N)=d_j-a_jN$ ($j=x,y$) and cost $C_r^m(N)=k_r^m+b_r^mN$ ($r=1,2$; $m=0,1$) functions. The base case parameters were set equal to $d_x=d_y=100$, $a_{sum}=0.015$ (where a_{sum} denotes the slope of the total inverse demand function, $a_{sum}=(a_ia_u)/(a_i+a_u)$), $k_1^0=k_1^1=k_2^0=k_2^1=20$, $b_1^0=b_2^0=0.015$, $b_1^1=b_2^1=0.04$, and $p_1=p_2=0.25$. Hence, it is assumed throughout the section that both links are identical and that the k-variables are all equal in all possible states, implying that information provision will always be welfare increasing, see Proposition 6.2.

Section 6.4.1 focuses on savings in expected travel costs due to the information provision, while Section 6.4.2 centres around the welfare implications of information in a two-link network.

6.4.1 Information provision and expected travel costs

Figure 6.2 and Figure 6.3 show the impact of the expected number of informed drivers on the expected travel costs. The x-axis shows the expected number of informed drivers using the two-link network (hence depending on the state probabilities p_1 and p_2) as a percentage of the total expected number of drivers using the network.

To create these figures, total demand for road usage (i.e., the horizontal summation of the inverse demand curves of the informed and uninformed potential road users) was kept constant, while the slope parameters of the respective inverse demand functions were varied in order to create different group sizes of informed and uninformed individuals, resulting in variations in the expected number of informed

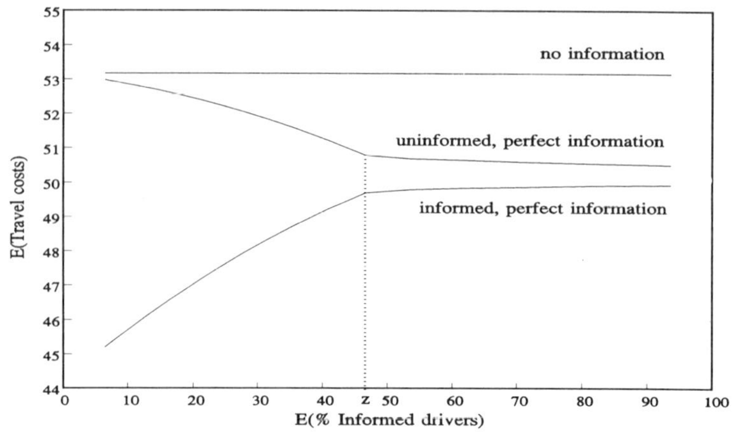

Figure 6.2 Expected travel costs as a function of the expected number of informed drivers.

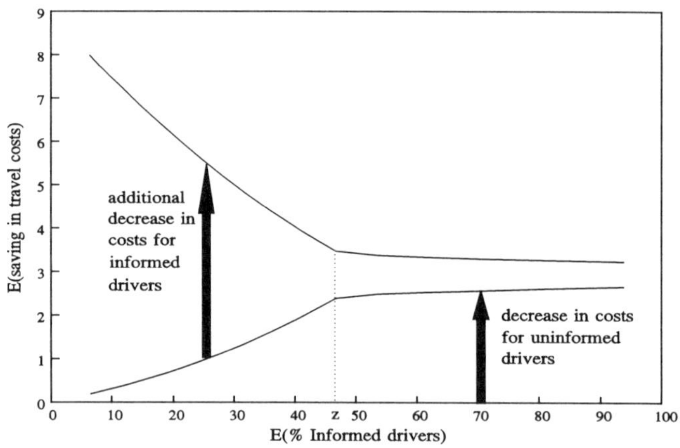

Figure 6.3 Expected saving in travel costs as a function of the expected number of informed drivers.

drivers using the network. Before interpreting the graphs it is important to note that the expected cost curve of informed drivers under perfect information is not defined at a zero level of market penetration. Clearly, the same holds true for the expected cost curve of uninformed drivers at 100 per cent market penetration. The shape of the figures is representative for the experiments conducted with the two-link model.

First, notice that the curves are kinked at a level of approximately 47 per cent of informed drivers (indicated in Figure 6.2 and Figure 6.3 by the point z).[6] On the right-hand side of z, the group-route-regularity condition applies to both groups of potential road users and both routes, while on the left-hand side of z this condition is violated for the group of informed drivers. In this case, informed drivers who are using the network will choose route 2 (route 1) if state s^{10} (s^{01}) prevails, and will use both routes if either state s^{00} or s^{11} prevails. The large benefits incurred to the informed drivers on the left-hand side of z are mainly caused by so-called route-split effects, that is, all informed drivers who are using the network choose the less congested (and therefore less costly) route, while the uninformed drivers split themselves equally between the two identical routes, since the *expected* route travel costs are the same on both identical routes. On the right-hand side of z, the group-route-regularity condition applies to both groups and both routes, which implies that link travel costs on both routes are in all states the same. Therefore, benefits to informed drivers are now caused by so-called mode-split effects, that is, some informed drivers decide not to use the network under high cost circumstances. Furthermore, Figure 6.2 shows that uninformed drivers are better off owing to the information provision to informed drivers. Their expected travel costs decrease quite substantially as the expected number of informed drivers increases.

Figure 6.3 is directly derived from Figure 6.2 by presenting Figure 6.2 upside down, and depicts the expected savings in travel costs corresponding to the experiments that generated Figure 6.2. A couple of points is worth noting. First, the size of the decrease in costs for uninformed drivers (indicated by the right arrow in Figure 6.3) - owing to the information provision to informed drivers - increases as more informed drivers are using the network. In Chapter 2, this phenomenon was already mentioned and referred to as a positive external effect to non-equipped drivers: the more informed drivers are using the network the lower the expected travel costs for uninformed drivers. Second, it is interesting to pay some attention to the difference between the expected travel cost curve of the informed and uninformed drivers (indicated by the left arrow in Figure 6.3). Given a certain expected number of informed drivers, the size of this difference represents the expected travel cost savings obtained from buying the motorist information system. These are referred to as *information benefits to equipped drivers*; see also Chapters 2, 11 and 12. The *information benefits to equipped drivers* curve is downward sloping, which means that a marginal equipped driver adversely affects the already equipped drivers. Hence, the marginal equipped driver generates a negative external effect to already equipped drivers, but since the information benefits are positive, it might still be beneficial for the marginal driver himself to buy the equipment.

To conclude, at low levels of market penetration, route-split effects will dominate, that is, benefits to informed drivers are mainly due to better route choice decision-making. At larger levels of market penetration, the route-split effects will disappear, while the mode-split effects will take over. Hence, at these levels *mode guidance* rather than *route guidance* is more important. Finally, the higher the level

---

[6]Obviously, the position of the kink is fully dependent on the parameters used.

of market penetration, the smaller the discrepancy between expected user costs for informed and uninformed drivers, and hence, the lower the expected net benefits of being equipped.

6.4.2 Information provision and welfare effects

The welfare implications of information provision in a two-link network will now be studied by using a performance indicator ω, which measures the relative welfare improvement of providing information to a group of drivers. The index ω, see also Chapter 5, is defined as:

$$\omega = \frac{Welfare(Information) - Welfare(No\ Information)}{Welfare(Optimum) - Welfare(No\ Information)} \tag{6.6}$$

The index ω gives the achievable welfare gains as a proportion of the theoretically possible welfare gains. Clearly, ω cannot exceed the value of one. In addition, under the base case parameters, ω cannot be smaller than zero, since the numerator of ω will always be greater than or equal to zero under the assumptions made. To prove this, two cases are distinguished between. First, if the group-route-regularity condition holds true for all groups and all routes, then it was shown in Proposition 6.1 that information provision will lead to a strict Pareto improvement, and hence the numerator of ω is positive. Second, if the group-route-regularity condition does not hold true for the group of informed drivers, then the numerator of ω will be positive as well under the sufficient condition that $k_1^0=k_1^1=k_2^0=k_2^1=20$, as was shown in Proposition 6.2.

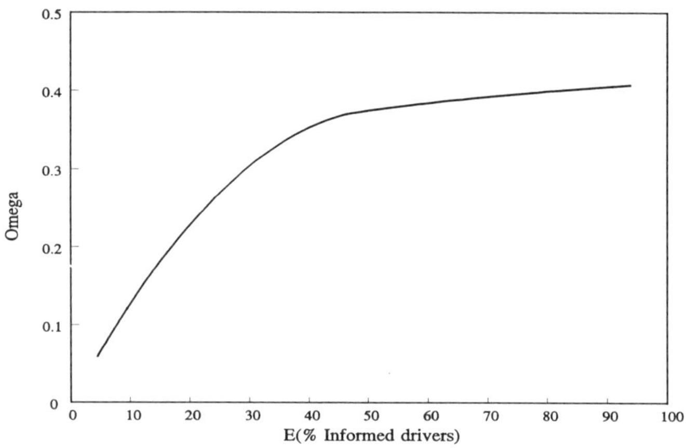

Figure 6.4 Relative welfare improvement as a function of the expected number of informed drivers.

Figure 6.4 depicts the relative welfare improvement performance indicator ω as a function of the expected number of informed drivers using the network, using the base case parameters. Four points are worth mentioning. In the first instance, ω increases as the expected number of informed drivers increases, and hence the more drivers have access to the information, the better the system performance. As was shown previously, this result depends on the assumption of identical k-parameters. Second, the curve in Figure 6.4 is concave, implying that the efficiency improvements of providing an additional driver with information increase at a decreasing rate. Third, the curve becomes much flatter when the group-route-regularity condition is satisfied. Finally, experiments with different b_r^1-values (r=1,2) have shown that the larger the shock, the greater the efficiency improving potential of information provision.

6.5 CONCLUSION

In this chapter the welfare economic effects of information provision to a group of drivers under stochastic cost functions in a two-link network have been studied. An economic equilibrium framework as proposed in Chapter 4 was used with elastic demand for road usage to account for the interaction between road usage and travel costs. Two groups of drivers were considered, informed and uninformed ones.

It was shown that if: (1) the level of market penetration of the information provision device is not too small, and; (2) the stochastic shock in the link travel cost function is not too extreme, then the two-link network can be modelled as a one-link network. In that case, beneficial route-split effects do not exist; information benefits arise from so-called mode-split effects. However, when these two conditions are not satisfied, then the information benefits to equipped drivers are substantially larger due to the existence of route-split benefits, that is, all informed drivers choose the less congested route. Finally, the analytical model showed that it is not necessarily always optimal to provide as many drivers as possible with information.

The next chapter further extends the model by addressing the issue of endogenous demand for information. The number of informed drivers in the network is then determined endogenously, that is, within the model, rather than exogenously, that is, outside the model.

APPENDIX 6.A1 PROOF OF PROPOSITION 6.1

Proposition 6.1: In a two-link network assuming linear inverse demand (D_x, D_y) and cost (C_r^0, C_r^1) functions, $C_r^0(N) \leq C_r^1(N)$, and $dC_r^0(N)/dN \leq dC_r^1(N)/dN$ (r=1,2), and assuming that the group-route-regularity condition holds for both groups and both routes, then the following results can be derived:

- *expected road usage increases due to information*;
- *expected link travel costs decrease due to information*;
- *the number of uninformed road users in the network increases due to information*;
- *the expected number of informed road users in the network increases due to information*;
- *expected welfare increases due to information.*

Proof:

Appendix 6.A1 consists of two parts. In the first part, it will be proven that expected road usage and expected welfare on a two-link network - satisfying the group-route-regularity condition for both groups and both routes, and with linear demand and cost functions - does not decrease (and generally increases) with perfect information provision to all drivers (public information) compared with no information provision at all. In the second part the same is shown to hold (both for the informed and for the uninformed drivers) when perfect information is only provided to a subset of drivers (club information).

Public information

Consider a two-link network with the following two stochastic link travel cost functions (subscripts denote routes; superscripts denote states), $C_1^j(N_1)=k_1^j+b_1^jN_1$ (j=0,1) and $C_2^k(N_2)=k_2^k+b_2^kN_2$ (k=0,1). The (single) demand curve is given by $D(N_1+N_2)=d-a(N_1+N_2)$.

For a group-route-regular network (with one group this is obviously the same as a route-regular network) with perfect information in each state $C_1^j=C_2^k$ and without information $\bar{C}_1=\bar{C}_2$ (bars denote expectations), this model is identical to a one-link four-state model with the following cost functions (where cost parameters in capitals denote the equivalent one-link four-state model):

$$C^{jk}(N) = \frac{b_1^j \cdot k_2^k + b_2^k \cdot k_1^j}{b_1^j + b_2^k} + \frac{b_1^j \cdot b_2^k}{b_1^j + b_2^k} \cdot N = K^{jk} + B^{jk} \cdot N \tag{6A.1}$$

Expected user costs without information are then given by:

$$\bar{C}(N) = \sum_{j=0}^{1} \sum_{k=0}^{1} \rho^{jk} \cdot C^{jk}(N) = \bar{K} + \bar{B} \cdot N \tag{6A.2}$$

Denote expected total usage with perfect information $\bar{N}_p$, and usage without information N_n. The following expression for N_p can be derived:

$$\bar{N}_p = N_p^{00} - \rho^{10}\cdot(N_p^{00}-N_p^{10}) - \rho^{01}\cdot(N_p^{00}-N_p^{01}) - \rho^{11}\cdot(N_p^{00}-N_p^{11})$$
$$= N_p^{00} - \sum_{\substack{j=0\\ \{j,k\}\neq\{0,0\}}}^{1}\sum_{k=0}^{1} \rho^{jk}\cdot\frac{C^{jk}(N_p^{00}) - C^{00}(N_p^{00})}{a+B^{jk}} \tag{6A.3}$$

where $C^{jk}(N_p^{00})$ denotes user costs in state s^{jk} evaluated at the perfect information state s^{00} usage level. It can also be shown that:

$$N_n = N_p^{00} - \sum_{\substack{j=0\\ \{j,k\}\neq\{0,0\}}}^{1}\sum_{k=0}^{1} \rho^{jk}\cdot\frac{C^{jk}(N_p^{00}) - C^{00}(N_p^{00})}{a+\bar{B}} \tag{6A.4}$$

Hence:

$$\bar{N}_p - N_n = \sum_{\substack{j=0\\ \{j,k\}\neq\{0,0\}}}^{1}\sum_{k=0}^{1} \rho^{jk}\cdot\delta^{jk}\cdot\left[\frac{1}{a+\bar{B}} - \frac{1}{a+B^{jk}}\right] \tag{6A.5}$$

with $\delta^{jk}=C^{jk}(N_p^{00})-C^{00}(N_p^{00})$.
Note that:

$$\frac{1}{a+\bar{B}} - \frac{1}{a+B^{jk}} = \frac{B^{jk}-\bar{B}}{(a+B^{jk})\cdot(a+\bar{B})} \tag{6A.6}$$

and that:

$$\begin{aligned} B^{10}-\bar{B} &= \rho^{00}\cdot(B^{10}-B^{00}) + \rho^{01}\cdot(B^{10}-B^{01}) + \rho^{11}\cdot(B^{10}-B^{11}) \\ B^{01}-\bar{B} &= \rho^{00}\cdot(B^{01}-B^{00}) + \rho^{10}\cdot(B^{01}-B^{10}) + \rho^{11}\cdot(B^{01}-B^{11}) \\ B^{11}-\bar{B} &= \rho^{00}\cdot(B^{11}-B^{00}) + \rho^{10}\cdot(B^{11}-B^{10}) + \rho^{01}\cdot(B^{11}-B^{01}) \end{aligned} \tag{6A.7}$$

After substituting (6A.7) into (6A.6), and then (6A.6) into (6A.5), and rearranging terms, one finds:

$$\begin{aligned} \bar{N}_p - N_n = \frac{1}{a+\bar{B}}\cdot\Bigg\{&\rho^{11}\cdot\rho^{00}\cdot(B^{11}-B^{00})\cdot\frac{\delta^{11}}{a+B^{11}} + \rho^{10}\cdot\rho^{00}\cdot(B^{10}-B^{00})\cdot\frac{\delta^{10}}{a+B^{10}} \\ &+\rho^{01}\cdot\rho^{00}\cdot(B^{01}-B^{00})\cdot\frac{\delta^{01}}{a+B^{01}} + \rho^{10}\cdot\rho^{11}\cdot(B^{11}-B^{10})\cdot\left[\frac{\delta^{11}}{a+B^{11}} - \frac{\delta^{10}}{a+B^{10}}\right] \\ &+\rho^{01}\cdot\rho^{11}\cdot(B^{11}-B^{01})\cdot\left[\frac{\delta^{11}}{a+B^{11}} - \frac{\delta^{01}}{a+B^{01}}\right] + \rho^{10}\cdot\rho^{01}\cdot(B^{10}-B^{01})\cdot\left[\frac{\delta^{10}}{a+B^{10}} - \frac{\delta^{01}}{a+B^{01}}\right]\Bigg\} \end{aligned} \tag{6A.8}$$

This can finally be rewritten as:

$$\begin{aligned}\bar{N}_p - N_n = \frac{1}{a+\bar{B}}\cdot\Big\{&\rho^{11}\cdot\rho^{00}\cdot(B^{11}-B^{00})\cdot(N_p^{00}-N_p^{11}) + \rho^{10}\cdot\rho^{00}\cdot(B^{10}-B^{00})\cdot(N_p^{00}-N_p^{10})\\ &+\rho^{01}\cdot\rho^{00}\cdot(B^{01}-B^{00})\cdot(N_p^{00}-N_p^{01}) + \rho^{10}\cdot\rho^{11}\cdot(B^{11}-B^{10})\cdot(N_p^{10}-N_p^{11})\\ &+\rho^{01}\cdot\rho^{11}\cdot(B^{11}-B^{01})\cdot(N_p^{01}-N_p^{11}) + \rho^{10}\cdot\rho^{01}\cdot(B^{10}-B^{01})\cdot(N_p^{01}-N_p^{10})\Big\} \geq 0\end{aligned} \tag{6A.9}$$

Six terms can be distinguished between the large parentheses, all of which, except perhaps the last one, are greater than (or equal to) zero. The last term between the large parentheses may be negative; however, since $\rho^{11}\cdot\rho^{00}=\rho^{10}\cdot\rho^{01}$, it can never (in absolute value) exceed the first term. Therefore, expression (6A.9) cannot be smaller than zero, and will exceed zero if there is uncertainty in congestion and if demand is not completely inelastic.

Next, using a similar methodology as in Appendix 5.A, one can show that expected welfare is directly related to the squared expectation of road usage. Therefore, it is then proven that information provision to all drivers is never welfare reducing and generally welfare improving.

Club information

In this section, it will be proven that expected usage also increases (or does at least not decrease) when perfect information is provided to only a subset of drivers (denoted group x), while group y remains behaving on the basis of no information. The proof will be set up in reverse order: it will be proven that expected usage will generally decline when perfect information is taken away from group x. A sub-model M will be used, where it is assumed that the demand of group y is completely inelastic, and remains at the usage consistent with the perfect information provision to group x. A group-route-regular network is considered.

In model M, where group y's D_y is perfectly inelastic, the proof in the previous section is then sufficient to secure that, in the non-trivial case where D_x is not completely inelastic, expected usage will decline when removing perfect information: $M_x \leq N_{p,x}$. The model is then identical to the one discussed before; be it that for group x now a vertical axis at D_y should be considered. Given the inelastic demand of group y, $M_y=N_{p,y}$, and hence overall expected usage will decline as well: $M_x+M_y \leq N_{p,x}+N_{p,y}$. Furthermore, expected user cost will increase when removing information. With perfect information for group x, expected user cost for group x is equal to $c^p=\rho^{00}\cdot D_x(N_{p,x}^{00})+\rho^{10}\cdot D_x(N_{p,x}^{10})+\rho^{01}\cdot D_x(N_{p,x}^{01})+\rho^{11}\cdot D_x(N_{p,x}^{11})$. Because of the assumed linearity of the demand function, this may be rewritten as $c^p=D_x(\bar{N}_{p,x})$. When removing information and keeping usage by group y fixed, expected user costs for group x increase to $c^M=D_x(M_x)$. Since $M_x \leq N_{p,x}$ and D_x is negatively sloped, it follows that $c^M \geq c^p$. Given the complete group-route-regularity of the network, group y faces the same expected user costs as does group x; hence, also for group y expected user costs increase from c^p to c^M when removing information.

Next, turn to model n (model with no information), where group y is allowed to have elastic demand. It will be shown that overall expected usage will decline even further in comparison to model M if group y's demand is indeed not completely inelastic. To do so, consider Figure 6A.1. From the left origin, group y's usage is measured, while group x's usage is measured from the right origin. Also from the left origin measured, curve $\bar{C}$ depicts expected user costs. Both groups are now uninformed, and therefore base their behaviour on $\bar{C}$. By horizontal substraction of D_x from $\bar{C}$, the curve S_y is obtained. This curve gives expected user costs for group y given the equilibrium behaviour of group x, who will enter the network up to the point where marginal benefits D_x are equal to expected marginal private cost as given by $\bar{C}$.

Now if D_y is indeed inelastic as assumed in model M and as depicted by D_y^M, without perfect information provision to group x equilibrium levels of usage are given by M_x and $M_y=N_{p,y}$, and the proof is complete. However, if D_y is not completely inelastic, as depicted by curve D_y, an equilibrium as given by $N_{n,y}$ and $N_{n,x}$ will arise, where for both groups marginal benefits are equal to expected marginal private costs c^n. In this case, it is known that at M_y, the equality $D_y(M_y)=c^p$ holds (otherwise, M_y would not be consistent with perfect information provision to group x). Furthermore, it is known that $S_y(M_y)=c^M$, as the equilibrium in sub-model M is derived under the assumption of an inelastic D_y.

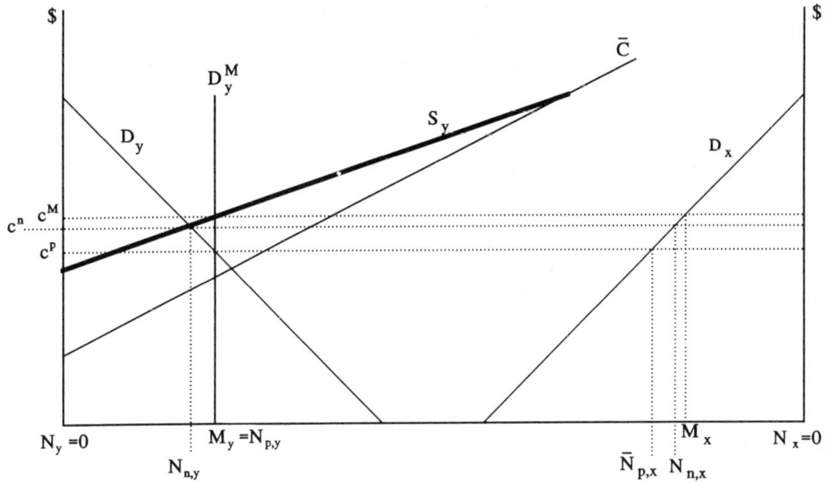

Figure 6A.1 Expected usage declines both for group y and for group x when perfect information is taken away from group x.

Hence, with elastic D_y, $N_{n,y}<M_y$, since $M_y-N_{n,y}=(c^M-c^p)/(a_y+s_y)$ (where a_y is the absolute value of the slope of D_y and s_y is the slope of S_y); and (c^M-c^p), a_y and s_y are all greater than zero. On the other hand, the inequality $N_{n,x}>M_x$ holds since $c^n<c^M$. Still, it is easy to show that $M_y+M_x>N_{n,y}+N_{n,x}$. To do so, notice that:

$$N_{n,y} - M_y = -\frac{c^M - c^n}{s^y} = -\frac{c^M - c^n}{a_x \cdot \dfrac{\bar{c}}{a_x + \bar{c}}} \qquad (6A.10)$$

(where $\bar{c}$ gives the slope of $\bar{C}$, and a_x gives the slope of D_x), and that:

$$N_{n,x} - M_x = \frac{c^M - c^n}{a_x} \qquad (6A.11)$$

As the denominator in (6A.11) is larger than the denominator in (6A.10), it is found that $N_{n,x}+N_{n,y} \le M_x+M_y \le N_{p,x}+N_{p,y}$: also with elastic D_y, expected usage is smaller without information than it is with perfect information provision to group x.

Finally, since $c^p \le c^n$, $N_{n,y} \le N_{p,y}$ and $N_{n,x} \le N_{p,x}$ and using the relationship between expected level of road usage and welfare (see Appendix 5.A) also expected welfare does not increase and generally decreases when perfect information is taken away from group x.

APPENDIX 6.A2 PROOF OF PROPOSITION 6.2

Proposition 6.2: In a two-link network assuming linear demand (D_x, D_y) and cost (C^0_r, C^1_r) functions, $C^0_r(N) \leq C^1_r(N)$, and $dC^0_r(N)/dN \leq dC^1_r(N)/dN$ (r=1,2), and assuming that $k^0_1=k^1_1=k^0_2=k^1_2$, then the following result can be derived:

- *expected welfare increases due to information.*

Proof:

The proposition will be proven by considering what happens to expected link travel costs if one informed road user switches to the route not used by informed road users. Assume that in a particular state s^{jk} the group-route-regularity condition does not hold for the group of informed drivers and route 1. Without loss of generality, it is assumed that j=1 and k=0. This implies that in state s^{10} informed drivers are using route 2 rather than route 1. This can only be the case when $C^0_2(N_{2,y}+N^{10}_{2,x})<C^1_1(N_{1,y})$. Total travel costs incurred by the road users in state s^{10} can be given by:

$$C^0_2(N_{2,y}+N^{10}_{2,x})\cdot(N_{2,y}+N^{10}_{2,x}) + C^1_1(N_{1,y})\cdot N_{1,y} \tag{6A.12}$$

Now consider what happens to the total travel costs in state s^{10} if one informed road user switches from route 2 to route 1. Then total travel costs are given by:

$$C^0_2(N_{2,y}+N^{10}_{2,x}-1)\cdot(N_{2,y}+N^{10}_{2,x}-1) + C^1_1(N_{1,y}+1)\cdot(N_{1,y}+1) \tag{6A.13}$$

After some rearranging and using the functional forms of the link travel cost functions C, the difference of expression (6A.12) and (6A.13) can be written as:

$$\left(b^0_2\cdot(N_{2,y}+N^{10}_{2,x}-1) - b^1_1\cdot N_{1,y}\right) + \left((k^0_2+b^0_2\cdot(N_{2,y}+N^{10}_{2,x})) - (k^1_1+b^1_1\cdot(N_{1,y}+1))\right) \tag{6A.14}$$

Expression (6A.14) has an intuitive explanation. The first term indicates the total increase in congestion costs to the other road users due to route switching by one informed driver. The second term represents the change in travel costs for the informed road users that switches route. Since in the non group-route-regular equilibrium all informed road users are travelling on link 2 it follows that the second term is negative. Hence, if it can be shown that the first term is negative as well, then route switching by one informed driver increases total travel costs in state s^{10}, and hence decreases expected welfare. Using the assumption of non group-route-regularity, it follows that travel costs on route 1 exceed those on route 2:

$$\begin{gathered} k^0_2 + b^0_2\cdot(N_{2,y}+N^{10}_{2,x}) < k^1_1 + b^1_1\cdot N_{1,y} \\ \Leftrightarrow \\ k^0_2 + b^0_2\cdot(N_{2,y}+N^{10}_{2,x}-1) < k^1_1 + b^1_1\cdot N_{1,y} \\ \Leftrightarrow \\ b^0_2\cdot(N_{2,y}+N^{10}_{2,x}-1) - b^1_1\cdot N_{1,y} < k^1_1 - k^0_2 \end{gathered} \tag{6A.15}$$

Hence, under the condition that $k^0_1=k^1_1=k^0_2=k^1_2$, the right-hand side of the last expression is equal to zero. Therefore, switching some informed drivers to the uninformed category will decrease welfare. Since the group-route-regularity will still not hold, this process can be repeated until all informed drivers have been switched to the uninformed category. Therefore, it follows that - under the condition that $k^0_1=k^1_1=k^0_2=k^1_2$ - providing information will increase welfare in the non group-route regular case.

7 ENDOGENOUS DEMAND FOR MOTORIST INFORMATION[1]

7.1 INTRODUCTION

In Chapters 5 and 6, it was assumed that the number of informed and uninformed drivers is *exogenously* determined. It is more realistic to assume that there are costs associated with motorist information. Costs, which may, for instance, reflect the price of the necessary information technology equipment. In this manner, the choice of being informed is modelled *endogenously* in the present chapter. Hence, an actor in the present chapter does not only decide on whether or not to use the transport network, but also decides upon whether or not to buy the information on the traffic situation. Clearly, an actor will buy the information only if the private benefits of being informed at least exceed the private costs of doing so.

The model will also be used to examine the efficiency improving properties of two types of government regulation. First, the possible implications of subsidising the costs of information for social welfare are considered. The idea is that owing to the external benefits generated by the information to uninformed actors, it may be attractive for the government (or an infrastructure authority) to subsidise information, see Chapters 2 and 5, and it will be analysed under which conditions it is socially desirable for the government to do so. Next, the link between fine congestion-pricing and endogenous provision of information is investigated. Fine (in the sense of fluctuating) congestion-pricing will yield its first-best characteristic only if the users of the system are perfectly aware of the prevailing fine congestion toll. Without perfect information on actual levels of congestion and tolls, users would base their behaviour on expected costs rather than on actual costs. Therefore, it seems logical to consider the efficiency of first-best congestion tolling in combination with the endogenous provision of perfect information. With respect to the second point, work by El Sanhouri (1994) and part of Chapter 10 is taken further here. In Chapter 10 it will be assumed that information is available for free, both for all actors and for the government.

The structure of the chapter is as follows. In Section 7.2, the model is presented. In Section 7.3, the above mentioned regulatory issues are treated. Section 7.4 contains the conclusions.

7.2 SNDUE MODEL WITH ENDOGENOUS DEMAND FOR INFORMATION

Thus far, it was assumed that the decision of being informed is *exogenous*, that is, this decision is not determined within the model. In this section a model is presented in

---

[1]This chapter is based on Emmerink, Verhoef, Nijkamp and Rietveld (1996b) published in *Annals of Regional Science*.

which the choice of being informed is *endogenous*, and dependent on the *internal* benefits that an informed individual derives from the information.

7.2.1 The model

In Chapter 5, Section 5.3, it was shown that there are two kinds of information benefits, *internal* and *external* ones. In the model to be presented here after, it is assumed that an individual decides to be informed when the *internal* benefits of information exceed the private costs of being informed. It is rational to consider the internal benefits only, because for an arbitrary individual the *external* benefits are (by definition) independent of whether or not that particular individual is himself informed. These external benefits (and costs) are caused by the fact that *other* road users are informed on the actual traffic situation. This observation implies that only drivers between N_p^1 and N_p^0 are potentially interested in being informed, since these are the ones that may incur internal benefits from the information.[2] Drivers at the left-hand side of N_p^1 are not interested, since they will use the road network regardless the prevailing state, whereas drivers to the right-hand side of N_p^0 are not interested since they will use the network in neither state.

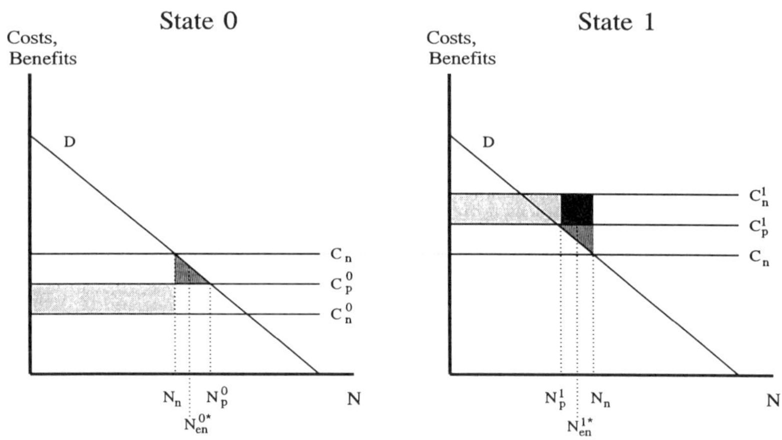

Figure 7.1 Welfare effects of information.

Now consider driver N_{en}^{0*} in the left-hand panel of Figure 7.1, where the subscript *en* (endogenous) refers to the model in which the demand for information is endogenous. If, for this driver, the internal benefits of being supplied with information exceed the

[2]The subscripts x and y can be dropped as the present model does not use *exogenously* determined group sizes.

costs, then the same holds for all drivers between N_n and N_{en}^{0*}, since for these drivers internal benefits are larger than those of driver N_{en}^{0*}. A similar reasoning holds for driver N_{en}^{1*} in the right-hand panel of Figure 7.1. If it is beneficial for this particular driver to be informed, then this is the case for all the drivers between N_{en}^{1*} and N_n. Therefore, if the marginally informed driver (the driver who is indifferent between being informed or not) on the left-hand side of N_n is denoted by N_{en}^{1}, and the marginally informed driver on the right-hand side of N_n with N_{en}^{0}, then in equilibrium N_{en}^{0}-N_{en}^{1} drivers are informed.

The resulting network situation in the model with endogenous demand can now be characterised as follows. First, the informed drivers (those between N_{en}^{1} and N_{en}^{0}) will only use the network when state 0 is prevailing, and will thus fully benefit from their *internal* decision-making benefits. Second, drivers on the left-hand side of N_{en}^{1} will not buy the motorist information equipment, and will always use the transport network. This can easily be seen by noting that they would do the same in case no information was available for equipped drivers. Finally, drivers to the right-hand side of N_{en}^{0} will not buy the motorist information equipment and will use the network in neither state.

In order to determine N_{en}^{0} and N_{en}^{1}, the price of information is of course needed. Due to the static equilibrium nature of the model, the price of information π to be considered below is short of any time dimension. In other words, whereas one would intuitively think of π as an individual investment, the internal benefits of which were to be reaped during a subsequent (large) number of travel decisions, such reasoning is not in the spirit of static equilibrium analysis. Hence, for the translation of the present model into more practical terms, one should either interpret π as the daily equivalent of some purchase price Π (for example leasing, where π reflects daily interest and depreciation), or one could consider π as the real purchase price and take D and C to be some discounted measures of the future stream of benefits and costs of road usage.

The mathematical formulation of the model is presented below. Since, by definition, the marginally informed driver is indifferent between being informed or not, the following two relationships should hold:

$$(1-p)\left(D(N_{en}^{0})-C^{0}(N_{en}^{0})\right) = \pi \tag{7.1}$$

$$p\left(C^{1}(N_{en}^{1})-D(N_{en}^{1})\right) = \pi \tag{7.2}$$

where π denotes the private costs of information. The term in the large parentheses in expression (7.1) gives the internal decision-making benefits for the marginally informed driver N_{en}^{0} in state 0. Hence, the *expected* internal decision-making benefits are given by multiplying this term with the probability that state 0 occurs. The left-hand side of expression (7.2) gives the expected internal decision-making benefits for a marginally informed driver N_{en}^{1}. Obviously, the expected internal decision-making benefits of the marginally informed driver should equal the private costs of information π.

Also, note that for both N^0_{en} and N^1_{en}, expressions (7.1) and (7.2) guarantee that the expected net private benefits of the marginally *uninformed* driver and marginally *informed* driver are the same, and equal to 0 for N^0_{en}, and equal to $(1\text{-}p)(D(N^1_{en})\text{-}D(N^0_{en}))$ for N^1_{en}. Therefore, both the marginally informed driver N^0_{en} and N^1_{en} are indifferent in buying the information.

To derive the properties of the model discussed above, it will be assumed that both the demand and cost curves are linear functions over the relevant ranges considered, that is, $D(N)=d\text{-}aN$ and $C^j(N)=k^j+b^jN$ for $j=0,1$. Then solving for N^0_{en} and N^1_{en} in expression (7.1) and (7.2) yields:

$$N^0_{en} = \frac{d-k^0}{a+b^0} - \frac{\pi}{(1-p)(a+b^0)} = N^0_p - \frac{\pi}{(1-p)(a+b^0)} \tag{7.3}$$

$$N^1_{en} = \frac{d-k^1}{a+b^1} + \frac{\pi}{p(a+b^1)} = N^1_p + \frac{\pi}{p(a+b^1)} \tag{7.4}$$

where π is in the interval $[0\ ,\ \pi_{max}]$ to ensure that N^1_{en} is smaller than N^0_{en}; for a derivation of π_{max}, see Appendix 7.A1. The first term on the right-hand side of expression (7.3) ((7.4)) gives the number of informed drivers using the network in state 0 (state 1), in case the information were provided for free, as in Chapter 5. The second term on the right-hand side of these expressions captures the effect of the price of information.

The proposition below presents some of the properties of the model with endogenous demand for information compared with the model in which no information is available.

Proposition 7.1: In a one-link network with endogenous demand for information, and assuming linear inverse demand (D) and cost (C^0, C^1) functions, $C^0(N) \leq C^1(N)$, and $dC^0(N)/dN \leq dC^1(N)/dN$, the following relationships hold:

1. *expected road usage is higher with information than without*;
2. *expected link travel costs are smaller with information than without*;
3. *expected welfare is higher with information than without.*

Proof: Appendix 7.A1

The net private benefits of *endogenous* provision of information compared with the situation in which no information is available are depicted in Figure 7.2. Drivers on the left-hand side of N^1_{en} benefit from an *external* decrease in congestion costs of the size C_n minus C_{en}, where C_m ($m=n$, en) denotes the expected link travel costs in model m. Drivers between N^1_{en} and N_n benefit, in addition, from *internal* decision-making benefits owing to the purchased information. Finally, drivers between N_n and N^0_{en} benefit from *internal* decision-making benefits solely.

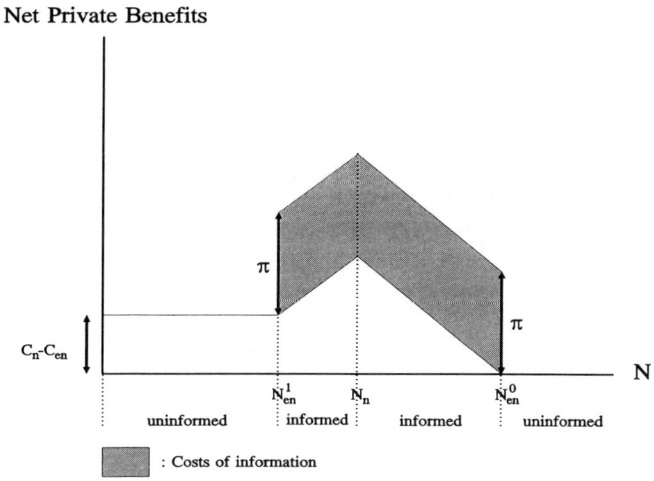

Figure 7.2 Difference in expected net private benefits of the model with endogenous demand for information and the model in which no information is available.

Using the equilibrium levels of road usage as given by expressions (7.3) and (7.4) when the private costs of being provided with information are equal to π, the demand curve for information can be derived. Given a price of information of π, N^0_{en} minus N^1_{en} drivers would like to be supplied with information. Therefore, the relationship between the number of informed drivers and the private costs of information is given by:

$$D^{-1}(\pi) = N_p^0 - N_p^1 - \pi\left(\frac{1}{(1-p)(a+b^0)} + \frac{1}{p(a+b^1)}\right) \tag{7.5}$$

The term in large parentheses multiplied by the costs of being equipped with a motorist information device π denotes the number of drivers for whom the costs of information exceed the internal benefits of being informed. The term N^0_p minus N^1_p reflects the number of drivers who are interested in being informed when the information is provided for free.

7.2.2 Welfare effects

Next, the welfare properties of the model with endogenous demand for information are explored. In Proposition 7.1 it was already shown that the *availability* of information will lead to an increase in social welfare, defined by total system benefits minus total system costs. The manner in which this is accomplished is graphically shown in Figure 7.3.

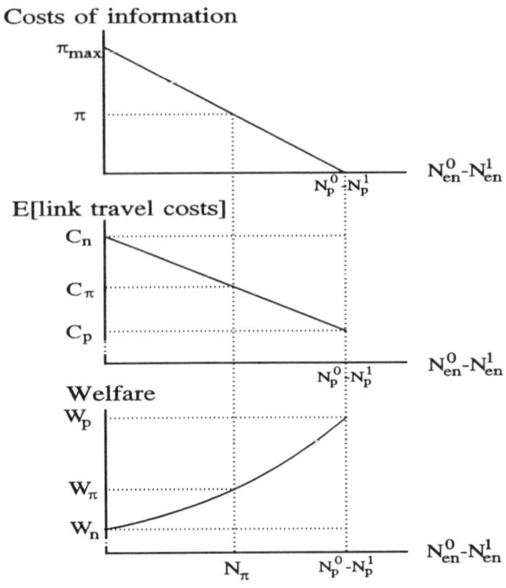

Figure 7.3 Welfare effects of endogenous information.

Figure 7.3 depicts the relationship between the private costs of information, the number of informed drivers, the expected link travel costs, and expected welfare; see Appendix 7.A2 for a derivation of the curves in Figure 7.3. For example, a price of information equal to π leads to N_π informed drivers; expected link travel costs will then be equal to C_π; and expected welfare equal to W_π. The upper panel of Figure 7.3 shows the linear demand function for information as given by expression (7.5), where π_{max} is the price for which demand is equal to zero. Then, the middle panel of Figure 7.3 gives the expected link travel costs (C_{en}=(1-p)$C^0(N^0_{en})$+p$C^1(N^1_{en})$) as a function of the number of informed drivers, which is of course a function of the price of information. With linear D and C^j (j=0,1) curves, it can be shown that this relationship is linear as well. Clearly, expected link travel costs run from a minimum of C_p (where information is free, so that every one is willing to be informed) to a maximum of C_n (where information is so expensive that no one is willing to be informed).

Finally, the lower panel of Figure 7.3 depicts expected welfare as a function of the number of informed drivers, which is a function of the price of information. Under the assumptions made it can be shown that this is an increasing convex quadratic function, running from a minimum when no one is informed (W_n), to a maximum when every one is informed (W_p). The convexity of the expected welfare curve implies that as the number of informed drivers increases, then so does social welfare, and even at an increasing rate. Three effects are playing a role here. First, as the number of informed drivers increases, welfare for the "newly" informed drivers increases. If this were not the case, then they would not have bought the information in the first place. Second, owing to more informed drivers in the network, the expected link travel costs decrease, leading to additional benefits for the uninformed drivers. Finally, in the model an increase in the number of informed drivers can only be realised when the private costs of information π decrease, thereby increasing total welfare.

7.3 REGULATORY ISSUES

In the previous sections it was shown that information is not only beneficial to the (endogenously) informed drivers, but in addition to the (endogenously) uninformed ones, because the expected link travel costs in the network will decrease (see Figure 7.2). From economic theory it is well known that such kind of external effects may distort the efficiency of the market system in economic processes. In these circumstances, the market outcome without government intervention will generally not coincide with the allocation that maximises social welfare. Hence, the existence of external effects to uninformed drivers renders the issue of subsidising motorist information systems relevant. In the following sections, the previously presented model is used to analyse this in more detail. First, in Section 7.3.1 the optimal level of the subsidy on motorist information equipment is derived; next (Section 7.3.2), the scope of the analysis is extended towards a situation in which a motorist information system exists in combination with fine congestion tolling. Finally, a comparison between these two types of policy instruments from an efficiency point of view is made in Section 7.3.3.

7.3.1 Subsidising information without tolling

The optimal subsidy s for the motorist information equipment, given the price of information π, can be found by maximising expected welfare subject to individual maximising utility behaviour, see expression (7.6). The two restrictions in (7.6) ensure that information is allocated in an individually rational manner. Furthermore, due to the subsidy s, the costs of information as paid for by the users of the information system have decreased from π to π-s. Also, note that the objective function is not explicitly dependent on the subsidy s, as, from a welfare economic point of view, the *redistributive* impact of the subsidy is not relevant for the policy's efficiency.

$$\max_{s} \ (1-p)\left(\int_0^{N_{en}^0} D(x)dx - C^0(N_{en}^0)N_{en}^0\right) + p\left(\int_0^{N_{en}^1} D(x)dx - C^1(N_{en}^1)N_{en}^1\right) - \pi\left(N_{en}^0 - N_{en}^1\right) \tag{7.6}$$

subject to

$$(1-p)\left(D(N_{en}^0) - C^0(N_{en}^0)\right) = \pi - s$$
$$p\left(C^1(N_{en}^1) - D(N_{en}^1)\right) = \pi - s$$

Using the technique of Lagrange, the optimal subsidy s is equal to:

$$s = \frac{\gamma^1 w^0 - \gamma^0 w^1}{w^0 + w^1} \tag{7.7}$$

where

$$\begin{aligned}
\gamma^0 &= (1-p)\cdot C^{0\prime}(N_{en}^0) N_{en}^0 \\
\gamma^1 &= p C^{1\prime}(N_{en}^1) N_{en}^1 \\
w^0 &= (1-p)\left(D'(N_{en}^0) - C^{0\prime}(N_{en}^0)\right) \\
w^1 &= p\left(D'(N_{en}^1) - C^{1\prime}(N_{en}^1)\right)
\end{aligned} \tag{7.8}$$

For linear demand and cost curves the optimal subsidy can be given explicitly by:

$$s = \frac{N_{en}^1(\pi)\left(\dfrac{b^1}{a+b^1}\right) - N_{en}^0(\pi)\left(\dfrac{b^0}{a+b^0}\right)}{\dfrac{a+2b^0}{(1-p)(a+b^0)^2} + \dfrac{a+2b^1}{p(a+b^1)^2}} \tag{7.9}$$

where $N_{en}^j(\pi)$ denotes the equilibrium levels of road usage with a price of information equal to π and no subsidy (j=0,1).

After substituting $N_{en}^j(\pi)$ (j=0,1; see equations (7.3) and (7.4)) in expression (7.9), it follows that in the model with linear demand and cost curves the optimal subsidy is an increasing linear function of the price of information π. Hence, the higher the price of information, the higher the optimal regulator's subsidy. This linear relationship, however, may also imply that the optimal value of the subsidy may be negative when the price of the information is relatively low. This then leads to the interesting result that it might be welfare increasing to *tax* rather than to *subsidise* information. At first sight, this may seem a counter intuitive result, since it was found in previous sections that information provision leads to positive external effects to uninformed drivers as well. However, it should be realised that information induces *more* travel in the network, while in order to achieve system optimal behaviour in case of congestion, road usage should be *reduced*. Apparently, when the optimal subsidy is negative, it is more efficient for the government to use the motorist information system as an instrument to price for congestion rather than to stimulate the use of information.

To conclude, the optimal subsidy captures two effects. On the one hand, a positive value of the subsidy will induce more drivers to be informed, and hence will generate a more efficient use of the network. On the other hand, a negative subsidy (a tax) is a kind of (second-best) instrument to price for congestion, and hence to reduce overall levels of road usage. The optimal subsidy is a compromise between these two effects, and turns out to be more in favour of the latter at a relatively low price of information, and more in favour of the former at relatively high prices.

Using the parameter values from Chapter 5, a typical situation is depicted in Figure 7.4.[3]

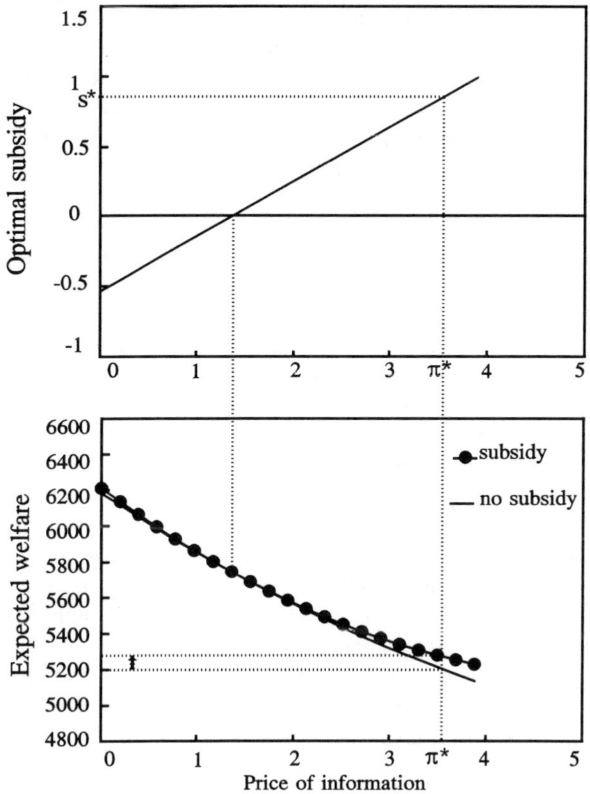

Figure 7.4 The relationship between the price of information, subsidising information, and expected welfare.

The linear increasing relationship between the price of information π and the optimal subsidy s is given in the upper panel of Figure 7.4. The bottom panel depicts the mapping between the price of information π and expected welfare. Here, expected welfare with and without government intervention to optimally subsidise motorist information systems are compared. Clearly, expected welfare with the imposition of the optimal (welfare maximising) subsidy is at least as high as without, and both are

[3]The following parameter values were used to generate Figure 7.4: d=50, a=0.015, $k^0=k^1=20$, $b^0=0.015$, $b^1=0.04$, and p=0.25. With these parameter values, π_{max} (the maximum possible price of information under which demand for information is still positive) is approximately equal to 3.88.

(of course) identical when the optimal subsidy is equal to zero. In the bottom panel of Figure 7.4 it is important to note that the difference in expected welfare under the two types of regulation is very small in the model. The difference is significant at only relatively low and high prices of information π. A conclusion regarding this point is that (for a wide range of parameter values) the model indicates that subsidising information as a tool to achieve a more efficient allocation of road usage only leads to small efficiency gains. In Section 7.3.3 this topic will be returned to; in that section a comparison of other types of government regulation is carried out. Before doing so, fine (fluctuating) congestion-pricing in combination with endogenous information is introduced first.

7.3.2 Combining fine congestion tolling and subsidising motorist information

The first-best solution in congested transport networks - the solution that is optimal from a social welfare point of view - can be achieved by implementing a fine (i.e. dependent on the level of congestion) congestion-pricing scheme. By doing so, marginal private costs will coincide with marginal social costs, so that individually optimal behaviour is also in line with overall optimality. In order to obtain these socially desired behavioural responses, it is necessary for all potential road users to be perfectly aware of the prevailing level of the fine tolls. Hence, some kind of information system is needed to ensure that this first-best policy will indeed achieve what it is intended to. Or, in other words, proper fine tolling cannot be implemented without the simultaneous implementation of some kind of information system.

In the present section, the optimal fine congestion tolls (f^0: in state 0; f^1: in state 1) and the optimal motorist information subsidy s are derived simultaneously by assuming that the cost price of information is equal to π.

In the model, the government will determine the fine fees f^0 and f^1, and the optimal level of the subsidy s in order to maximise expected social welfare. Hence, the regulator faces the following maximisation problem:

$$\max_{f^0, f^1, s} \; (1-p)\left(\int_0^{N_{en}^0} D(x)dx - C^0(N_{en}^0)N_{en}^0\right) + p\left(\int_0^{N_{en}^1} D(x)dx - C^1(N_{en}^1)N_{en}^1\right) - \pi\left(N_{en}^0 - N_{en}^1\right) \tag{7.10}$$

subject to

$$(1-p)\left(D(N_{en}^0) - C^0(N_{en}^0) - f^0\right) = \pi - s$$
$$p\left(C^1(N_{en}^1) + f^1 - D(N_{en}^1)\right) = \pi - s$$

The two restrictions in (7.10) ensure that individual behaviour is rational, that is, no individual can increase expected net private benefits by changing behaviour, and hence the solution is an equilibrium. Clearly, these restrictions follow directly from expressions (7.1) and (7.2) after taking proper account of the introduction of fine congestion fees f^0 and f^1, and the subsidy for the motorist information system s.

The optimal fine fees and subsidy for information can be found using the technique of Lagrange; for details, see Emmerink, Verhoef, Nijkamp and Rietveld (1996b).

$$\begin{aligned} f^0 - \frac{s}{1-p} &= C^{0\prime}(N_{en}^0)\,N_{en}^0 \\ f^1 + \frac{s}{p} &= C^{1\prime}(N_{en}^1)\,N_{en}^1 \end{aligned} \tag{7.11}$$

There are two equations to solve for three variables, implying that one can fix one variable and then solve for the other two. For example, fixing the subsidy at 0, leads to:

$$\begin{aligned} f^0 &= C^{0\prime}(N_{en}^0)\,N_{en}^0 \\ f^1 &= C^{1\prime}(N_{en}^1)\,N_{en}^1 \end{aligned} \tag{7.12}$$

The expressions in (7.12) show that once fine congestion tolling equal to marginal external congestion cost takes place, there is no need for subsidising motorist information systems. In this case, individuals face the optimal incentives for the decision whether or not to be informed. Furthermore, it can be seen that these optimal fine fees are given by the traditional expression reflecting the external costs of congestion, that is, the external costs that the marginal driver imposes on the other drivers.

However, the expressions in (7.11) also show that there are more (an infinite number of) ways to reach the socially optimal level of road usage. For example, when information is subsidised, the corresponding optimal fee in state 0 is somewhat larger, and the fee in state 1 is smaller. This can intuitively be explained by noting that when information is subsidised, then more drivers are willing to be informed. This in turn will imply that more people will be inclined to use the network when state 0 prevails, and hence, the fine fee in state 0 will be set at a higher level. Similarly, less people will be inclined to use the network in state 1, so that the fine fee in state 1 will be smaller.

An important implication of the above reasoning is that in the model presented above, flat congestion-pricing (which means that $f^0=f^1$, so that the fine fee is independent of the level of congestion in a particular state) in combination with the optimal subsidy for motorist information equipment (see expression (7.13) for the optimal flat fee and subsidy), will lead to the social welfare maximising solution. Under the additional, and rather weak, assumption that the free flow travel costs are identical in both states - $C^0(0)$ is equal to $C^1(0)$ -, it can easily be proven that the optimal subsidy *s* corresponding to the flat fee scheme is always positive, since under this assumption $C^{1\prime}(N_{en}^1)N_{en}^1$ minus $C^{0\prime}(N_{en}^0)N_{en}^0$ is always greater than zero. Therefore, the subsidy will never be a tax. Moreover, the expression of the flat fee in (7.13) shows that it can be interpreted as the expected external congestion costs.
The reason that the combination of a flat fee and a subsidy is as efficient as fine tolling is of course closely related to the fact that the subsidy is in equilibrium only

$$f^0 = f^1 = (1-p)C^{0'}(N^0_{en})N^0_{en} + pC^{1'}(N^1_{en})N^1_{en}$$
$$s = p(1-p)\left(C^{1'}(N^1_{en})N^1_{en} - C^{0'}(N^0_{en})N^0_{en}\right) \tag{7.13}$$

enjoyed by actors using the network in state 0. It therefore enables the regulator to discriminate perfectly between actors for whom it is efficient to use the network only in state 0, and those for whom it is efficient to use it in both states.

Furthermore, it can be shown that all the available welfare maximising strategies, as given by the expressions in (7.11), lead to the same expected net private benefits for the actors (the uninformed drivers, the informed drivers, and the regulator) in the system. This follows from the fact that the expected net private benefits for all actors in the system are independent of the optimal subsidy s. Hence, the subsidy on information cannot be used by the regulator to influence the equity distribution among actors under a first-best policy.

The results obtained above cannot easily be generalised to more complex networks and more states of nature. For instance, in a one-link three state transport network, the optimal flat fee and optimal subsidy on information will not yield the system optimum, since in that particular situation there are only *two* instruments (the flat fee and the subsidy) to deal with *three* different traffic situations. However, in Chapter 10 in a two-link network, it will be shown that the combination of information provision and flat tolling is a rather robust instrument, and works out to be almost as good as the theoretically first-best option of fine tolling. Moreover, given the public and political opposition to fine congestion-pricing, flat pricing in combination with subsidising information may be an attractive policy option.[4]

In the next section, the welfare effects of the first-best regulatory pricing and information scheme discussed above are compared with the second-best policy as presented in Section 7.3.1.

7.3.3 Efficiency of regulation

The welfare effects of four different policies will be compared. As previously, welfare is defined as total system benefits minus total system costs. The four policies to be compared are:

1. first-best: the regulator sets optimal fine tolls; the price of information is π.
2. second best: there are no tolls, but the regulator sets optimal subsidy on information; the price of information is π.
3. non intervention: the regulator does not intervene, but information is available at a price π.

---

[4]See Chapter 3 for arguments pro and contra congestion-pricing. For empirical evidence that suggests public aversion against pricing in a fluctuating manner, see the arguments in Frey and Pommerehne (1993). These authors found that a price rise due to a non-recurrent shock (such as a sudden excess in demand) is considered unfair by 80 per cent of the respondents in their survey.

4. no information: the regulator does not intervene, and information is not available.

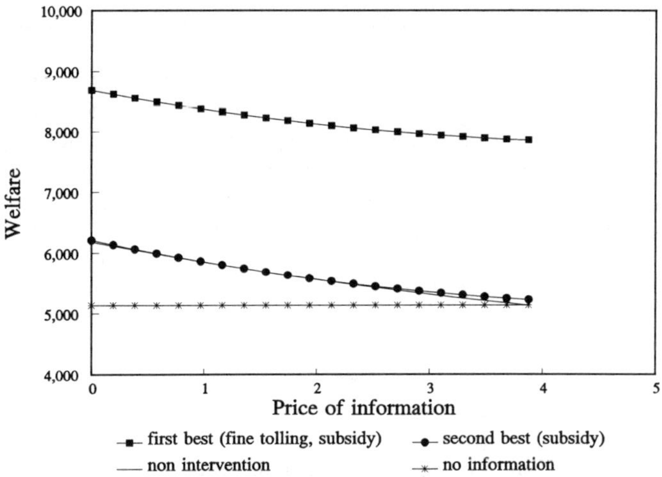

Figure 7.5 Relationship between the price of information and welfare.

In Figure 7.5, the impacts of these four policy options on welfare are depicted as a function of the price of information, using the same parameter values as in Section 7.3.1.

Clearly, the curve labelled *no information* is independent of the price of information, and is therefore a horizontal line. The *non intervention* curve crosses the *no information* one at $\pi_{max}(\approx 3.88)$, since there are no informed actors at this price. Next, notice that as in Figure 7.4, the impact of subsidising information (without fine tolling) is negligible. Finally, it is important to see that welfare can be increased quite substantially under *first-best* transport policy (optimal fine tolls and no subsidy, or optimal flat toll and optimal subsidy). As will be argued in Chapter 10, information provision without some kind of congestion-pricing will in most cases not direct the traffic flows towards a level that is close to the system optimum.

7.4 CONCLUSION

In this chapter, the impact of the endogenous provision of information to potential road users in transport networks was analysed. To do so, the static economic equilibrium framework, as discussed in Chapters 4 and 5 was adjusted in order to allow potential road users to buy information on the prevailing stochastic traffic situation. An individual driver will acquire information, only if the private benefits of being informed exceed the private costs. Then, uninformed potential road users base their trip-making behaviour on the expected costs, while informed ones use the actual costs.

By comparing this model with the one in which no information is available, it was proven that the endogenous provision of information leads to a strict Pareto improvement. The size of this increase in social welfare depends strongly on the private costs of information. When the private costs of information decrease, then more actors are willing to be informed, which in turn decreases expected link travel costs and increases social welfare.

Next, given a certain price of information, the optimal subsidy was derived, that is, the subsidy that maximises social welfare. It was found that the optimal subsidy increases with the price of information. However, the optimal subsidy can be negative when the information is very cheap. Then, it is from a government point of view more efficient to use the motorist information system as an instrument to price for congestion rather than to stimulate a more efficient use of the network. In both situations, however, it was shown that the potential welfare improvement of subsidising (or taxing) information is relatively small.

The analysis of the relation between fine congestion-pricing and subsidising motorist information revealed that there is no rationale to subsidise information as long as the government implements a proper fine congestion-pricing scheme. However, subsidising information may become an attractive (and necessary) policy option when a flat pricing scheme is adopted.

In the present chapter it was assumed that the price of information is independent of the number of informed drivers. However, it may be more realistic to describe the costs of a motorist information system by a fixed cost component, reflecting the necessary equipment needed to provide information (for instance, road-side equipment or a central computer system), and a variable cost component, representing the costs of the in-vehicle unit (for instance, the on-board computer system). Such a cost structure would imply the existence of *economies of scale*, see also Chapter 2.

In the next chapter, the model is further extended to include costs of uncertainty into the driver's decision-making process.

APPENDIX 7.A1 PROOF OF PROPOSITION 7.1

Proposition 7.1: In a one-link network with endogenous demand for information, and assuming linear inverse demand (D) and cost (C^0, C^1) functions, $C^0(N) \leq C^1(N)$, and $dC^0(N)/dN \leq dC^1(N)/dN$, the following relationships hold:

1. *expected road usage is higher with information than without;*
2. *expected link travel costs are smaller with information than without;*
3. *expected welfare is higher with information than without.*

Proof:

In the following it will be assumed that the linear demand curve D can be written as $D(N)=d-aN$, and the linear cost curves as $C^j(N)=k^j+b^jN$ for $j=0,1$.

First, the maximum possible price of information under which demand is non-negative is derived. In state 0, this price (π^0_{max}) can be found by equating N^0_{en} and N_n. Using expression [7] in Section 3.1 it follows that:

$$\pi^0_{max} = (N^0_p - N_n) \cdot (1-p) \cdot (a+b^0) \tag{7A.1}$$

A similar reasoning for state 1 yields:

$$\pi^1_{max} = (N_n - N^1_p) \cdot p \cdot (a+b^0) \tag{7A.2}$$

Using the explicit formulas for N_n, N^0_p, and N^1_p it can be easily shown that $\pi^0_{max}=\pi^1_{max}(=\pi_{max})$.

Next, the first point in Proposition 7.1 will be proven. First, using the equilibrium condition of model *n* (no information available), it follows that:

$$N_n = D^{-1}\big((1-p)\cdot C^0(N_n) + p\cdot C^1(N_n)\big) \tag{7A.3}$$

Then, using the equilibrium conditions of model *en* (endogenous demand for information) and the linearity of the demand function of road usage D, it can be shown that:

$$E(N_{en}) = (1-p)\cdot N^0_{en} + p\cdot N^1_{en} = D^{-1}\big((1-p)\cdot C^0(N^0_{en}) + p\cdot C^1(N^1_{en}) + (1-p)\cdot p\cdot (N^0_{en} - N^1_{en})\cdot (b^0 - b^1)\big) \tag{7A.4}$$

Since $N^0_{en}-N^1_{en}$ is positive, and b^0-b^1 negative, it follows that:

$$\Delta = (1-p)\cdot p\cdot (N^0_{en} - N^1_{en})\cdot (b^0 - b^1) < 0 \tag{7A.5}$$

Writing $s=E(N_{en})$ and $t=N_n$, the following relationships are identical to (7A.4) and (7A.3):

$$\begin{aligned} s &= D^{-1}\big((1-p)\cdot C^0(s) + p\cdot C^1(s) + \Delta\big) \\ t &= D^{-1}\big((1-p)\cdot C^0(t) + p\cdot C^1(t)\big) \end{aligned} \tag{7A.6}$$

Since, Δ is smaller than zero, and D^{-1} is a decreasing function, it follows that t is smaller than s. Hence, the first point in the proposition holds true.

Next, since Δ is smaller than zero, the second point follows directly. Finally, the last point in Proposition 7.1 will be proven. It will be shown that a similar relationship between expected level of road usage and welfare will hold as was derived in Chapter 5. For the model without information (model *n*) this was proven in Appendix 5.A. Next, rewriting the expressions of the model in which information is endogenised (model *en*) one finds:

$$C^0(N_{en}^0) = D(N_{en}^0) - \frac{\pi}{1-p} \tag{7A.7}$$

and

$$C^1(N_{en}^1) = D(N_{en}^1) + \frac{\pi}{p} \tag{7A.8}$$

Expected welfare is given by:

$$\begin{aligned} E(Welfare) &= (1-p)\cdot\left(\int_0^{N_{en}^0} D(x)dx - C^0(N_{en}^0)\cdot N_{en}^0\right) + p\cdot\left(\int_0^{N_{en}^1} D(x)dx - C^1(N_{en}^1)\cdot N_{en}^1\right) \\ &\quad - \pi\cdot(N_{en}^0 - N_{en}^1) = \\ &(1-p)\cdot\left(\frac{1}{2}\cdot a\cdot(N_{en}^0)^2 + N_{en}^0\cdot\frac{\pi}{1-p}\right) + p\cdot\left(\frac{1}{2}\cdot a\cdot(N_{en}^1)^2 - N_{en}^1\cdot\frac{\pi}{p}\right) - \pi\cdot(N_{en}^0 - N_{en}^1) = \frac{1}{2}\cdot a\cdot E(N_{en}^2) \end{aligned} \tag{7A.9}$$

where the last equality follows from the substitution of expressions (7A.7) and (7A.8) in expression (7A.9). The final point in Proposition 7.1 then follows from applying the first point in Proposition 7.1.

APPENDIX 7.A2 DERIVATION OF FIGURE 7.3

Under the assumptions of linear demand for travel (D(N)=d-aN) and linear link travel cost functions ($C^j(N)=k^j+b^jN$; j=0,1), it was shown in Section 3.1 that demand for information is linearly dependent on the private costs of being informed. Here, it will be shown that under these conditions expected link travel costs are also a linear function of demand for information. Expected link travel costs in the model with endogenous demand for information are given by:

$$(1-p)\cdot C^0(N_{en}^0)+p\cdot C^1(N_{en}^1) = (1-p)\cdot C^0\left(N_p^0-\frac{\pi}{(1-p)\cdot(a+b^0)}\right)+p\cdot C^1\left(N_p^1+\frac{\pi}{p\cdot(a+b^1)}\right) =$$
$$Constant + \pi\cdot\left(\frac{a\cdot(b^1-b^0)}{(a+b^0)\cdot(a+b^1)}\right), \text{ where} \tag{7A.10}$$
$$Constant = (1-p)\cdot k^0+p\cdot k^1+(1-p)\cdot b^0\cdot N_p^0+p\cdot b^1\cdot N_p^1$$

Hence, the private costs of information π and the expected link travel costs are positively linearly dependent. Now, since demand for information and private costs of information π are negatively linearly dependent, it follows that the expected link travel costs are negatively linearly dependent on demand for information.

Finally, it will be shown that social welfare, as defined by total system benefits minus total system costs, is a convex quadratic increasing function of demand for information. This will be done by proving that the first derivative of welfare with respect to private costs of information π is negative, while the second derivative of welfare with respect to private costs of information π is positive, and the third derivative is equal to zero.

$$\frac{\partial Welfare}{\partial\pi} = \frac{\partial Welfare}{\partial N_{en}^0}\cdot\frac{\partial N_{en}^0}{\partial\pi}+\frac{\partial Welfare}{\partial N_{en}^1}\cdot\frac{\partial N_{en}^1}{\partial\pi} = \frac{a\left(N_{en}^1\cdot(a+b^0)-N_{en}^0\cdot(a+b^1)\right)}{(a+b^0)\cdot(a+b^1)} < 0 \tag{7A.11}$$

This expression is negative because N_{en}^1 is smaller than N_{en}^0, and b^0 is smaller than b^1.

$$\frac{\partial^2 Welfare}{\partial\pi^2} = a\cdot\left(\frac{1}{(1-p)\cdot(a+b^0)^2}+\frac{1}{p\cdot(a+b^1)^2}\right) > 0 \tag{7A.12}$$

$$\frac{\partial^3 Welfare}{\partial\pi^3} = 0 \tag{7A.13}$$

Hence, welfare is a convex quadratic decreasing function of the private costs of being informed π. Now, since the demand function for information is negatively linearly dependent on π, it follows that welfare is a convex quadratic increasing function of the number of informed drivers.

8 INCLUDING THE COSTS OF UNCERTAINTY[1]

8.1 INTRODUCTION

This chapter complements the previous chapters by including a term related to the costs of travel time uncertainty into the generalised travel cost function. By doing so, the model presented hereafter gains much in realism, as the importance of uncertainty costs has frequently been stressed in the literature (Arnott et al., 1990b; Hendrickson and Kocur, 1981). Furthermore, the work in this chapter advances Arnott et al. (1996) in two respects. First, besides *public* information also the case of *club* information is considered. Club information deals with the situation in which only a specific group of travellers is provided with information. Second, as in the previous chapter, attention will be paid to endogenous information provision, implying that travellers will acquire information only if the private benefits of the information exceed the private costs. Arnott et al. (1991, 1996) assumed that information is available for free for the whole population.

The chapter is organised in the following manner. In Section 8.2, the properties of four stochastic network deterministic user equilibrium (SNDUE) models are analysed. In Section 8.3, the efficiency impacts of driver information systems are studied. Section 8.4 contains the conclusions.

8.2 SNDUE MODELS WITH COSTS OF UNCERTAINTY

In this section, four models for road use in which drivers take costs of uncertainty into account are presented. Previous work (see, for example, Arnott et al. (1996) and Chapters 5, 6, and 7) considered the case in which travel costs were solely determined by stochastic costs related to travel time. Costs resulting from uncertainty per se were not explicitly considered. Here, it is assumed that travellers base their behaviour on the following generalised cost function:

$$E(\textit{travel costs}) = \alpha\, E(\textit{travel time}) + \beta\, Sd(\textit{travel time}) \qquad (8.1)$$

where *Sd* indicates the standard deviation of the stochastic variable *travel time*, and *E* denotes the expectation operator. The parameters α and β can be interpreted as the monetary *value-of-time* and *value-of-uncertainty*, respectively. This type of generalised cost functions has previously been considered in the transportation literature; see for example, Arnott et al. (1990b), Hendrickson and Kocur (1981), Noland and Small (1995) and Small (1982). These authors stressed the importance of including costs related to travel time uncertainty in a generalised travel cost function. Empirical work

[1]This chapter is based on Emmerink, Verhoef, Nijkamp and Rietveld (1996e) forthcoming in *International Economic Review*.

revealed the important role played by uncertainty in travel behaviour. More precisely, the term *ßSd(travel time)* in equation (8.1) refers to planning costs that uncertainty in travel time imposes on individuals. For instance, one could think of costs associated with disruption of meetings with coworkers, stress etc. (Noland et al., 1994). Such costs make individuals effectively risk averse with respect to travel time. In economic terms, the model assumes that travellers have *rationally expected* first and second moments of the travel time distribution. For peak-hour travellers it seems reasonable to assume that drivers have no misperceptions of either expected travel time or the variance.

In order to confine the analysis to relevant situations the parameter ß is restricted to values within the interval $[0,\alpha]$. First, negative ß-values would imply risk-loving travelling behaviour; something which is rather unrealistic in the context of travel time valuation. Second, ß-values exceeding α would imply that reducing the uncertainty by a number of minutes is worth more than a reduction in travel time by the same amount. In the sequel, values of ß greater than α will only be used for purely illustrative purposes.[2]

Four models will be described, representing four different starting points concerning the availability of information. In each of these, travellers are seeking to optimise the (expected) generalised costs of travelling. In the first model, indicated as model N, there is no information available to the road users, and travellers base their trip-making decisions on expected travel costs. In the second model, model I, information on the actual traffic costs is available to all road users. Consequently, these will consider *actual* rather than *expected* costs in their trip-making decision. In the third model, denoted as model P, information is available to an exogenously determined fraction of the road users. Finally, in model E, information is modelled endogenously. In this model, the traveller's choice of being informed depends on the private benefits and private costs associated with the information. Drivers will then acquire traffic information when the internal private benefits derived from the information exceed the costs of information, see also Chapter 7.

As in the previous chapters, uncertainty originates from stochastic shocks in the cost function. Information, then, provides travellers with realisations of these random variables. It is assumed that the stochastic travel time functions follow a Bernoulli distribution. With probability (1-p) the travel time function is given by $C^0(N)$, while with probability p travel time is given by $C^1(N)$, where N denotes the level of road usage. As both $C^0(N)$ and $C^1(N)$ reflect travel time, these are increasing functions in N. The two realisations of the stochastic travel time variables are referred to as state 0 ($C^0(N)$) and state 1 ($C^1(N)$). State 1 reflects the situation with low capacity (e.g. due to traffic accidents, road works etc.); in state 0 capacity is relatively high.

The relation between the travel time in state 0 (high capacity) and state 1 (low capacity) can be written as:

---

[2]In fact, travellers with extremely large penalty costs for lateness might have a value of ß that exceeds α. However, this seems to be an exceptional case. Furthermore, the value of ß is restricted if the assumption is introduced that travel costs including uncertainty costs may never exceed travel costs under the least favourable congestion regime.

$$C^0(N) < C^1(N) \quad \text{and} \quad \frac{\partial C^0(N)}{\partial N} < \frac{\partial C^1(N)}{\partial N} \quad \text{for all N>0} \tag{8.2}$$

8.2.1 Model N: No information available

In order to implement the model based on individual optimising behaviour following equation (8.1), an expression for the standard deviation of the stochastic travel times is needed. By defining $\bar{C}=(1\text{-}p)C^0+pC^1$, it can easily be shown that:

$$Var(travel\ time) = (1-p)\left(C^0-\bar{C}\right)^2 + p\left(C^1-\bar{C}\right)^2 = p(1-p)\left(C^1-C^0\right)^2 \tag{8.3}$$

where *Var* indicates the variance operator. It follows that the standard deviation *Sd* is given by:

$$Sd(travel\ time) = \sqrt{p(1-p)}\left(C^1-C^0\right) \tag{8.4}$$

because, by construction, $C^1>C^0$; see expression (8.2).

The equilibrium condition for model N, where no information is available, is then given by:

$$\alpha\left((1-p)C^0(N_N)+pC^1(N_N)\right) + \beta\sqrt{p(1-p)}\left(C^1(N_N)-C^0(N_N)\right) = D(N_N) \tag{8.5}$$

where N_N denotes the equilibrium level of road usage of model N. The subscript is used to distinguish the equilibrium levels of road usage from this model with the ones to be presented hereafter. The left-hand side of expression (8.5) gives the expected travel costs, while the right-hand side denotes the willingness-to-pay for using the transport network. Clearly, for the marginal network user N_N, expected private costs should equal private benefits: he or she is indifferent between using the network or not.

Using the following explicit linear functions for $C^j(N)$ and D(N):

1. $C^j(N)=k^j+b^jN$ (j=0,1) and
2. D(N)=d-aN,

the equilibrium level of road usage N_N is given by:

$$N_N = \frac{d - \alpha\bar{k} - \beta\sqrt{p(1-p)}(k^1-k^0)}{\alpha\bar{b} + \beta\sqrt{p(1-p)}(b^1-b^0) + a} \tag{8.6}$$

where a bar indicates an expected value. As expected, an increase in the value-of-time (α) and the value-of-uncertainty (β) leads to a decrease of N_N.

8.2.2 Model I: Information available for all travellers

In model I it is assumed that all potential road users are perfectly informed on the actual traffic situation. This implies that the travellers base their trip-making decision on *actual* rather than *expected* costs, and do not face any costs related to travel time uncertainty. The two equilibrium conditions that describe the model are given by:

$$\alpha C^0(N_I^0) = D(N_I^0) \tag{8.7}$$

$$\alpha C^1(N_I^1) = D(N_I^1) \tag{8.8}$$

These conditions reveal that for the marginal network users N_I^0 (when state 0 occurs) and N_I^1 (when state 1 occurs) private costs are equal to private benefits in both states. Marginal user N_I^1 is indifferent between using the network in both states or just in state 0, while marginal user N_I^0 is indifferent between using the network in state 0 or not using it at all. Clearly, the parameter ß plays no role in these expressions, since informed drivers do not face uncertainty costs. When comparing models N and I, it can be shown that $N_N<N_I^0$. Regarding N_I^1, two regimes can be distinguished:

1. $N_I^1<N_N<N_I^0$: for relatively small ß;
2. $N_N<N_I^1<N_I^0$: for relatively large ß.

An analysis of the properties of model N and model I, under the condition of linear demand and travel time functions, leads to Proposition 8.1, containing the most important results of a welfare theoretical comparison of the two models.

Proposition 8.1: Due to information to all travellers:

1. expected road usage increases;
2. expected network travel costs decrease;
3. none of the road users is worse off and consequently, the system welfare in model I does not fall short of system welfare in model N.

Proof: Appendix 8.A1

Some implications of Proposition 8.1 are illustrated in Figure 8.1 and Figure 8.2. On the x-axis, the travellers are ranked according to decreasing willingness-to-pay, while on the y-axis the additional expected net private benefits (owing to information provision) are shown. The additional expected net private benefits are defined as the expected net private benefits for model I minus the expected net private benefits for model N.

Some observations are in order now. First, depending on whether N_I^1 is smaller or larger than N_N, most benefits from information accrue to either travellers in segment II_1 and some of those in III_1 (in Figure 8.1), or those in segment I_2 (in Figure 8.2). Second, the benefits depicted in Figure 8.1 and Figure 8.2 are the sum of the so-called *internal* and *external* individual benefits. For example, drivers in segment I_1 are not necessarily prepared to pay the whole amount of their information benefits - that is, $D(N_N)-D(E(N_I))$ - to become informed. Part of their information benefits is external in nature, i.e. is caused by changes in behaviour by other road users. In fact, drivers in segment I_1 are willing to pay a maximum of

$$\beta \sqrt{p(1-p)}\left(C^1(N_I^1) - C^0(N_I^0)\right) \tag{8.9}$$

to become informed. This amount represents the monetary costs of uncertainty which they are faced with. Furthermore, as theoretically shown in the previous chapters and Haltiwanger and Waldman (1985), the marginal effect of information decreases with the fraction of users who are informed. Simulation experiments by Mahmassani and Jayakrishnan (1991) led to even stronger results. They showed that information benefits might actually become negative as the fraction of informed road users exceeded a particular threshold value. The theoretical model presented here, however, shows that the system welfare (measured as the sum of the individual benefits minus the sum of the individual costs) in model I does not fall short of the system welfare in model N, see Figure 8.1 and Figure 8.2. In other words, information increases network efficiency.

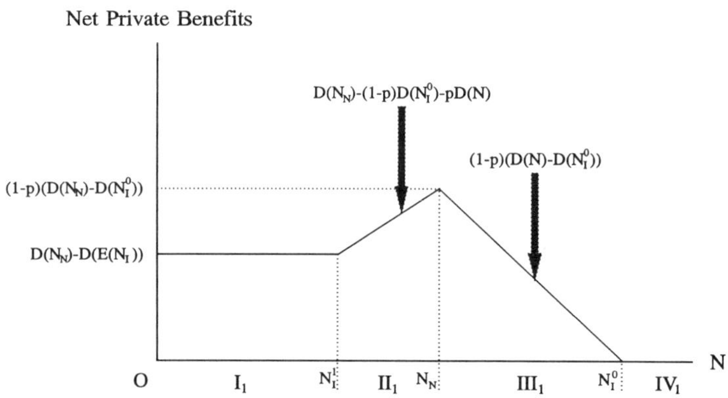

Figure 8.1 Model N and I: Case $N_I^1 < N_N < N_I^0$.

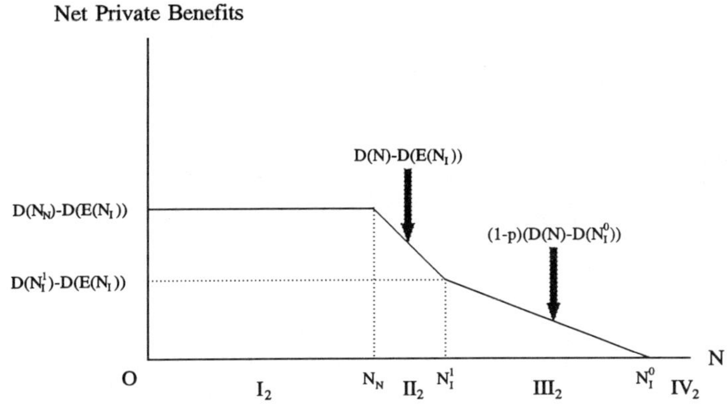

Figure 8.2 Model N and I: Case $N_N<N_I^1<N_I^0$.

Figure 8.1 was also discussed in Chapter 5, where no costs of uncertainty were considered (ß=0). When including monetary costs of uncertainty (ß>0) however, the pattern of benefit distribution may change to the one shown in Figure 8.2 for sufficiently high values of ß.

8.2.3 Model P: Information available for an exogenously determined group of travellers

Thus far, it was assumed that either none (model N) or all (model I) of the potential road users are supplied with information on the actual traffic situation. Next, in model P, it is assumed that information is provided to an *exogenously* determined group of potential road users denoted by subscript *x* (referring to *informed*), while *uninformed* travellers are denoted by subscript *y*. Respective inverse demand functions for these two groups of potential road users are given by $D_x(N_{P,x})$ and $D_y(N_{P,y})$, where as before a capital N indicates the level of road usage and the subscript P refers to the model under consideration. From these two groups, informed road users are perfectly aware of the prevailing (*actual*) travel costs and consequently do not face any uncertainty regarding the prevailing travel time; the cost component with the standard deviation is equal to zero. On the other hand, uninformed potential road users are not aware of any day-specific travel costs and base their behaviour on *expected* rather than *actual* costs. The three equilibrium conditions that fully describe model P are given in expressions (8.10) to (8.12).

$$\alpha C^0(N^0_{P,x}+N_{P,y}) = D_x(N^0_{P,x}) \tag{8.10}$$

$$\alpha C^1(N^1_{P,x}+N_{P,y}) = D_x(N^1_{P,x}) \tag{8.11}$$

$$\begin{aligned} &\alpha\left((1-p)\,C^0(N^0_{P,x}+N_{P,y}) + p\,C^1(N^1_{P,x}+N_{P,y})\right) \\ &+ \beta\sqrt{p(1-p)}\left(C^1(N^1_{P,x}+N_{P,y}) - C^0(N^0_{P,x}+N_{P,y})\right) \\ &= D_y(N_{P,y}) \end{aligned} \tag{8.12}$$

Equilibrium conditions (8.10) and (8.11) state the equality between private costs and private benefits for the informed marginal network users $N^0_{P,x}$ (when state 0 occurs) and $N^1_{P,x}$ (when state 1 occurs). The third equilibrium condition reflects the equality between expected private costs and private benefits for the uninformed marginal network user $N_{P,y}$; network user $N_{P,y}$ is indifferent between using the network or not.

In order to assess the welfare economic properties of model P, its features are compared with model N. To do so, model N is first presented for two groups of drivers. The equilibrium conditions of this model are straightforward:

$$\begin{aligned} &\alpha\left((1-p)\,C^0(N_{N,x}+N_{N,y}) + p\,C^1(N_{N,x}+N_{N,y})\right) \\ &+ \beta\sqrt{p(1-p)}\left(C^1(N_{N,x}+N_{N,y}) - C^0(N_{N,x}+N_{N,y})\right) \\ &= D_x(N_{N,x}) \end{aligned} \tag{8.13}$$

$$\begin{aligned} &\alpha\left((1-p)\,C^0(N_{N,x}+N_{N,y}) + p\,C^1(N_{N,x}+N_{N,y})\right) \\ &+ \beta\sqrt{p(1-p)}\left(C^1(N_{N,x}+N_{N,y}) - C^0(N_{N,x}+N_{N,y})\right) \\ &= D_y(N_{N,y}) \end{aligned} \tag{8.14}$$

Clearly, the x-travellers are then basing their behaviour on expected costs; see expression (8.13). Using the linear demand and stochastic link travel time functions, the properties of model P and model N are given by Proposition 8.2.

Proposition 8.2: Due to information to an exogenously determined group of travellers:

1. total expected road usage increases;
2. expected road usage of the group of informed travellers increases;
3. expected network travel costs of the group of informed travellers decreases;
4. none of the travellers of the informed group is worse off and consequently, welfare for the informed travellers does not decrease.

Proof: Appendix 8.A2

There are many similarities between Proposition 8.1 and Proposition 8.2. However, there is one important difference. Proposition 8.2 does not address the welfare economic effects of the information to the uninformed travellers. In fact, in model P it is impossible to prove that welfare of uninformed travellers will not decrease. The illustration in Figure 8.3, which gives the results of a simulation experiment with model N and P, shows that under certain conditions welfare of uninformed drivers decreases; in particular, if the value of uncertainty ß is sufficiently high.

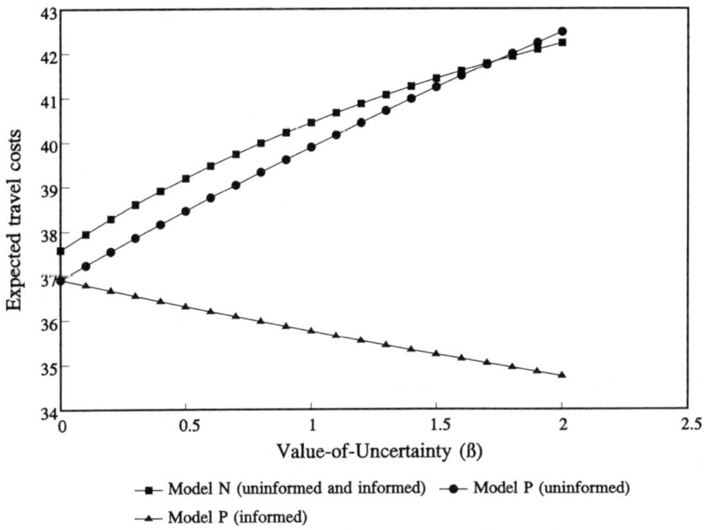

Figure 8.3 Expected network travel costs in model N and P.

In Figure 8.3, the expected network travel costs are depicted as a function of the value-of-uncertainty parameter ß.[3] As proven in Proposition 8.2, informed road users are better off due to information. However, this is not necessarily true for the uninformed ones. As can be seen in Figure 8.3, with the given parameter values, when ß gets beyond ß≈1.8 (which exceeds α), the uninformed road users are worse off due to the information. This can be explained as follows. First, notice that the value-of-time and value-of-uncertainty parameters are assumed to be identical for all potential road users, both informed and uninformed. Clearly, as the informed drivers do not face travel time uncertainty, the uncertainty component becomes zero in their expected travel cost function; see expressions (8.10) and (8.11). This, in turn, will lead to a

[3]Based on Chapter 5, the following parameters were used to produce Figure 8.3: $d_x=d_y=50$, $a_x=a_y=0.03$, $p=0.25$, $k^0=k^1=20$, $b^0=0.015$, $b^1=0.04$, $\alpha=1$.

larger number of informed travellers using the network. The increase will be larger with high values of ß. As a consequence, network travel times will increase, which leads to larger expected travel costs for the uninformed road users. On the left-hand side of Figure 8.3 at ß=0, however, uninformed drivers are absolutely certain to benefit from the information provision to informed drivers, results which have been derived in Chapter 5.

In the literature, several authors have called attention for potential negative effects of information provision for uninformed drivers; see for example Bonsall and Parry (1990). However, those assertions were often based on qualitative analyses. The formal analysis indicates that under the provision of perfect information in model P *the uninformed road users might be worse off, when the costs of uncertainty (ß) are very high.* With a more realistic value-of-uncertainty parameter, however, this is rather unlikely.

8.2.4 Model E: Information available for an endogenously determined group of travellers

In model E it is assumed that the choice of being informed depends on the internal private benefits derived from the information and the private costs of being informed. In contrast to the previous two models, it is now assumed that information is a commodity that can be purchased for a fixed price π, see Chapter 7. In this model, potential road users are faced with two decisions: either to use the network or not, and either to buy the information or not. Clearly, these decisions are interdependent.

Without loss of generality, only parameter values leading to at least a few travellers acquiring information are considered. Under this assumption the equilibrium conditions of the model can be written in terms of equalities. If none of the travellers is willing to buy the information (either because the price π is too high or the stochasticity in the network is too low, i.e. p-values close to zero or one) then the model simply collapses to model N, see Section 8.2.1.

It is not straightforward to derive the equilibrium conditions for this model. First, split the group of potential road users into four parts, see Figure 8.1. Road users in segment I_1 will always use the network, independent of the state of the network. For these drivers, information reduces their uncertainty costs to zero. Since the expected costs of the uncertainty are identical for these drivers, either all or none of the road users in segment I_1 will acquire information depending on the price of the information.

Next, consider the road users in segment II_1. Without being provided with information, they will always use the network, and therefore face travel costs related to both travel time and uncertainty. However, when provided with information, they will only use the network in state 0; their private benefits in state 1 fall short of their private costs in this state. Therefore, information will allow these drivers to reduce the costs of uncertainty to zero, and, in addition, will provide them with decision making benefits in state 1. In fact, the benefits in state 1 are stemming from the possibility to avoid using the network when capacity is low. A comparison of the information benefits of drivers from segment I_1 and II_1 shows that segment II_1 drivers benefit more than drivers from segment I_1 in Figure 8.1.

Next, attention is turned to the road users in segment III_1. When they base their trip-making behaviour on expected costs, these will exceed the private benefits of using the network. Therefore, they will not use the network when they are not provided with information. This also implies that drivers in segment III_1 do not face any costs related to uncertainty. When road users from segment III_1 are informed on the actual network conditions, they will decide to use the network only when state 0 prevails. These drivers will thus obtain individual benefits owing to improved decision making. Finally, the hypothetical road users of segment IV_1 will never use the network, nor will they therefore buy information.

To derive the equilibrium conditions of model E, one has to distinguish between two different situations. First, assume that for drivers in segment I_1 and I_2 it is not beneficial to be equipped with the information device. This implies that:

$$\beta\sqrt{p(1-p)}\left(C^1(N_E^1) - C^0(N_E^0)\right) < \pi \tag{8.15}$$

The above inequality shows that the benefits of being informed do not offset the costs of acquiring the information for drivers who will always use the network, and therefore, it ensures that drivers in segment I_1 and I_2 will not buy the information. The equilibrium conditions of model E under the assumption that inequality (8.15) holds can now be given:

$$(1-p)\left(D(N_E^0) - \alpha C^0(N_E^0)\right) = \pi \tag{8.16}$$

$$p\left(\alpha C^1(N_E^1) - D(N_E^1)\right) + \beta\sqrt{p(1-p)}\left(C^1(N_E^1) - C^0(N_E^0)\right) = \pi \tag{8.17}$$

First, the term in the big parentheses in expression (8.16) denotes the net *internal* private benefits to driver N_E^0 from segment III_1 in state 0. Multiplied by the probability of occurrence, 1-p, the left-hand side of expression (8.16) gives the net private benefits of informed road usage experienced by the informed driver N_E^0. To ensure that driver N_E^0 is the *marginal informed driver*, the net benefits should equal the costs of acquiring the information π.

Next, the first term in the big parentheses of expression (8.17) denotes the net private benefits to informed drivers in segment II_1, accruing from the fact that they are not using the network in state 1 when provided with information. Remember that without being provided with information, drivers from segment II_1 would always decide to use the network. The expected value of these benefits is obtained by multiplying this term with the probability of occurrence of state 1, p. The second term of expression (8.17) gives the benefits of avoiding the costs of uncertainty. Without being provided with information, these drivers would always use the network and therefore suffer from uncertainty costs. To ensure that driver N_E^1 is the *marginal informed driver* of segment II_1, the internal private benefits from being provided with information should equal the costs of information π. This then leads to expression (8.17). Notice that in expression (8.17) the equilibrium levels of road usage N_E^0 and N_E^1 interact in the term that relates to the costs of uncertainty.

Now, consider the case that inequality (8.15) does not hold, and hence:

$$\beta\sqrt{p(1-p)}\left(C^1(N_E^1) - C^0(N_E^0)\right) > \pi \tag{8.18}$$

This implies that all drivers in segment I_1 and I_2 will acquire information, since the benefits owing to the information exceed the costs of buying the information. Under this assumption, the *marginal informed driver* N_E^1 in segment I_2 is now marginal in the sense that he or she is indifferent between using the network in both states or just in state 0 when provided with information. Remember that under inequality (8.15) to hold, the marginal informed driver N_E^1 is indifferent between either using the network in both states and not buying the information or using the network in state 0 and buying the information. Under inequality (8.18), however, marginal driver N_E^1 is *not* indifferent between buying the information or not. He or she will buy the information as the benefits of avoiding the travel time uncertainty already offset the costs of acquiring the information; see inequality (8.18). The two equilibrium conditions that fully determine N_E^0 and N_E^1 are now given by expressions (8.19) and (8.20):

$$(1-p)\left(D(N_E^0) - \alpha C^0(N_E^0)\right) = \pi \tag{8.19}$$

$$\alpha C^1(N_E^1) = D(N_E^1) \tag{8.20}$$

Expression (8.19) is similar to expression (8.16). Expression (8.20) however, is different from expression (8.17). Equilibrium condition (8.20) ensures that the informed traveller N_E^1 is indifferent between either using the network in both states or in state 0 solely: the left-hand side represents the costs of using the network in state 1, while the right-hand side denotes the costs of not using the network in state 1.[4]

A comparison of the properties of model N and model E, under the assumption of linear demand and stochastic travel time functions, leads to Proposition 8.3:

Proposition 8.3: Due to endogenous provision of information:

1. expected road usage increases;
2. for all road users expected network travel costs do not increase;
3. none of the travellers is worse off and consequently, the system welfare in model E does not fall short of system welfare in model N.

Proof: Appendix 8.A3

---

[4]Notice that equation (8.17) collapses to equation (8.20) if

$$\beta\sqrt{p(1-p)}\left(C^1(N_E^1) - C^0(N_E^0)\right) = \pi$$

Hence, in this situation the two cases discussed are identical.

With respect to the third point of Proposition 8.3, it can be shown that the additional net expected private benefits due to the endogenous provision of information (net expected private benefits of model E minus net expected private benefits of model N) have a similar pattern as the curves in Figure 8.1 and Figure 8.2. Of course, subscript I has to be replaced by subscript E.

It is important to note that the negative external effects to uninformed road users are now not taking place. Even for very large values of ß, uninformed road users are benefiting from the endogenous provision of information. An obvious explanation for this essential difference between the equilibrium values of model P and model E is that the negatively affected uninformed road users of model P would buy the information in model E to ensure that they would also benefit from the information.

To conclude, in the previous sections various stochastic network equilibrium models were analysed. It was proven that information leads to a Pareto improvement for the informed drivers, that is, none of the informed drivers is worse off due to information. In model P, however, uninformed travellers might be negatively affected by the information, but this is unlikely to happen with realistic parameter values. Finally, in model E, where the choice of being informed is endogenous, it was shown that travellers who decide not to acquire information will always be better off. Therefore, a preliminary conclusion of the analysis is that information provision will lead to a strict Pareto improvement.

8.3 INFORMATION PROVISION AND SYSTEM OPTIMAL BEHAVIOUR

Information provision is a *second-best* policy, that is, provision of information does not necessarily direct the network performance to system optimal (*first-best*) levels. In the present section, the analysis is focused on the efficiency of endogenous information provision relative to the optimal policy. The level of welfare generated by endogenous information provision is compared with welfare of the *optimal* policy and welfare of the *non-intervention* policy, respectively. The non-intervention policy considered is given by model N. The optimal (welfare maximising) policy is presented in Section 8.3.1. The welfare effects of endogenous information provision are then investigated in Section 8.3.2.

8.3.1 System optimum (first-best) policy

The theoretical concept of road pricing, stemming from Pigou (1920) and Knight (1924), and further explored by Walters (1961), Smeed (1964), Sharp (1966) and Vickrey (1969, 1971), is based on levying a tax equal to the difference of the marginal social costs and marginal private costs (Pigouvian tax), see Chapter 3. In doing so, the negative external effects on other road users when joining a road are accounted for in the user's decision-making process: the system optimum will coincide with the user equilibrium (Wardrop, 1952). In economic terms, the road price ensures that only economic efficient trips are undertaken.

In this section, the theory of road pricing is expanded to situations in which road users suffer from costs related to uncertainty. A system optimal policy can be implemented by means of a fluctuating (depending on the state) pricing scheme; fluctuating in the sense that the road price is equal to f^0 (f^1) when state 0 (state 1) prevails. To analytically present the welfare maximising model - where welfare is measured as the sum of the individuals' private benefits minus the sum of their private costs - three situations have to be considered:

1. some of the travellers acquire information;
2. all travellers acquire information;
3. no information is available.

System optimum; Situation 1: some of the travellers acquire information
First, assume that inequality (8.15) holds true. This implies that travellers in the interval $[0,N_E^1]$ are not acquiring information. Only, for drivers in segment $[N_E^1,N_E^0]$ it is beneficial to buy information. In order to realise the welfare maximising traffic scheme, the regulator is faced with the following maximisation problem:

$$\max_{f^0, f^1} \; (1-p)\left(\int_0^{N_E^0} D(x)dx \; - \; \left(\alpha\, C^0(N_E^0) N_E^0\right)\right) + p\left(\int_0^{N_E^1} D(x)dx \; - \; \left(\alpha\, C^1(N_E^1) N_E^1\right)\right)$$

$$- \; \beta\sqrt{p(1-p)}\left(C^1(N_E^1) - C^0(N_E^0)\right) N_E^1 \; - \; \pi\left(N_E^0 - N_E^1\right) \tag{8.21}$$

subject to

$$(1-p)\left(D(N_E^0) - \alpha\, C^0(N_E^0) - f^0\right) \; = \; \pi$$

$$p\left(\alpha\, C^1(N_E^1) + f^1 - D(N_E^1)\right) \; + \; \beta\sqrt{p(1-p)}\left(C^1(N_E^1) - C^0(N_E^0)\right) \; = \; \pi$$

The term $\pi(N_E^0-N_E^1)$ in the objective function gives the costs of information, while the uncertainty cost component is multiplied by the number of road users faced with uncertainty: N_E^1. The two restrictions ensure that information is allocated in an individually rational manner. Apart from the incorporation of the fluctuating road price, the restrictions are similar to expressions (8.16) and (8.17). The second restriction implies that driver N_E^1 is indifferent between either always using the network or buying information and using the network in state 0 only; the first restriction implies that driver N_E^0 is indifferent between either never using the network or buying information and using the network in state 0 only. The optimal fluctuating fees f^0 and f^1 can be solved for by using the technique of Lagrange. Then, it follows that:

$$f^0 = \alpha C^{0'}(N_E^0) N_E^0 - \beta \sqrt{\frac{p}{1-p}} C^{0'}(N_E^0) N_E^1$$

$$f^1 = \alpha C^{1'}(N_E^1) N_E^1 + \beta \sqrt{\frac{1-p}{p}} C^{1'}(N_E^1) N_E^1 \tag{8.22}$$

The first terms on the right-hand side of (8.22) reflect the traditional external congestion costs of road traffic. The second term then represents the external costs of uncertainty imposed on the road users in the interval $[0, N_E^1]$. This additional term decreases the value of f^0 and increases the value of f^1 in order to diminish the fluctuations in travel time, and hence the external costs of uncertainty.

System optimum; Situation 2: all travellers acquire information

Next, assume that inequality (8.18) holds true, implying that all users of the network are acquiring information. The regulator then faces maximisation problem (8.23):

$$\max_{f^0, f^1} (1-p)\left(\int_0^{N_E^0} D(x)dx - \left(\alpha C^0(N_E^0) N_E^0\right)\right) + p\left(\int_0^{N_E^1} D(x)dx \quad \left(\alpha C^1(N_E^1) N_E^1\right)\right) - \pi N_E^0 \tag{8.23}$$

subject to

$$(1-p)\left(D(N_E^0) - \alpha C^0(N_E^0) - f^0\right) = \pi$$
$$\alpha C^1(N_E^1) + f^1 = D(N_E^1)$$

The term πN_E^0 in the objective function reflects the costs of information, while the restrictions ensure individually rational behaviour. Maximisation problem (8.23) can be solved using the technique of Lagrange. The optimal fees yield:

$$f^0 = \alpha C^{0'}(N_E^0) N_E^0$$
$$f^1 = \alpha C^{1'}(N_E^1) N_E^1 \tag{8.24}$$

which is the traditional expression for the optimal (first-best) congestion prices. Clearly, uncertainty does not play a role in these expressions, since none of the drivers in the network faces uncertainty costs as everyone is acquiring information.

System optimum; Situation 3: no information is available
Finally, assume that the costs of information (π) are relatively high. Then it might be optimal from a system point of view to supply none of the drivers with information. In such a situation, the network regulator faces the problem of finding the system optimum of model N (see also Section 8.2.1):

$$\max_f \int_0^{N_N} D(x)dx - (1-p)\,\alpha\, C^0(N_N) N_N - p\,\alpha\, C^1(N_N) N_N$$
$$- \beta\sqrt{p(1-p)}\left(C^1(N_N) - C^0(N_N)\right) N_N \tag{8.25}$$

subject to

$$\alpha\left((1-p)\,C^0(N_N) + p\,C^1(N_N)\right) + \beta\sqrt{p(1-p)}\left(C^1(N_N) - C^0(N_N)\right) + f = D(N_N)$$

Using the Lagrangrian technique, the optimal flat fee f has the following form:

$$f = \alpha\left\{(1-p)\cdot C^{0'}(N_N) + p\cdot C^{1'}(N_N)\right\}\cdot N_N + \beta\cdot\sqrt{p(1-p)}\left(C^{1'}(N_N) - C^{0'}(N_N)\right)\cdot N_N \tag{8.26}$$

The first term on the right-hand side of (8.26) denotes the expected external congestion costs; the second term reflects the external uncertainty costs.

Depending both on the functional forms of the demand and travel time functions, and on the values of the parameters, one of the three situations discussed above yields the system optimum for the transport network. In the next section, the welfare improving properties of information provision are compared with welfare levels under system optimum (first-best) behaviour.

8.3.2 Relative welfare effects of endogenous information provision

The impact of information on the relative efficiency of road usage will be measured using the performance measure ω, which has also been used in the previous chapters:

$$\omega = \frac{\textit{Welfare(Endogenous Information)} - \textit{Welfare(Non Intervention)}}{\textit{Welfare(System Optimum)} - \textit{Welfare(Non Intervention)}} \quad (8.27)$$

Welfare(Endogenous Information) denotes the amount of welfare generated by model E (Section 8.2.4), while *Welfare(Non Intervention)* represents welfare under model N (no information available, Section 8.2.1). ω then provides the achievable welfare gains as a proportion of the theoretically possible welfare gains. Clearly, ω cannot exceed the value of one. In addition, ω cannot be smaller than zero, since it was shown that endogenous provision of information leads to a strict Pareto improvement, see Proposition 8.3; the numerator of ω cannot take on negative values.

Various experiments have been conducted to assess the impact of key parameters of the model on the relative efficiency of information provision. Similar to the specifications underlying the propositions in Section 8.2, the functional form of the demand function used in the experiments is given by D(N)=d-aN; the link travel time functions are specified as $C^j(N)=k^j+b^jN$ (j=0,1).

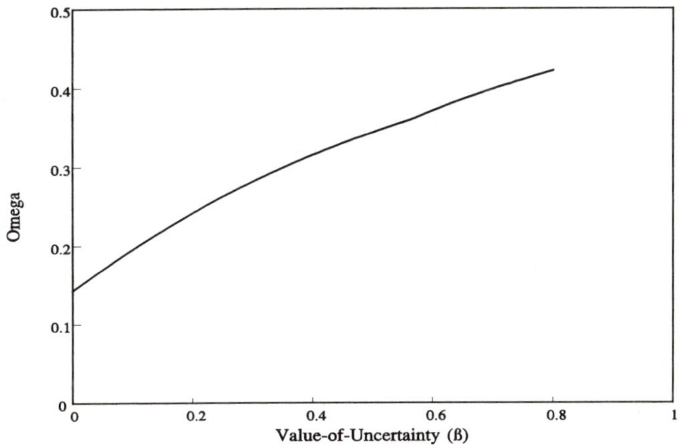

Figure 8.4 Relative welfare improvement as a function of the value-of-uncertainty (ß).

In previous chapters, costs related to uncertainty were not included in the generalised cost function, that is, ß was assumed to be equal to zero. Figure 8.4 however, shows that the inclusion of uncertainty costs strongly influences the relative efficiency of information provision.[5] The more important the uncertainty costs component, the higher the efficiency of information provision. Hence, the results obtained in previous

[5]The following parameters were used to produce Figure 8.4: d=50, a=0.015, p=0.25, $k^0=k^1=20$, $b^0=0.015$, $b^1=0.04$, $\alpha=1$, $\pi=2$.

chapters provide a lower bound for the potential efficiency gains of information; the positive impacts of information provision are underestimated when costs of uncertainty are ignored. Of course, it should be envisaged here that the information provided is perfect.

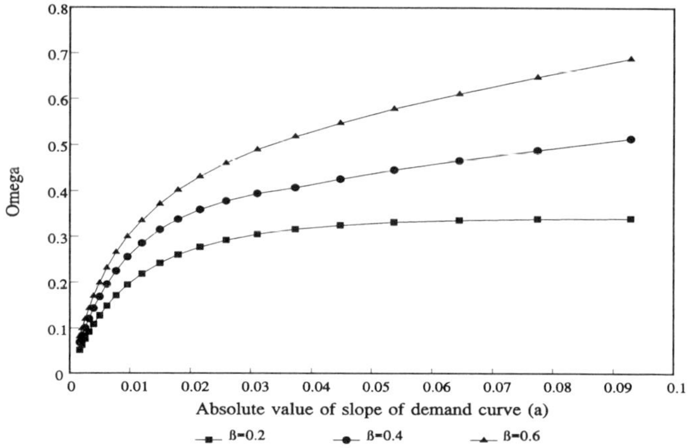

Figure 8.5 Relative welfare improvement as a function of the absolute value of the slope of the demand function.

Next, Figure 8.5 shows the impact of the slope of the demand curve on the relative efficiency of information for three values of ß. The slope of the demand curve is related to the elasticity of demand for road usage. In order to carry out a useful comparison, the level of road usage under the non-intervention case (model N) is kept constant.[6] On the left-hand side of Figure 8.5, demand is relatively elastic; on the right-hand side it is rather inelastic. Other researchers investigating the impact of information generally confined themselves to the case of inelastic demand. In the model, inelastic demand is just a limiting case, see the right-hand side of Figure 8.5. As shown in Figure 8.5, when only considering the case of inelastic demand, the impacts of information are likely to be overestimated. The more elastic demand, the less favourable the welfare improving properties of information.

8.4 CONCLUSION

In this chapter, the impacts of information provision to travellers were analysed using a stochastic equilibrium concept. Work in previous chapters was advanced by including an additional term related to the costs of travel time uncertainty into the

[6]The following parameters were used to produce Figure 8.5: p=0.25, $k^0=k^1=20$, $b^0=0.015$, $b^1=0.04$, $\alpha=1$, $\pi=2$. The non-intervention (model N) level of road usage is calculated with the demand function parameters d=50 and a=0.015.

generalised cost function. Costs of uncertainty have frequently been mentioned as being important in the travel behaviour literature, particularly with respect to driver information systems.

Four models were discussed. It was proven that for realistic values of the value-of-uncertainty parameter (ß), information is beneficial to both the informed and uninformed drivers, i.e., information will lead to a strict Pareto improvement. For exceptionally large ß-values however, *exogenous* provision of information (model P) might negatively affect the uninformed drivers.

The relative efficiency of endogenous information provision (model E) was analysed, using the system optimal welfare level as a benchmark. System optimal welfare can be achieved by means of a fluctuating (depending on the level of congestion) road pricing scheme. With the inclusion of costs related to uncertainty, the optimal fluctuating road price depends on both the *external* congestion costs and the *external* uncertainty costs. Experiments revealed that the larger the costs of uncertainty, the more attractive information as a policy option to optimise traffic flows. Finally, it was demonstrated that with inelastic demand information is more efficient. Therefore, other researchers who confined themselves to inelastic demand are likely to overestimate the impacts of driver information.

In the next chapter, the analysis is extended to a network structure with more than one origin-destination pair.

APPENDIX 8.A1 PROOF OF PROPOSITION 8.1

Proposition 8.1: Due to information to all travellers:
(1) expected road usage increases;
(2) expected network travel costs decrease;
(3) none of the road users is worse off and consequently, the system welfare in model I does not fall short of system welfare in model N.

Proof:

Proposition 8.1 (1): It has to be shown that $E(N_I)=(1-p)N_I^0+pN_I^1>N_N$. Using the equilibrium conditions of model I and the linearity of D, it follows that:

$$E(N_I) = D^{-1}\big(\alpha\cdot\big((1-p)\cdot C^0(N_I^0) + p\cdot C^1(N_I^1)\big)\big) \tag{8A.1}$$

Now substituting

$$\begin{aligned} N_I^0 &= E(N_I) + p\cdot(N_I^0-N_I^1) \\ N_I^1 &= E(N_I) + (1-p)\cdot(N_I^1-N_I^0) \end{aligned} \tag{8A.2}$$

it follows that:

$$E(N_I) = D^{-1}\big(\alpha\cdot\big((1-p)\cdot C^0(E(N_I)) + p\cdot C^1(E(N_I))\big) + \alpha\cdot p\cdot(1-p)\cdot(b^1-b^0)\cdot(N_I^1-N_I^0)\big) \tag{8A.3}$$

Since $b^1>b^0$ and $N_I^1<N_I^0$

$$\Delta_y = \alpha\cdot p\cdot(1-p)\cdot(b^1-b^0)\cdot(N_I^1-N_I^0) < 0 \tag{8A.4}$$

Rewriting the equilibrium condition of model N gives:

$$N_N = D^{-1}\big(\alpha\cdot\big((1-p)\cdot C^0(N_N) + p\cdot C^1(N_N)\big) + \beta\cdot\sqrt{p(1-p)}\cdot\big(C^1(N_N)-C^0(N_N)\big)\big) \tag{8A.5}$$

Obviously,

$$\Delta_x = \beta\cdot\sqrt{p(1-p)}\cdot\big(C^1(N_N)-C^0(N_N)\big) > 0 \tag{8A.6}$$

Now define $x=N_N$ and $y=E(N_I)$. Then, the reasoning above shows that:

$$x = D^{-1}\big(\alpha\cdot\big((1-p)\cdot C^0(x) + p\cdot C^1(x)\big) + \Delta_x\big) = D^{-1}\big(\alpha\cdot\big((1-p)\cdot C^0(x) + p\cdot C^1(x)\big)\big) + \Delta_{xx} \tag{8A.7}$$

and

$$y = D^{-1}\big(\alpha\cdot\big((1-p)\cdot C^0(y) + p\cdot C^1(y)\big) + \Delta_y\big) = D^{-1}\big(\alpha\cdot\big((1-p)\cdot C^0(y) + p\cdot C^1(y)\big)\big) + \Delta_{yy} \tag{8A.8}$$

for some $\Delta_{xx}<0$ and $\Delta_{yy}>0$, since D and thus also D^{-1} is a non increasing function, $\Delta_x>0$ and $\Delta_y<0$. Again using the non increasing property of D^{-1} it follows that $N_N=x<y=E(N_I)$.

Proposition 8.1 (2): Follows directly from Proposition 8.1 (1), since the expected travel costs in model N are given by $D(N_N)$, and in model I by $D(E(N_I))$. Applying Proposition 8.1 (1) and using the non increasing property of D leads to the desired result.

Proposition 8.1 (3): Next, it will be proven that the expected net private benefits in Model I do not fall short of those in Model N; hence, due to information none of the road users will be worse off.

First, consider case 1: $N_I^1<N_N<N_I^0$. The notation $npc_M(N)$ will be used, referring to the expected net private costs of traveller N in model M (M=N,I) (ranked from high to low willingness-to-pay). It can be shown that $npc_I(N) \leq npc_N(N)$ for all N, then the expected net private benefits in Model I do not fall short of those in Model N.

It can easily be proven that:

- $npc_N(N)=D(N_N) \geq npc_I(N)=D(E(N_I))$ for all $n \in [0,N_I^1]$, since $N_N<E(N_I))$.
- $npc_N(N)=D(N_N) \geq npc_I(N)=(1-p)D(N_I^0)+pD(N)$ for all $N \in [N_I^1,N_N]$, since $N \geq N_I^1$.
- $npc_N(N)=D(N) \geq npc_I(N)=(1-p)D(N_I^0)+pD(N)$ for all $N \in [N_N,N_I^0]$, since $N \leq N_I^0$.
- $npc_N(N)=D(N) = npc_I(N)=D(N)$ for all $N \in [N_I^0,\infty]$.

This completes the proof for case 1: $N_I^1<N_N<N_I^0$.

A similar reasoning holds for case 2: $N_N<N_I^1<N_I^0$:

- $npc_N(N)=D(N_N) \geq npc_I(N)=D(E(N_I))$ for all $n \in [0,N_N]$, since $N_N<E(N_I))$.
- $npc_N(N)=D(N) \geq npc_I(N)=D(E(N_I))$ for all $N \in [N_N,N_I^1]$, since $N_N \leq N_I^1 \leq E(N_I))$.
- $npc_N(N)=D(N) \geq npc_I(N)=(1-p)D(N_I^0)+pD(N)$ for all $N \in [N_I^1,N_I^0]$, since $N \leq N_I^0$.
- $npc_N(N)=D(N) = npc_I(N)=D(N)$ for all $N \in [N_I^0,\infty]$.

This completes the proof for case 2.

A direct consequence of the reasoning above is that system welfare - measured by the sum of the benefits of all the individuals minus the sum of the costs of all the individuals - in model I exceeds system welfare in model N.

APPENDIX 8.A2 PROOF OF PROPOSITION 8.2

Proposition 8.2: Due to information to an exogenously determined group of travellers:
(1) total expected road usage increases;
(2) expected road usage of the group of informed travellers increases;
(3) expected network travel costs of the group of informed travellers decreases;
(4) none of the travellers of the informed group is worse off and consequently, welfare for the informed travellers does not decrease.

Proof:

First, the equilibrium conditions of both models are rewritten. Model P can be rewritten to:

$$\alpha \cdot C^0(N^0_{P,x}+N_{P,y}) = D_x(N^0_{P,x}) \tag{8A.9}$$

$$\alpha \cdot C^1(N^1_{P,x}+N_{P,y}) = D_x(N^1_{P,x}) \tag{8A.10}$$

$$\gamma^0 \cdot C^0(N^0_{P,x}+N_{P,y}) + \gamma^1 \cdot C^1(N^1_{P,x}+N_{P,y}) = D_y(N_{P,y}) \tag{8A.11}$$

with

$$\begin{aligned} \gamma^0 &= \alpha(1-p) - \beta\sqrt{p(1-p)} \\ \gamma^1 &= \alpha p + \beta\sqrt{p(1-p)} \end{aligned} \tag{8A.12}$$

Model N for two groups of potential travellers reads as:

$$\gamma^0 \cdot C^0(N_{N,x}+N_{N,y}) + \gamma^1 \cdot C^1(N_{N,x}+N_{N,y}) = D_x(N_{N,x}) \tag{8A.13}$$

$$\gamma^0 \cdot C^0(N_{N,x}+N_{N,y}) + \gamma^1 \cdot C^1(N_{N,x}+N_{N,y}) = D_y(N_{N,y}) \tag{8A.14}$$

Proposition 8.2 (1): First, it will be proven that total expected road usage will increase due to information to group x, i.e. $N_{P,y}+E(N_{P,x})>N_{N,y}+N_{N,x}$. Using the equilibrium conditions of model N, total road usage in model N can be expressed as:

$$\begin{gathered} N_{N,y}+N_{N,x} = \\ D_y^{-1}\left(\gamma^0 \cdot C^0(N_{N,x}+N_{N,y}) + \gamma^1 \cdot C^1(N_{N,x}+N_{N,y})\right) + D_x^{-1}\left(\gamma^0 \cdot C^0(N_{N,x}+N_{N,y}) + \gamma^1 \cdot C^1(N_{N,x}+N_{N,y})\right) \end{gathered} \tag{8A.15}$$

Total expected road usage in model P can be written as:

$$\begin{aligned}
N_{P,y} + E(N_{P,x}) &= D_y^{-1}\Big(\gamma^0 \cdot C^0(N_{P,x}^0 + N_{P,y}) + \gamma^1 \cdot C^1(N_{P,x}^1 + N_{P,y})\Big) + D_x^{-1}\Big(\alpha \cdot \big((1-p) \cdot C^0(N_{P,x}^0 + N_{P,y}) + \\
&\quad p \cdot C^1(N_{P,x}^1 + N_{P,y})\big)\Big) \\
&= D_y^{-1}\Big(\gamma^0 \cdot C^0\big(N_{P,y} + E(N_{P,x}) + p \cdot (N_{P,x}^0 - N_{P,x}^1)\big) + \gamma^1 \cdot C^1\big(N_{P,y} + E(N_{P,x}) + (1-p) \cdot (N_{P,x}^1 - N_{P,x}^0)\big)\Big) + \\
&\quad D_x^{-1}\Big(\alpha \cdot \big((1-p) \cdot C^0\big(N_{P,y} + E(N_{P,x}) + p \cdot (N_{P,x}^0 - N_{P,x}^1)\big) + p \cdot C^1\big(N_{P,y} + E(N_{P,x}) + (1-p) \cdot (N_{P,x}^1 - N_{P,x}^0)\big)\big)\Big) \\
&= D_y^{-1}\Big(\gamma^0 \cdot C^0\big(N_{P,y} + E(N_{P,x})\big) + \gamma^1 \cdot C^1\big(N_{P,y} + E(N_{P,x})\big) + (N_{P,x}^1 - N_{P,x}^0)\big(\gamma^1 b^1 (1-p) - \gamma^0 b^0 p\big)\Big) + \\
&\quad D_x^{-1}\Big(\alpha\big((1-p) \cdot C^0\big(N_{P,y} + E(N_{P,x})\big) + p \cdot C^1\big(N_{P,y} + E(N_{P,x})\big)\big) + (N_{P,x}^1 - N_{P,x}^0)\alpha p(1-p)(b^1 - b^0)\Big) \qquad \text{(8A.16)} \\
&= D_y^{-1}\Big(\gamma^0 \cdot C^0\big(N_{P,y} + E(N_{P,x})\big) + \gamma^1 \cdot C^1\big(N_{P,y} + E(N_{P,x})\big) + (N_{P,x}^1 - N_{P,x}^0)\big(\gamma^1 b^1 (1-p) - \gamma^0 b^0 p\big)\Big) + \\
&\quad D_x^{-1}\Big(\gamma^0 \cdot C^0\big(N_{P,y} + E(N_{P,x})\big) + \gamma^1 \cdot C^1\big(N_{P,y} + E(N_{P,x})\big) + \\
&\quad (b^1 - b^0)\big((N_{P,y}^1 - N_{P,x}^0)\alpha p(1-p) - (N_{P,y} + E(N_{P,x}))\beta\sqrt{p(1-p)}\big) + k^n\Big)
\end{aligned}$$

$$\textit{with } k^n = \beta\sqrt{p(1-p)}(k^0 - k^1) \le 0$$

where the first equality follows from the equilibrium conditions of model P and the linearity of D_x. Now define $s=N_{N,y}+N_{N,x}$ and $t=N_{P,y}+E(N_{P,x})$. Then, the above expressions can be written as:

$$\begin{aligned}
s &= D_y^{-1}\big(\gamma^0 \cdot C^0(s) + \gamma^1 \cdot C^1(s)\big) + D_x^{-1}\big(\gamma^0 \cdot C^0(s) + \gamma^1 \cdot C^1(s)\big) \\
t &= D_y^{-1}\big(\gamma^0 \cdot C^0(t) + \gamma^1 \cdot C^1(t) + \Delta_{y,t}\big) + D_x^{-1}\big(\gamma^0 \cdot C^0(t) + \gamma^1 \cdot C^1(t) + \Delta_{x,t}\big)
\end{aligned}$$

$$\begin{aligned}
\textit{with } \Delta_{y,t} &= (N_{P,x}^1 - N_{P,x}^0) \cdot \big(\gamma^1 b^1 (1-p) - \gamma^0 b^0 p\big) \\
&= (N_{P,x}^1 - N_{P,x}^0) \cdot \big(\alpha p(1-p)(b^1 - b^0) + \beta\sqrt{p(1-p)}(b^1(1-p) + b^0 p)\big) \qquad \text{(8A.17)} \\
&< 0 \\
\Delta_{x,t} &= (b^1 - b^0) \cdot \big((N_{P,x}^1 - N_{P,x}^0)\alpha p(1-p) - (N_{P,y} + E(N_{P,x}))\beta\sqrt{p(1-p)}\big) + k^n \\
&< 0
\end{aligned}$$

Because D_y^{-1} and D_x^{-1} are non increasing functions, and $\Delta_{y,t}<0$ and $\Delta_{x,t}<0$ for all t there exists a $\Delta>0$ so that:

$$t = D_y^{-1}\big(\gamma^0 \cdot C^0(t) + \gamma^1 \cdot C^1(t)\big) + D_x^{-1}\big(\gamma^0 \cdot C^0(t) + \gamma^1 \cdot C^1(t)\big) + \Delta \qquad \text{(8A.18)}$$

$$\textit{with } \Delta > 0$$

Again using the non increasing property of D_y^{-1} and D_x^{1} it follows that t>s, i.e. $N_{P,y}+E(N_{P,x})>N_{N,y}+N_{N,x}$. This completes the proof of Proposition 8.2 (1).

Proposition 8.2 (2): First, define $s=N_{N,x}$ and $t=E(N_{P,x})$. Using the equilibrium conditions of model N and P and the reasoning above it follows that:

$$\begin{aligned}
s &= D_x^{-1}\big(\gamma^0 \cdot C^0(s + N_{N,y}) + \gamma^1 \cdot C^1(s + N_{N,y})\big) \\
t &= D_x^{-1}\big(\gamma^0 \cdot C^0(t + N_{P,y}) + \gamma^1 \cdot C^1(t + N_{P,y}) + \Delta_{x,t}\big)
\end{aligned} \qquad \text{(8A.19)}$$

Define Δ_N so that $N_{N,y}+\Delta_N=N_{P,y}$. Then, the above expressions can be written as:

$$\begin{aligned} s &= D_x^{-1}\left(\gamma \cdot C^0(s+N_{N,y}) + \gamma^1 \cdot C^1(s+N_{N,y})\right) \\ t &= D_x^{-1}\left(\gamma^0 \cdot C^0(t+N_{N,y}+\Delta_N) + \gamma^1 \cdot C^1(t+N_{N,y}+\Delta_N) + \Delta_{x,t}\right) \\ &= D_x^{-1}\left(\gamma^0 \cdot C^0(t+N_{N,y}) + \gamma^1 \cdot C^1(t+N_{N,y}) + \gamma^0 b^0 \Delta_N + \gamma^1 b^1 \Delta_N + \Delta_{x,t}\right) \\ &= D_x^{-1}\left(\gamma^0 \cdot C^0(t+N_{N,y}) + \gamma^1 \cdot C^1(t+N_{N,y}) + \theta\right) \end{aligned} \quad (8A.20)$$

$$with\ \theta = \alpha p \Delta_N (b^1-b^0) + \alpha b^0 \Delta_N - (b^1-b^0)(t+N_{N,y})\beta\sqrt{p(1-p)} + (b^1-b^0)(N^1_{P,x}-N^0_{P,x})\alpha p(1-p) + k^n$$

From the non increasing property of D_x it follows that s is smaller than t if $\theta<0$. To prove that $\theta<0$ three cases have to be distinguished between:

1: $\Delta_N=0$: It can easily be seen that θ is smaller than 0.
2: $\Delta_N<0$: It can easily be seen that θ is smaller than 0.
3: $\Delta_N>0$: Using the equilibrium conditions of the model N and P, and $\Delta_N>0$ it follows that:

$$\begin{aligned} N_{P,y} &= D_y^{-1}\left(\gamma^0 \cdot C^0(N^0_{P,x}+N_{P,y}) + \gamma^1 \cdot C^1(N^1_{P,x}+N_{P,y})\right) \\ &> D_y^{-1}\left(\gamma^0 \cdot C^0(N_{N,x}+N_{N,y}) + \gamma^1 \cdot C^1(N_{N,x}+N_{N,y})\right) \\ &= N_{N,y} \end{aligned} \quad (8A.21)$$

This implies that:

$$\begin{aligned} &\gamma^0 \cdot C^0(N^0_{P,x}+N_{P,y}) + \gamma^1 \cdot C^1(N^1_{P,x}+N_{P,y}) < \gamma^0 \cdot C^0(N_{N,x}+N_{N,y}) + \gamma^1 \cdot C^1(N_{N,x}+N_{N,y}) \\ &\Leftrightarrow \\ &\alpha(1-p)C^0(N^0_{P,x}+N_{P,y}) + \alpha p C^1(N^1_{P,x}+N_{P,y}) + \beta\sqrt{p(1-p)}\cdot\left(C^1(N^1_{P,x}+N_{P,y}) - C^0(N^0_{P,x}+N_{P,y})\right) < \\ &\gamma^0 \cdot C^0(N_{N,x}+N_{N,y}) + \gamma^1 \cdot C^1(N+_{N,x}+N_{N,y}) \end{aligned} \quad (8A.22)$$

Since, $C^1(N^1_{P,x}+N_{P,y})>C^0(N^0_{P,x}+N_{P,y})$, it follows that:

$$\alpha(1-p)C^0(N^0_{P,x}+N_{P,y}) + \alpha p C^1(N^1_{P,x}+N_{P,y}) < \gamma^0 \cdot C^0(N_{N,x}+N_{N,y}) + \gamma^1 \cdot C^1(N+_{N,x}+N_{N,y}) \quad (8A.23)$$

Due to the non increasing property of D_x this means that:

$$\begin{aligned} E(N_{P,x}) &= D_x^{-1}\left(\alpha(1-p)C^0(N^0_{P,x}+N_{P,y}) + \alpha p C^1(N^1_{P,x}+N_{P,y})\right) > \\ &D_x^{-1}\left(\gamma^0 \cdot C^0(N_{N,x}+N_{N,y}) + \gamma^1 \cdot C^1(N+_{N,x}+N_{N,y})\right) = N_{N,x} \end{aligned} \quad (8A.24)$$

This completes the proof of Proposition 8.2 (2).

<u>Proposition 8.2 (3):</u> Proposition 8.2 (3) follows directly from Proposition 8.2 (1), since the expected travel costs in model N are given by $D_x(N_{N,x})$, and in model P by $D_x(E(N_{P,x}))$. Applying Proposition 8.2 (1) and using the non increasing property of D_x leads to the desired result.

<u>Proposition 8.2 (4):</u> Finally, it will be proven that all informed road users are better off due to the information. Two cases will be distinguished between: $N^1_{P,x}<N_{N,x}<N^0_{P,x}$ and $N_{N,x}<N^1_{P,x}<N^0_{P,x}$:

Case 1: $N^1_{P,x}<N_{N,x}<N^0_{P,x}$:

1. Road users in $[0,N^1_{P,x}]$ always use the network. They benefit from a decline in expected network travel costs, see Proposition 8.2 (3).
2. Road users in $<N^1_{P,x},N_{N,x}]$: Travel costs without information: $D_x(N_{N,x})$. Travel costs with information: $(1-p)D_x(N^0_{P,x})+pD_x(N) < D_x(E(N_{P,x})) < D_x(N_{N,x})$ (because of Proposition 8.2 (1)).

3. Road users in $[N_{N,x}, N^0_{P,x}]$: Travel costs without information: $D_x(N)$. Travel costs with information: $(1\text{-}p)D_x(N^0_{P,x})+pD_x(N) < D_x(N)$, since $N^0_{P,x}>N$.

Case 2: $N_{N,x}<N^1_{P,x}<N^0_{P,x}$:

1. Road users in $[0,N_{N,x}]$ always use the network. They benefit from a decline in expected network travel costs, see Proposition 8.2 (3).
2. Road users in $<N_{N,x},N^1_{P,x}]$: Travel costs without information $D(N)$. Travel costs with information: $D(E(N_{P,x})) < D(N)$, because $N<E(N_{P,x})$.
3. Road users in $[N^1_{P,x},N^0_{P,x}]$: Travel costs without information $D(N)$. Travel costs with information: $(1\text{-}p)D(N^0_{P,x})+pD(N) < D(N)$, because $N<N^0_{P,x}$.

This completes the proof of Proposition 8.2.

APPENDIX 8.A3 PROOF OF PROPOSITION 8.3

Proposition 8.3: Due to endogenous provision of information:

(1) expected road usage increases;

(2) for all road users expected network travel costs do not increase;

(3) none of the travellers is worse off and consequently, the system welfare in model E does not fall short of system welfare in model N.

Proof:

Proposition 8.3 (1): It has to be proven that $N_N<E(N_E)$. The case that $N_N<N_E^1<N_E^0$ is obvious. It will therefore be assumed that $N_E^1<N_N<N_E^0$. Rewriting the equilibrium condition of model N gives:

$$N_N = D^{-1}\left(\alpha\cdot\left((1-p)\cdot C^0(N_N) + p\cdot C^1(N_N)\right) + \beta\cdot\sqrt{p(1-p)}\cdot\left(C^1(N_N)-C^0(N_N)\right)\right) \quad (8A.25)$$

Further, consider first the situation that inequality (8.15) holds, implying that road users in the interval $[0,N_E^1]$ are not acquiring information. Using the equilibrium conditions of N_E^0 and N_E^1, the expected level of road usage in model E can be written as:

$$\begin{aligned}
E(N_E) &= (1-p)\cdot N_E^0 + p\cdot N_E^1\\
&= (1-p)\cdot D^{-1}\left(\alpha\cdot C^0(N_E^0) + \frac{\pi}{1-p}\right) + p\cdot D^{-1}\left(\alpha\cdot C^1(N_E^1) + \beta\sqrt{\frac{1-p}{p}}\left(C^1(N_E^1) - C^0(N_E^0)\right) - \frac{\pi}{p}\right)\\
&= D^{-1}\left(\alpha\left((1-p)\cdot C^0(N_E^0) + p\cdot C^1(N_E^1)\right) + \beta\sqrt{p(1-p)}\left(C^1(N_E^1) - C^0(N_E^0)\right)\right)\\
&= D^{-1}(\alpha\left((1-p)\cdot C^0\left(E(N_E)+p(N_E^0-N_E^1)\right) + p\cdot C^1\left(E(N_E)+(1-p)(N_E^1-N_E^0)\right)\right) +\\
&\quad \beta\sqrt{p(1-p)}\left(C^1\left(E(N_E)+(1-p)(N_E^1-N_E^0)\right) - C^0\left(E(N_E)+p(N_E^0-N_E^1)\right)\right))\\
&= D^{-1}\left(\alpha\left((1-p)\cdot C^0(E(N_E)) + p\cdot C^1(E(N_E))\right) + \beta\sqrt{p(1-p)}\left(C^1(E(N_E)) - C^0(E(N_E))\right) + \Delta_t\right)
\end{aligned} \quad (8A.26)$$

$$\text{with } \Delta_t = \alpha p(1-p)(b^0-b^1)(N_E^0-N_E^1) + \beta\sqrt{p(1-p)}(N_E^1-N_E^0)((1-p)b^1+pb^0) < 0$$

Writing $s=N_N$ and $t=E(N_E)$, then

$$s = D^{-1}\left(\alpha\cdot\left((1-p)\cdot C^0(s) + p\cdot C^1(s)\right) + \beta\cdot\sqrt{p(1-p)}\cdot\left(C^1(s)-C^0(s)\right)\right)$$

$$t = D^{-1}\left(\alpha\cdot\left((1-p)\cdot C^0(t) + p\cdot C^1(t)\right) + \beta\cdot\sqrt{p(1-p}\cdot\left(C^1(t)-C^0(t)\right) + \Delta_t\right) \quad (8A.27)$$

$$\text{with } \Delta_t < 0$$

Using the non increasing property of D, it follows that s<t.

Next, consider the case that inequality (8.18) holds, implying that all travellers in the interval $[0,N_E^0]$ are buying information. Using the equilibrium conditions for this situation, expected road usage in model E can be written as:

$$\begin{aligned}
E(N_E) &= (1-p)\cdot N_E^0 + p\cdot N_E^1 \\
&= (1-p)\cdot D^{-1}\left(\alpha C^0(N_E^0) + \frac{\pi}{1-p}\right) + p\cdot D^{-1}\left(\alpha C^1(N_E^1)\right) \\
&= D^{-1}\left(\alpha\left((1-p)\cdot C^0(N_E^0) + p\cdot C^1(N_E^1)\right) + \pi\right) \\
&= D^{-1}\left(\alpha\left((1-p)\cdot C^0(E(N_E)) + p\cdot C^1(E(N_E))\right) + \alpha(1-p)p(b^0-b^1)(N_E^0-N_E^1) + \pi\right) \\
&> D^{-1}\left(\alpha\left((1-p)\cdot C^0(E(N_E)) + p\cdot C^1(E(N_E))\right) + \alpha(1-p)p(b^0-b^1)(N_E^0-N_E^1) + \beta\sqrt{p(1-p)}\left(C^1(N_E^1) - C^0(N_E^0)\right)\right) \\
&= D^{-1}\left(\alpha\left((1-p)\cdot C^0(E(N_E)) + p\cdot C^1(E(N_E))\right) + \beta\sqrt{p(1-p)}\left(C^1(E(N_E)) - C^0(E(N_E))\right) + \Delta_t\right)
\end{aligned}$$

$$\text{with } \Delta_t = \alpha p(1-p)(b^0-b^1)(N_E^0-N_E^1) + \beta\sqrt{p(1-p)}(N_E^1-N_E^0)((1-p)b^1+pb^0) < 0 \tag{8A.28}$$

where the ">" sign follows from inequality (8.18) and the non increasing property of D. Hence, again using the non increasing property of D this implies that $N_N<E(N_E)$.

Proposition 8.3 (2): Four cases have to be distinguished between: First, either inequality (8.15) or (8.18) holds. Second, either $N_E^1<N_N<N_E^0$ or $N_N<N_E^1<N_E^0$. Below, the detailed proof is not provided, but rather an indication of how it is derived. Using the equilibrium conditions of the respective models, and using the property that $N_N<E(N_E)$ it can easily be shown that independent of whether inequality (8.15) or (8.18) holds true, the difference in expected private costs of model N and model E is given by the pattern in Figure 1 when $N_E^1<N_N<N_E^0$ holds true, and by the pattern in Figure 2 when $N_N<N_E^1<N_E^0$ holds true (where obviously subscript I has to be replaced by subscript E).

Proposition 8.3 (3): This is a direct consequence of Proposition 8.3 (2) and completes the proof of Proposition 8.3.

9 MOTORIST INFORMATION IN NETWORKS WITH MULTIPLE OD-PAIRS[1]

9.1 INTRODUCTION

The models presented in the previous chapters have yielded much insight into the key factors that influence the usefulness of information provision. However, an often mentioned drawback of these type of theoretical models is their lack of realism. This lack of realism has been taken for granted by the desire to answer policy questions related to information provision by using transparent analytical models rather than simulation models of more realistic networks that generate "black-box" type of answers. Nevertheless, one of the more serious restrictions in the theoretical models is the assumption of a single origin-destination (OD) pair and very simple network structures (either one or two-link networks); more complicated network structures are mathematically difficult to handle. The simplicity of these network structures might, however, seriously cloud the impacts of information provision in realistic networks.

In the present chapter the theoretical model proposed in Chapter 4 is extended in order to deal with a network consisting of two OD-pairs. The results obtained with this network are indicative for the impact of information in more realistic networks with more than two OD-pairs.

The chapter is organised as follows. In Section 9.2 the network structure to be analysed is discussed, while Section 9.3 formalises the models under consideration. In Section 9.4 the model experiments are presented. The conclusions are contained in Section 9.5.

9.2 THE NETWORK STRUCTURE

The network that will be used in the remainder of this chapter is depicted in Figure 9.1.[2] The network consists of two OD-pairs (O_1D_1 and O_2D_2) connected by seven links numbered 1, 2, 3, h1, h2, h3, and h4. The capacity of the network is assumed to be stochastic. It is assumed that with a probability p an incident occurs on link 1, thereby reducing its capacity. Informed drivers are informed about the occurrence of such an incident and they are therefore able to base their behaviour on *actual* costs; uninformed drivers, in contrast, base their behaviour on *expected* costs, see also Chapter 4.

---

[1]This chapter is based on Emmerink, Verhoef, Nijkamp and Rietveld (1996c) forthcoming in *Regional Science and Urban Economics*.

[2]The idea of this network originated from fruitful discussions with Peter Bonsall (University of Leeds) and Piet Bovy (Delft Institute of Technology) at the 7*th* World Conference on Transport Research, Sydney, Australia, July 16-21, 1995.

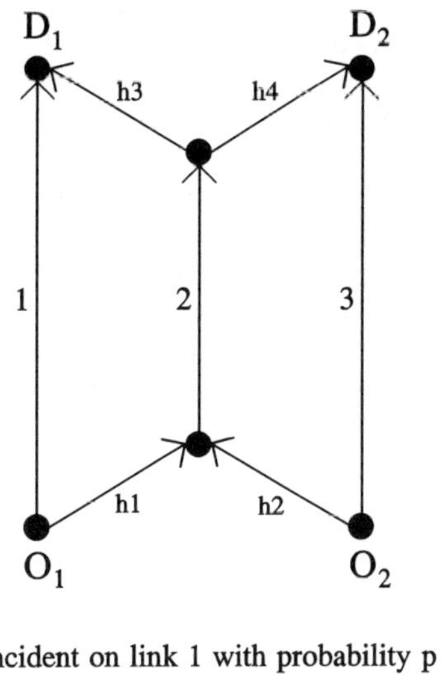

Figure 9.1 Network structure.

As an incident is only assumed to take place on link 1, the network under consideration is rather unbalanced. In fact, this network is the extreme example of a situation where the incident costs incurred by one group of travellers (the O_1D_1-travellers) can be passed on to another group of travellers (the O_2D_2-travellers). Consequently, this network is probably close to a worst case analysis for studying the welfare impacts of information provision on uninformed O_2D_2-travellers.

As all links indicated in Figure 9.1 represent one-way traffic, drivers travelling from origin O_1 to destination D_1 can choose between using route 1 or route 2. On the other hand, drivers travelling from origin O_2 to destination D_2 have the option to use either route 2 or route 3. In addition, as it is assumed that demand is elastic (implying that the number of drivers using the network is dependent on the costs of using the network), drivers from O_1D_1 and O_2D_2 have a third option of not using the network at all (for instance, to turn to an alternative mode). The elasticity of demand might also refer to travel time shifts, i.e., there is a demand for peak-hour travel which is elastic. If peak-hour travel is too congested people shift to travel times before or after the peak.

The route choice decisions of drivers from O_1D_1 and O_2D_2 interact on route 2, as this is the alternative that is available for both. Consequently, incidents on route 1 indirectly affect drivers travelling from O_2 to D_2, because O_1D_1-drivers diverting from route 1 to route 2 increase the travel time on the joint alternative. This, in turn, might also affect the travel time on route 3, as more O_2D_2-drivers might choose route 3 when the travel time on route 2 is relatively high.

The model experiments in Section 9.4 consist of two parts. First, in Section 9.4.1 it is assumed that the travel time on route 3 is infinitely high. Hence, in the case of an accident on route 1, drivers willing to travel from O_2 to D_2 have the option either to undertake the trip by using route 2 and consequently incur higher travel costs or to refrain from using the network (elastic demand). On the other hand, drivers willing to travel from O_1 to D_1 have three alternatives: (1) continue using route 1; (2) diverting to route 2; (3) abstaining from using the network. Second, in Section 9.4.2 it is assumed that drivers travelling from O_2 to D_2 can also choose route 3. Therefore, in case of an incident on route 1, O_2D_2-drivers then have the choice between route 2, route 3 or refraining from using the network.

In the chapter, it is assumed that the travel times on the links h1, h2, h3 and h4 are equal to zero and therefore independent of the number of drivers using these links. Without making this assumption, the complexity of the analysis and the number of parameters in the model would unnecessarily be raised. Furthermore, for notational convenience, the origin-destination pair O_1D_1 is abbreviated with x and the pair O_2D_2 with y in the remainder of the chapter.

9.3 THE MODELS

In the experiments in Section 9.4 the performance of three models is compared. First, model N is described; a model where no information is available. Travellers base their decision-making on expected travel costs. Second, model P_x is discussed; a model where information is available for all drivers willing to travel from O_1 to D_1. Third, model P_{x+y} is portrayed; a model where information is available for all the travellers: both those travelling from O_1 to D_1 and those travelling from O_2 to D_2.

A comparison between model N and model P_x yields insight into the impact of information on the uninformed drivers travelling from O_2 to D_2. In single OD-networks it was shown in previous chapters that information is both beneficial to informed and uninformed drivers. The present analysis will detect whether this still holds true for a network structure with multiple OD-pairs.

Next, it is worthwhile to compare model P_x and model P_{x+y} in order to find out whether:

1. information to y-drivers (travelling from O_2 to D_2) will benefit y-drivers;
2. information to y-drivers will also be beneficial to x-drivers;
3. further information provision is beneficial from a system perspective.

First, Section 9.3.1 is devoted to clarifying the model's principles as described in Chapter 4. Next Section 9.3.2 provides a detailed description of the models N, P_x and P_{x+y}, while in Section 9.3.3 the concept of *complete regularity* is introduced. A concept that will prove useful in the experiments in Section 9.4.

9.3.1 Principles of the models

Demand for using the network depicted in Figure 9.1 is origin-destination specific. The *inverse demand function* for x-travellers is given by $D_x(N_x)$, where N_x denotes the level of road usage of group x. Clearly, $D_x(N_x)$ is a decreasing function in N_x and reflects the willingness-to-pay (or private benefits) of the population of x-travellers. A similar relationship holds for y-travellers: their inverse demand function is given by $D_y(N_y)$.

The link travel cost functions for route 2 and route 3 are given by $C_r(N_r)$ (r=2, 3), where N_r denotes the number of drivers using route r. On route 1, an incident occurs with a fixed probability p and leads to a decrease in capacity. It is assumed that the low capacity link travel cost function of route 1 is given by $C_1^1(N_1)$ and occurs with probability p. The high capacity link travel cost function has a probability of occurrence equal to 1-p and is denoted by $C_1^0(N_1)$. Superscript 1 indicates low capacity, superscript 0 denotes high capacity; superscript 0 and 1 are called state 0 and state 1, respectively. The relationship between $C_1^0(N_1)$ and $C_1^1(N_1)$ is given below:

$$C_1^0(N_1) < C_1^1(N_1) \quad \text{and} \quad \frac{\partial C_1^0(N_1)}{\partial N_1} < \frac{\partial C_1^1(N_1)}{\partial N_1} \quad \text{for all } N_1>0 \qquad (9.1)$$

If drivers are informed, they are assumed to know the actual state of route 1 (either state 0 or 1). Consequently, by making their trip decision informed drivers base themselves on the actual costs of using a route in the network. On the other hand, uninformed drivers do not know the state of route 1, and therefore base their trip decision on the expected costs of using a route. The equilibrium principle can now be phrased in terms of private benefits and private costs. An *informed* driver uses a particular route only if the *actual* costs of using that route do not exceed the private benefits of the informed driver. An *uninformed* driver uses a particular route only if the *expected* costs of using that route do not exceed the private benefits of the uninformed driver. The model thus follows the SNDUE principles.

9.3.2 Formulation of the models

For reasons of space, the three models are formulated for the situation in which the travel costs on route 3 are infinitely high, implying that route 3 is not a feasible alternative for y-drivers. It is straightforward to formulate the models for the situation in which route 3 is a viable alternative.

Model N

In model N it is assumed that no information is available. Therefore, all travellers base themselves on expected costs. Model N is given in expressions (9.2) to (9.4):

$$D_x(N_{1,x}+N_{2,x}) \leq (1-p)\,C_1^0(N_{1,x}) + p\,C_1^1(N_{1,x}),\quad N_1^x \geq 0 \quad \text{and}$$
$$N_{1,x}\Big(D_x(N_{1,x}+N_{2,x}) - \big((1-p)\,C_1^0(N_{1,x}) + p\,C_1^1(N_{1,x})\big)\Big) = 0 \tag{9.2}$$

$$D_x(N_{1,x}+N_{2,x}) \leq C_2(N_{2,x}+N_{2,y}),\quad N_{2,x} \geq 0 \quad \text{and}$$
$$N_{2,x}\big(D_x(N_{1,x}+N_{2,x}) - C_2(N_{2,x}+N_{2,y})\big) = 0 \tag{9.3}$$

$$D_y(N_{2,y}) \leq C_2(N_{2,x}+N_{2,y}),\quad N_{2,y} \geq 0 \quad \text{and}$$
$$N_{2,y}\big(D_y(N_{2,y}) - C_2(N_{2,x}+N_{2,y})\big) = 0 \tag{9.4}$$

The equilibrium levels of road usage are denoted by $N_{1,x}$, $N_{2,x}$ and $N_{2,y}$, where the first subscript denotes the route (1 or 2), and the second the origin-destination pair (x or y). Hence, the number of drivers using route 1 is equal to $N_{1,x}$, while $N_{2,x}+N_{2,y}$ drivers use route 2.

First, the expressions ensure that levels of road usage are nonnegative. Second, if levels of road usage are positive, then this implies the equality between private benefits and private costs for the marginal road user. In addition, if all levels of road usage are positive, then it can be easily seen that expected travel costs are identical on both routes, thereby satisfying Wardrop's user equilibrium principle (Wardrop, 1952).

In fact, expressions (9.2) to (9.4) form a complementarity problem and can be rewritten in terms of a variational inequality problem (Nagurney, 1993). This might help in solving this type of equilibrium models in very complex networks.

Model P_x

In model P_x it is assumed that information is available to the travellers with origin O_1 and destination D_1. These travellers base themselves on actual travel costs, while y-travellers use expected travel costs. Model P_x is given in the expressions (9.5) to (9.7):

$$D_x(N_{1,x}^s+N_{2,x}^s) \leq C_1^s(N_{1,x}^s),\quad N_{1,x}^s \geq 0 \quad \text{and}$$
$$N_{1,x}^s\big(D_x(N_{1,x}^s+N_{2,x}^s) - C_1^s(N_{1,x}^s)\big) = 0 \qquad s=0,1 \tag{9.5}$$

$$D_x(N_{1,x}^s+N_{2,x}^s) \leq C_2(N_{2,x}^s+N_{2,y}),\quad N_{2,x}^s \geq 0 \quad \text{and}$$
$$N_{2,x}^s\big(D_x(N_{1,x}^s+N_{2,x}^s) - C_2(N_{2,x}^s+N_{2,y})\big) = 0 \qquad s=0,1 \tag{9.6}$$

$$D_y(N_{2,y}) \leq (1-p)\,C_2(N_{2,x}^0+N_{2,y}) \;+\; p\,C_2(N_{2,x}^1+N_{2,y}),\quad N_{2,y} \geq 0$$
$$\text{and}$$
$$N_{2,y}\Big(D_y(N_{2,y}) \;-\; \big((1-p)\,C_2(N_{2,x}^0+N_{2,y}) \;+\; p\,C_2(N_{2,x}^1+N_{2,y})\big)\Big) = 0 \tag{9.7}$$

The equilibrium levels of road usage are given by $N^0_{1,x}$, $N^0_{2,x}$, $N^1_{1,x}$, $N^1_{2,x}$ and $N_{2,y}$. The subscripts denotes route (1 or 2) and OD-pair, respectively; the superscripts denote state (0 or 1). Clearly, equilibrium values for drivers with origin O_1 and destination D_1 are state dependent: the informed drivers are able to adapt their behaviour according to the actual travel costs on the routes. Uninformed drivers (those with origin O_2 and destination D_2) do not know the actual travel costs and therefore base their behaviour on expected travel costs. Therefore, equilibrium levels of road usage of y-drivers are independent of the state. To conclude, if state 0 occurs then $N^0_{1,x}$ drivers use route 1, while $N^0_{2,x}+N_{2,y}$ drivers use route 2. On the other hand, if state 1 takes place then $N^1_{1,x}$ drivers use route 1, while $N^1_{2,x}+N_{2,y}$ drivers use route 2.

Model P_{x+y}

In model P_{x+y} it is assumed that information is available to all travellers, both those travelling from O_1 to D_1 and those travelling from O_2 to D_2. Behaviour for both groups is therefore based on actual travel costs. The equilibrium conditions are given in expressions (9.8) to (9.10):

$$\begin{aligned} &D_x(N^s_{1,x}+N^s_{2,x}) \leq C^s_1(N^s_{1,x}), \quad N^s_{1,x} \geq 0 \quad \text{and} \\ &N^s_{1,x}\left(D_x(N^s_{1,x}+N^s_{2,x}) - C^s_1(N^s_{1,x})\right) = 0 \quad s=0,1 \end{aligned} \tag{9.8}$$

$$\begin{aligned} &D_x(N^s_{1,x}+N^s_{2,x}) \leq C_2(N^s_{2,x}+N^s_{2,y}), \quad N^s_{2,x} \geq 0 \quad \text{and} \\ &N^s_{2,x}\left(D_x(N^s_{1,x}+N^s_{2,x}) - C_2(N^s_{2,x}+N^s_{2,y})\right) = 0 \quad s=0,1 \end{aligned} \tag{9.9}$$

$$\begin{aligned} &D_y(N^s_{2,y}) \leq C_2(N^s_{2,x}+N^s_{2,y}), \quad N^s_{2,y} \geq 0 \quad \text{and} \\ &N^s_{2,y}\left(D_y(N^s_{2,y}) - C_2(N^s_{2,x}+N^s_{2,y})\right) = 0 \quad s=0,1 \end{aligned} \tag{9.10}$$

The equilibrium levels of road usage are given by $N^0_{1,x}$, $N^0_{2,x}$, $N^1_{1,x}$, $N^1_{2,x}$ and $N^0_{2,y}$, $N^1_{2,y}$. Compared with model P_x, y-drivers now also have the opportunity to adapt levels of road usage of route 2 to the actual costs: equilibrium levels of y-travellers are state-dependent.

9.3.3 Complete regularity

The concept of *complete regularity* that will be widely used in the subsequent sections is defined as:

> An equilibrium of the models presented above is said to satisfy the *complete regularity* condition if and only if all levels of road usage are positive.

Of course, whether the complete regularity condition holds is dependent on the parameter values in the model. Compared with Chapter 6, Section 6.2, the complete regularity condition holds if the group-route-regularity condition holds for all groups and all routes. It can easily be derived that if the *complete regularity* condition is satisfied in model N, then expected link travel costs of route 1 and 2 are the same;

otherwise uninformed drivers would either use route 1 or route 2. Similarly, if the *complete regularity* condition is met in model P_x or P_{x+y} it implies that for both states travel costs on route 1 and 2 are equal.

9.4 MODEL EXPERIMENTS

9.4.1 Route 1 and 2 available

An analysis of the network without the availability of route 3 for y-travellers enables to determine whether information provision to x-travellers has a negative impact on y-travellers. In fact, since route 2 is the only route choice alternative available to y-travellers this network forms an extreme example to examine whether information benefits to one group (the informed x-travellers) are realised at the expense of the other group, the uninformed y-travellers.

It is assumed that both the link travel cost functions and the inverse demand functions are linear over the relevant ranges of road usage, that is, $D_g(N_g)=d_g-a_gN_g$ (g=x, y) and $C_1^0(N_1)=k_1^0+b_1^0N_1$, $C_1^1(N_1)=k_1^1+b_1^1N_1$, $C_2(N_2)=k_2+b_2N_2$, where d_g, a_g, k_1^0, b_1^0, k_1^1, b_1^1, k_2 and b_2 are the parameters of these functions. Given these linear functional forms, the equilibrium values of road usage follow from a system of linear equations with nonnegativity constraints imposed on the road usage values.

A regularity condition is imposed to restrict the experiments to relevant situations. It is assumed that for at least one state (either state 0 or state 1) some x-travellers use the joint route choice alternative (route 2), and that for at least one state (not necessarily the same state!) some y-travellers use route 2. If this assumption were not satisfied, then the x- and y-travellers do not interact (do not compete for road space on route 2), and the model could be split into two disjoint models: one for each origin-destination pair. This situation was analysed in previous chapters.

The base case parameter-values are chosen in the following fashion: $d_x=d_y=50$, $a_x=0.02$, $a_y=0.03$, p=0.25, $k_1^0=k_1^1=k_2=20$, $b_1^0=0.015$, $b_1^1=0.04$, $b_2=0.015$.

Experiments with models N and P_x

The key-parameters for testing whether uninformed y-travellers are negatively affected by the information provided to x-travellers are the characteristics of the joint alternative (route 2) in combination with the size of the stochastic capacity change of route 1.

Figure 9.2 and Figure 9.3 show the net welfare gain of model P_x compared to model N as a function of the stochastic capacity shock. More precisely, vertically the difference between total expected net private benefits (total private benefits minus total expected private costs) generated by model P_x and the total expected net private benefits generated by model N are shown. In Figure 9.2 the congestion cost parameter of route 2 (b_2) is equal to 0.015, while in Figure 9.3 this parameter takes on the value of 0.03, implying that the capacity of the joint route 2 is lower in Figure 9.3. Clearly, both in Figure 9.2 and Figure 9.3 the net welfare gain is equal to zero if the network is deterministic, that is, the size of the stochastic shock is zero.

Figure 9.2 reveals that both the informed drivers from x and the uninformed drivers from y benefit from the information provision to x-travellers. Consequently,

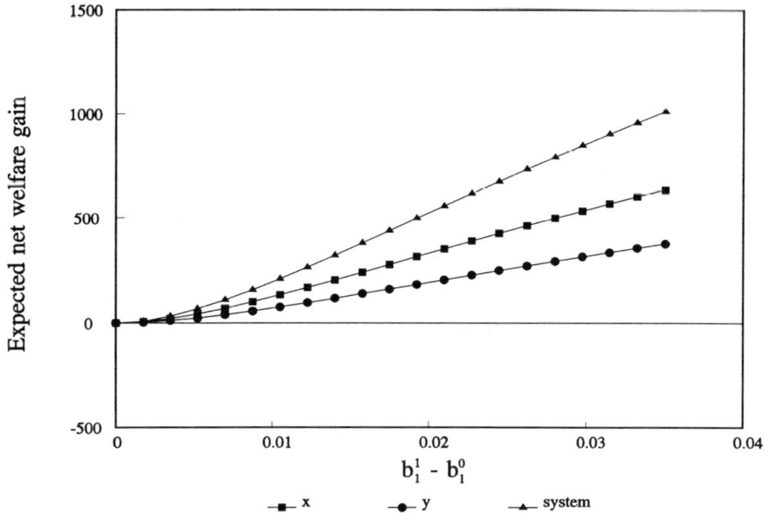

Figure 9.2 Expected net welfare gain due to information provision to x-travellers as a function of the capacity shock b_1^1-b_1^0 (Model P_x - Model N); b_2=0.015.

system welfare - given by the sum of the welfare of group x and the welfare of group y - increases with information provision to group x. Both in model N and in model P_x the equilibrium levels of road usage for all routes and all states are positive in Figure 9.2. Hence, the *complete regularity* condition is met in both models. For model N this implies that the expected link travel costs for route 1 and 2 are identical, while it implies that for model P_x the link travel costs for both routes are identical in each state. The equality between link travel costs on both routes is guaranteed by the provision of information to x-travellers, who use both routes in both states.

Before proceeding to Figure 9.3 it is useful to stress that the net welfare gain plotted on the y-axis is an expected value, that is, the figure is averaged over state 0 (with probability 1-p) and state 1 (with probability p). Net welfare gain due to information - welfare in model P_x minus welfare in model N - of either group x or group y might actually be negative in a particular state. However, in terms of expectations welfare in model P_x exceeds welfare in model N for both group x and group y in Figure 9.2.

Next, consider Figure 9.3 in which the congestion cost parameter of route 2 (b_2) is set to 0.03, implying that the capacity of route 2 is lower than in the experiments depicted in Figure 9.2. For values of b_1^1-b_1^0 less than 0.02 information turns out to negatively affect the y-travellers, while the slope of this trend is reversed for values greater than 0.02. For values less than 0.02, x-travellers never choose route 2 in model N (that is, $N_{2,x}$=0). On the other hand, if information is given to x-travellers then some of these

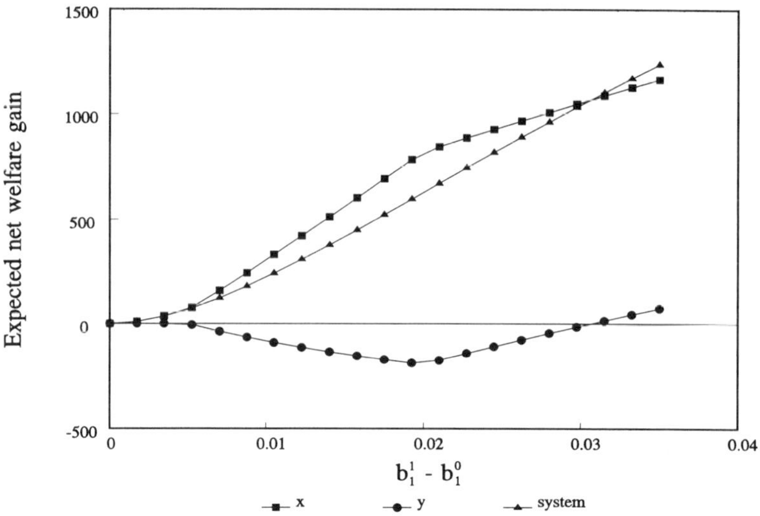

Figure 9.3 Expected net welfare gain due to information provision to x-travellers as a function of the capacity shock b_1^1-b_1^0 (Model P_x - Model N); b_2=0.03.

will use route 2 in state 1 (that is, $N_{2,x}^1>0$). This clearly has a negative impact on the uninformed drivers travelling from O_2 to D_2. For values greater than 0.02, x-travellers use both routes in model N (i.e., $N_{2,x}>0$). The stochastic capacity shock in state 1 is now large enough to render it efficient for x-travellers to also use route 2 in model N. The figure shows that as long as b_1^1-b_1^0<0.03, y-travellers are negatively affected by information provision.

It is interesting to notice that the information benefits to informed drivers still increase beyond 0.02, but at a decreasing rate. Apparently, the welfare gains to uninformed y-travellers to some extent adversely affect informed x-travellers. Furthermore, the curve showing the expected net welfare gain of system benefits is very smooth, that is, this curve does not contain any kinks.

To conclude, in the experiments discussed so far information provision:

1. increases system welfare;
2. increases welfare of the informed travellers;
3. increases or decreases welfare of the uninformed travellers.
4. hence, leads to a *potential* Pareto improvement, but not necessarily to a *strict* Pareto improvement.

The last one of the points mentioned summarizes the main differences with the results obtained with single OD-pairs. In models with single OD-pairs information was found to be useful to both informed and uninformed drivers. In this model however, welfare of the uninformed travellers is likely to decrease when the capacity of their route is relatively low. In all other circumstances, uninformed travellers obtain benefits from the information provision to informed travellers.

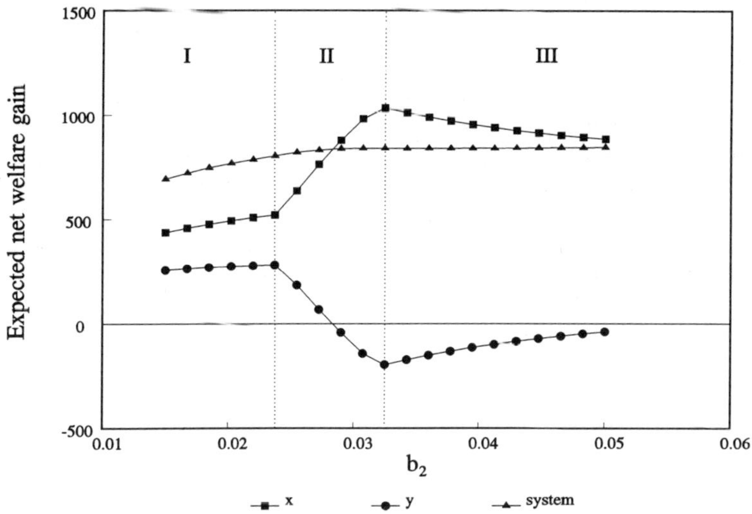

Figure 9.4 Expected net welfare gain due to information provision to x-travellers as a function of b_2 (Model P_x - Model N).

The impact of the capacity of route 2 (the only alternative available to y-travellers) on the expected welfare effects of information is depicted in Figure 9.4. The x-axis of this figure can be divided into 3 intervals, I, II and III, respectively.

1. interval I: *complete regularity* in both model N and model P_x;
2. interval II: $N^0_{2,x}$=0, other equilibrium levels of road usage are positive; *complete regularity* in model N;
3. interval III: $N^0_{2,x}$=0, $N_{2,x}$=0, other equilibrium levels of road usage are positive.

In interval I the *complete regularity* condition holds. Figure 9.4 shows that in this case information provision to x-travellers has a positive impact on all travellers.[3] In interval II, the informed x-travellers do not use route 2 in state 0. Hence, due to the provision of information x-travellers incur route-split benefits in state 0. Apparently, this has a negative impact on uninformed y-travellers, see Figure 9.4. It can be concluded that a relative increase in welfare benefits to one group implies a relative decrease in welfare benefits to the other group and even becomes negative. Finally, in interval III $N_{2,x}$ is equal to zero in model N and $N^0_{2,x}$ is equal to zero in model P_x. Hence, there is no change of behaviour which implies that the information is in fact of less use to the informed x-travellers in comparison with interval II. Consequently, in interval III the information benefits to x-travellers decrease. However, the

[3]Notice that the same result was witnessed in Figure 9.2.

uninformed y-travellers benefit from this decrease in welfare: welfare of y-travellers increases, but they are still negatively affected by the information.

Experiments with models P_x and P_{x+y}
Next, it is examined to what extent information provision to y-travellers influences the results derived above. In order to do so, equilibrium levels of road usage in model P_{x+y} (information available to both x- and y-travellers) are compared with model P_x (information available to x-travellers). Similar plots as in Figure 9.2, Figure 9.3 and Figure 9.4 are depicted in Figure 9.5, Figure 9.6 and Figure 9.7.

First, it is remarkable that the expected net welfare differences between model P_x and model P_{x+y} are much smaller than those between model N and model P_x. Apparently, providing information to a broader group does not affect the net welfare positions of the different groups much. However, it is noteworthy that system performance in model P_{x+y} exceeds system performance in model P_x in all figures, implying that providing information to more people does not negatively affect system performance.

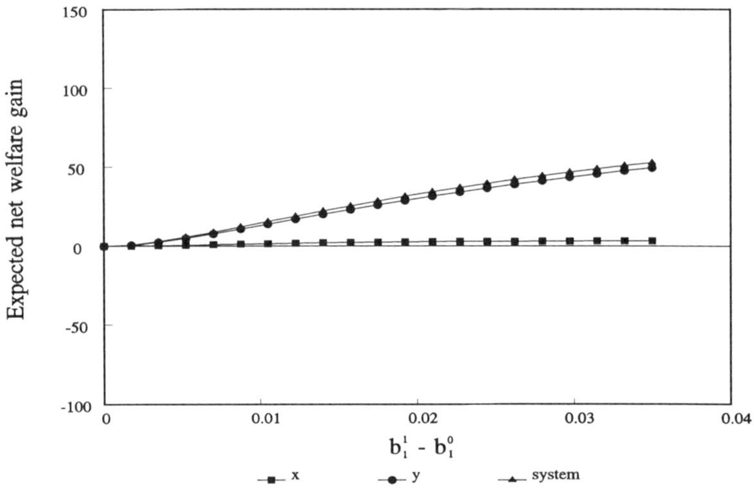

Figure 9.5 Expected net welfare gain due to information to y-travellers as a function of the capacity shock b_1^1-b_1^0 (Model P_{x+y} - Model P_x); b_2=0.015.

A closer look at the figures separately leads to the following analysis. In Figure 9.5 the *complete regularity* condition is satisfied in both model P_x and model P_{x+y}. Here, adding more information is beneficial to all groups involved.

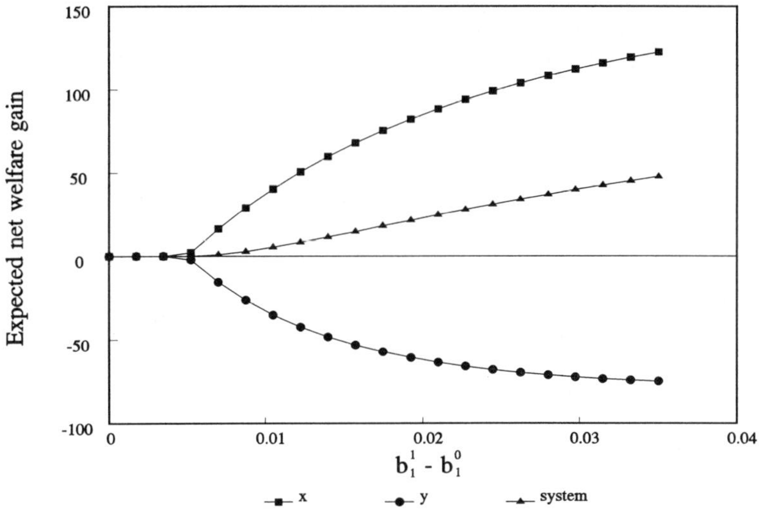

Figure 9.6 Expected net welfare gain due to information to y-travellers as a function of the capacity shock b_1^1-b_1^0 (Model P_{x+y} - Model P_x); b_2=0.03.

Figure 9.6 reveals an interesting pattern.[4] Here, it is shown that providing information to y-travellers is negatively affecting this group, while it is beneficial to the already informed x-travellers. This requires some explanation. As the capacity shock increases, more x-drivers shift to the relatively low capacity alternative route 2 (b_2=0.03) in state 1. This forces some of the y-travellers to not use the transport network, thereby decreasing their welfare.

If in Figure 9.6 the y-travellers would cooperate to maximise their total welfare as a group, then y-travellers would benefit from the information provided. In fact, y-travellers would never be worse off in model P_{x+y} compared with model P_x if the individuals in this group would cooperate. This is easy to see, since y-travellers could always behave as if no information were provided, that is, stick to their equilibrium levels of road usage in model P_x. Information provision to cooperating y-travellers would lead to a potential Pareto improvement within this group; a strict Pareto improvement could be realised by redistributing the revenues obtained.[5] However, the levels of road usage corresponding with this potential Pareto improvement are not

[4]If the capacity shock is very small in Figure 9.6 (smaller than 0.005), then x-travellers never use route 2, neither in state 0 nor in state 1. This implies that the network considered can be split into two one-link networks: A one-link network for x-travellers and a one-link network for y-travellers. This situation has been dealt with in previous chapters.

[5]Of course, cooperation of y-travellers also affects welfare of x-travellers.

optimal from an individual point of view for all y-travellers. Consequently, these levels of road usage are unstable, and therefore do not coincide with the *equilibrium* levels of road usage in model P_{x+y}.[6] It is socially optimal for y-travellers to cooperate, however, the social optimum does not coincide with the stable (sub-optimal) user equilibrium.

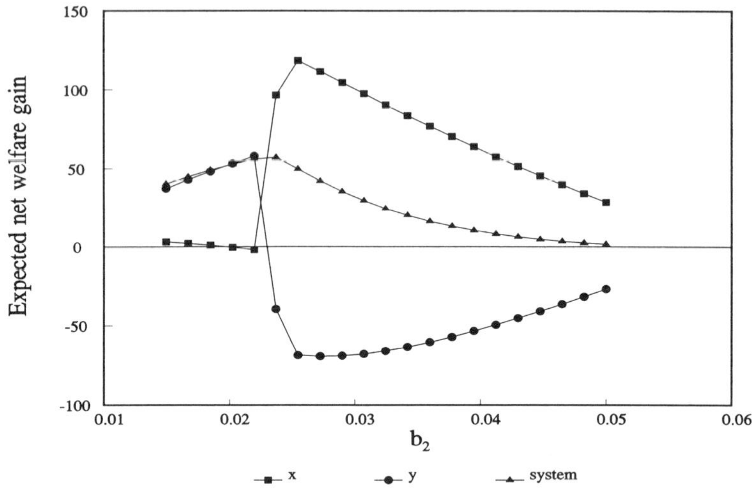

Figure 9.7 Expected net welfare gain due to information to y-travellers as a function of b_2 (Model P_{x+y} - Model P_x).

In Figure 9.7 a similar pattern as in Figure 9.6 is witnessed. Information provision adversely affects the informed y-travellers if b_2 increases beyond approximately 0.023.

Finally, if the intervals in which y-travellers in model P_{x+y} are worse off than in model P_x are compared with the intervals in which y-travellers in model P_x are worse off than in model N, it follows that the intervals are very similar but larger in the present section. This might indicate that if information provision to a particular group negatively affects another uninformed group (model P_x), then the negatively affected group prefers no information availability (model N) to information availability for all travellers (model P_{x+y}).

Summary

Information provision increases network efficiency in terms of system welfare also with multiple OD-pairs. Roughly speaking, the informed group of x-travellers is better off, and so is the group of uninformed y-travellers in most of the experiments.

[6]State-dependent tolling within the group of y-travellers would be needed to render these levels of road usage optimal from an individual point of view.

However, if the capacity of the route choice alternative available to y-travellers (route 2) is relatively low then information provision to x-travellers might harm the uninformed y-travellers. Moreover, in these circumstances additional provision of information to the group of y-travellers would render them even worse off. Note however, that the negative welfare impacts on y-travellers only occur in situations where the joint route has a, relatively speaking, rather low capacity. Consequently, in a situation with one large motorway (available for one group) and a small capacity residential route (available for both groups), information provision might hurt the uninformed travellers.

9.4.2 Route 1, 2 and 3 available

In the previous section it was shown that the provision of information is in most cases welfare improving for the uninformed travellers. In the present section it is investigated whether the availability of another route choice alternative for uninformed travellers changes the results obtained - in particular the results related to the negative welfare effects. The choice set of the y-travellers now consists of three elements: two routes (route 2 and route 3) and not using the network (to abstain).

The parameters used in the experiments are mostly identical to those in the previous sections. However, parameter a_y is decreased in size to 0.02 in order to generate more y-travellers in the network. The reason being the increase in capacity of the routes available to y-travellers. To summarise, the base case parameter-values are equal to: $d_x=d_y=50$, $a_x=a_y=0.02$, $p=0.25$, $k_1^0=k_1^1=k_2=k_3=20$, $b_1^0=0.015$, $b_1^1=0.04$, $b_2=b_3=0.015$.

The key-parameter for testing the efficiency impacts of information is the capacity of the newly added route, route 3. If the congestion cost parameter b_3 of this route is relatively low, it implies that capacity is relatively high.

Experiments with models N and model P_x

By comparing model N and model P_x, the impact of information provision to x-travellers on the uninformed y-travellers and the network efficiency is investigated. In Figure 9.8 and Figure 9.9 the net welfare gain (net welfare in model P_x minus net welfare in model N) is depicted as a function of the congestion cost parameter of route 3 (b_3). The congestion cost parameter of route b_2 is equal to 0.02 in Figure 9.8 and equal to 0.04 in Figure 9.9.

In Figure 9.8 it is shown that information provision to x-travellers is beneficial to all groups: the informed and the uninformed. Here, the *complete regularity* condition is satisfied in both model N and model P_x.

Next, consider Figure 9.9 where the congestion costs of the joint alternative are relatively high ($b_2=0.04$). The x-axis of this figure can be divided into three disjoint intervals, I, II and III:

1. interval I: *complete regularity* in both model N and model P_x;
2. interval II: $N_{2,x}^0=0$, other equilibrium levels of road usage are positive; *complete regularity* in model N;
3. interval III: $N_{2,x}^0=0$, $N_{2,x}=0$, other equilibrium levels of road usage are positive.

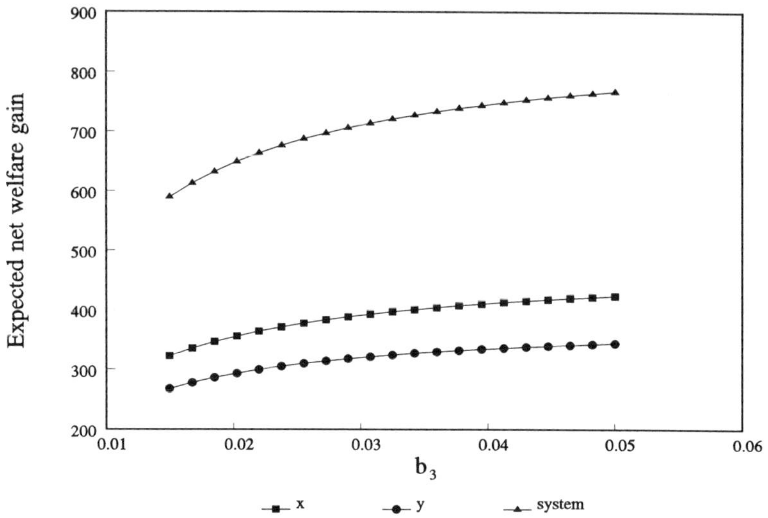

Figure 9.8 Expected net welfare gain due to information provision to x-travellers as a function of b_3 (Model P^x - Model N); b_2=0.02.

In interval I *complete regularity* in both model N and model P_x applies. As before, model P_x outperforms model N in terms of welfare for both informed x- and uninformed y-travellers. In the second interval, $N^0_{2,x}$ is equal to zero. b_3 is now so high that many uninformed y-travellers are using the joint alternative (route 2) so that it becomes beneficial for x-travellers to only use route 1 in state 0. Consequently, x-travellers incur route-split benefits when information is available that state 0 applies, which is reflected by their net welfare increase in interval II in Figure 9.9. In interval III, the equilibrium level of road usage $N_{2,x}$ in model N is also equal to zero. This implies that the information is in fact of less use to the informed x-travellers in comparison with interval II. Therefore, their net welfare gain slightly decreases. Notice that information provision to x-travellers is negatively affecting y-travellers in a part of interval II and in interval III in Figure 9.9.

As in the previous sections, it was shown that in some cases a reduction in capacity (increase in b) leads to an increase in net welfare (owing to information supply) for one group, and a decrease for the other group of travellers. The network as a whole however, is never worse off due to information: The provision of information increases system welfare.

Experiments with models P_x and P_{x+y}

A comparison of the models P_x (information available to x-travellers) and P_{x+y} (information available to x- and y-travellers) provides an indication whether additional information to y-travellers might be harmful to any of the two groups involved.

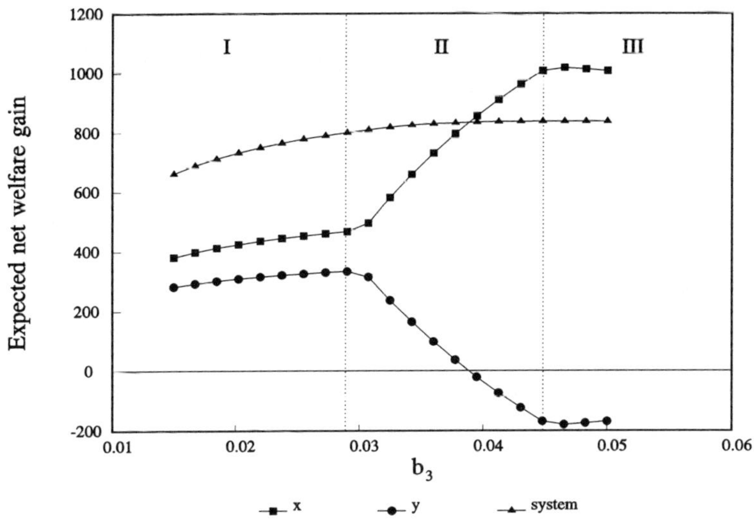

Figure 9.9 Expected net welfare gain due to information provision to x-travellers as a function of b_3 (Model P^x - Model N); b_2=0.04.

In Figure 9.10 and Figure 9.11 the net welfare gain (net welfare in model P_{x+y} minus net welfare in model P_x) is depicted as a function of the congestion cost parameter of route 3 (b_3). As in the previous section, the congestion cost parameter of route b_2 is equal to 0.02 in Figure 9.10 and equal to 0.04 in Figure 9.11.

A couple of points are worth noting. First, from a system point of view, giving information to more users of the system increases network efficiency in terms of welfare: system welfare in model P_{x+y} exceeds system welfare in model P_x.

Second, the welfare gains to be realised when moving from model P_x to model P_{x+y} are significantly smaller than the welfare gains when moving from model N to model P_x. The marginal benefits of information are a decreasing function of the number of travellers informed.

Third, in Figure 9.10 welfare gains to the newly informed y-travellers are rather small. These minor gains are partly realised at the expense of the already informed x-travellers.

Fourth, Figure 9.11 reveals an interesting pattern. The x-axis of this figure can be divided into four disjoint intervals, I, II, III and IV:

1. interval I: $N^1_{2,y}$ in model P_{x+y} is equal to zero, other equilibrium levels of road usage are positive; *complete regularity* in model P_x;
2. interval II: *complete regularity* in both model P_x and model P_{x+y};
3. interval III: $N^0_{2,x}$ in model P_{x+y} is equal to zero, other equilibrium levels of road usage are positive; *complete regularity* in model P_x;
4. interval IV: $N^0_{2,x}$ in model P_x is equal to zero, $N^0_{2,x}$ in model P_{x+y} is equal to zero, other equilibrium levels of road usage are positive.

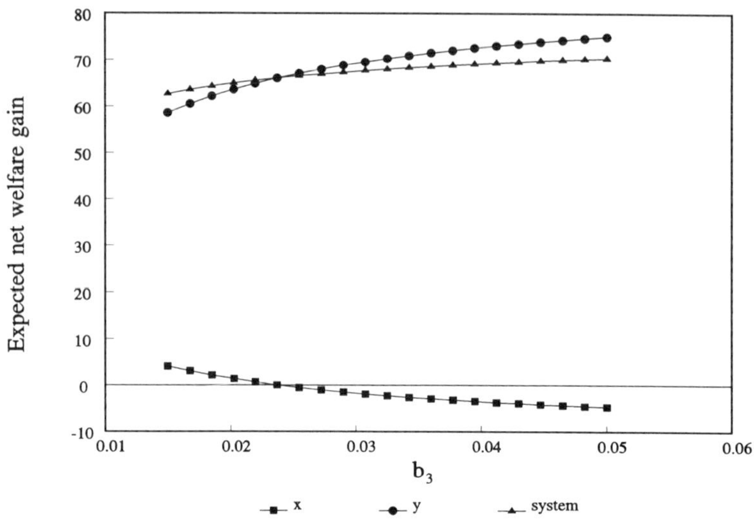

Figure 9.10 Expected net welfare gain due to information provision to y-travellers as a function of b_3 (Model P^{x+y} - Model P^x); b_2=0.02.

In interval I, y-travellers incur route-split benefits in model P_{x+y} by not using route 2 in state 1. In state 1, capacity on route 1 is relatively low implying that many x-travellers shift to the joint alternative route 2. Given the small value of b_3 (high capacity) in this interval it is beneficial for y-travellers to refrain from using route 2 in state 1. The informed x-travellers, on the other hand, are worse off compared with model P_x. In interval II, the *complete regularity* condition is satisfied in both model P_x and model P_{x+y}, and the situation is very much identical to the one depicted in Figure 9.10. In interval III, informed x-travellers incur route-split benefits in model P_{x+y} by not taking route 2 in state 0. In state 0 route 2 is relatively unattractive to x-travellers as more y-travellers are inclined to choose route 2 because the costs of route 3 are relatively high (b_3 increases). The benefits acquired by the x-travellers are obtained at the expense of the informed y-travellers. In interval IV x-travellers incur route-split benefits in state 0 both in model P_x and model P_{x+y}. As a consequence, their net welfare benefits in model P_{x+y} minus the net welfare benefits in model P_x decrease. This is beneficial to the informed y-travellers.

It is noteworthy that the net welfare gains are positive for both x- and y-travellers in only a very small interval in Figure 9.11. Moreover, the redistributional effects of adding information to the system are of greater importance than the efficiency improving impacts. Finally, notice again that the levels of net welfare gains for a specific group in model P_{x+y} might be smaller than in model P_x for particular parameter-values. This is particulary surprising when this concerns the group of y-travellers, that is, the travellers who were uninformed in model P_x.

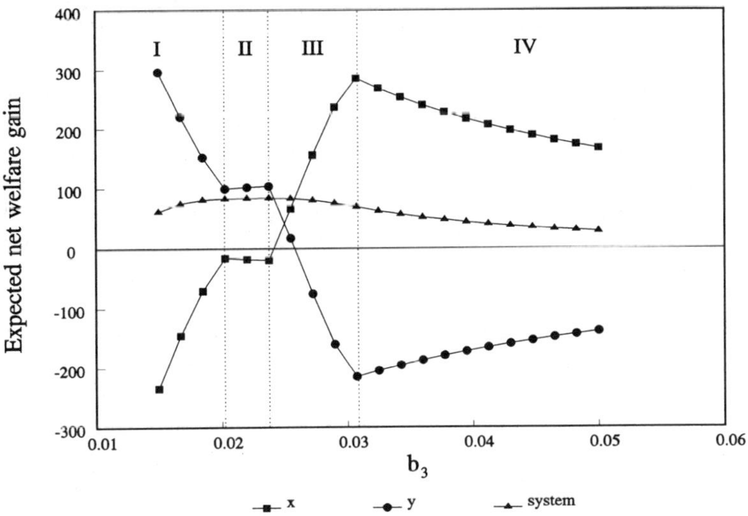

Figure 9.11 Expected net welfare gain due to information provision to y-travellers as a function of b_3 (Model P^{x+y} - Model P^x); b_2=0.04.

Summary

The results of the model experiments have demonstrated that information increases network efficiency in terms of system welfare. In addition, information provided to more travellers increases network efficiency even further, however, the redistributional effects of providing information to more travellers are relatively large. If information is only provided to x-travellers then these drivers always benefit, while the uninformed y-travellers in most cases also benefit. In fact, if the *complete regularity* condition in model N and model P_x is satisfied, then the experiments have demonstrated that information provision is beneficial to both informed and uninformed travellers, i.e., it leads to a strict Pareto improvement. If the y-travellers are worse off due to information provision to x-travellers it is mainly due to a relatively low capacity on their two route choice alternatives (route 2 and route 3). Finally, additional information provided to y-travellers might harm either the x- or y-travellers. If it harms the first group, then it is beneficial to the second, and vice versa.

9.4.3 Relative efficiency of information

In the previous sections it was shown that the provision of information increases network efficiency in terms of system welfare, although some groups might be negatively affected. In the present section the efficiency of the policy instrument information provision is briefly compared with the efficiency of the optimal (first-best) policy. To do so, the performance indicator ω can be used, see also Chapters 5 to 9. ω is defined as:

$$\omega = \frac{Welfare(2nd\ best) - Welfare(non\ intervention)}{Welfare(1st\ best) - Welfare(non\ intervention)} \tag{9.11}$$

Hence, ω gives the achievable welfare gains as a proportion of the theoretically possible welfare gains. Clearly, ω cannot exceed the value of one. In addition, the previous experiments indicated that ω is larger than zero, that is, information provision increases system welfare.

In the calculation of ω the non intervention policy is defined as the level of system welfare in model N, while the second-best policy is defined either as system welfare in model P_x or in model P_{x+y}.

It turns out that the efficiency of information in this network with two OD-pairs is very similar to the efficiency of information in a network with a single OD-pair, see chapters 5 to 7. In both cases, the index ω ranges between zero and 0.3. Clearly, ω is equal to zero if there is no stochasticity in the network, i.e., $b_1^0=b_1^1$ and $k_1^0=k_1^1$ or if either p=0 or p=1. With stochasticity the value of ω is mainly dependent on the size of the stochastic shock: information becomes more useful (ω increases) if the stochastic shock increases in size. For reasons of space, the results are not depicted graphically.

9.5 CONCLUSION

The models studied in this chapter were meant to analyse whether the results obtained with single OD-pair networks, indicating strict Pareto improvement due to information provision, can be generalised to more complex networks. The impression is that this is indeed, in general, the case. Although the results clearly demonstrate that with multiple OD-pairs information provision to a subset of drivers need not necessarily lead to a *strict* Pareto improvement, *potential* Pareto improvements always result. Moreover, the results reveal that when the condition of complete regularity is satisfied then information provision turns out to be beneficial to both informed and uninformed travellers.

In the next chapter, the interdependency between *first-best congestion-pricing* and *second-best information provision* is analysed.

10 SIMULTANEOUS CONGESTION-PRICING AND INFORMATION PROVISION[1]

10.1 INTRODUCTION

Chapters 2 and 3 discussed two main classes of instruments for congestion regulation: (1; Chapter 3) the *traditional* congestion tolls, essentially based on the works of Pigou (1920) and Knight (1924), and (2; Chapter 2) *modern* telematics solutions, using relatively new, user-oriented information technologies, Chapter 2. These two types of instruments are generally analysed in isolation. However, they do in fact bear a close similarity. As first-best congestion charges depend on the actual level of congestion, they clearly contain quite some information, and even imply perfect information provided road users are perfectly aware of their private costs. At the same time, fluctuating congestion tolling without proper pre-trip and on-route information provision is likely to have only a limited impact on user behaviour, as many choices (mode choice, departure time, route choice) will then be based on expected rather than actual tolls and congestion levels. In addition, such information provision to a considerable extent determines the social acceptability of fluctuating road pricing (imagine the emotions of a commuter not only ending up in unexpectedly severe congestion, but facing an unexpectedly high toll on top of that...). As is the case for any market, efficient pricing of road usage only yields its desired optimal effects if individual choices are based on perfect knowledge of the prevailing price and quality of the good to be purchased.

In response to the apparent social and political reluctance to introducing first-best pricing systems (see Chapter 3), second-best alternatives such as cordon pricing, regulatory parking policies, or fuel taxes have recently received much political and academic attention. These second-best alternatives often suffer from the impossibility of perfect fee differentiation (Verhoef, Nijkamp and Rietveld, 1995a). In such cases, regulatory taxes do not reflect perfect information. It is on the other hand evident that non-fluctuating tolls do bear some specific advantages, especially in terms of user-friendliness (transparency, predictability) and costs of toll collection.

Therefore, both with perfect and with imperfect tolling, information provision and tolling may be expected to be complementary measures for theoretical reasons. In addition, one may envisage technical complementarity of pricing and information systems. Nevertheless, most of the literature addresses the implications of congestion-pricing and information provision in isolation [see, among many others, for the first topic: Else (1981), Arnott et al. (1990a), Alan Evans (1992), Andrew Evans (1992), Verhoef, Nijkamp and Rietveld (1995a, 1995b, 1996) and Chapter 3; and for the

[1]This chapter is based on Verhoef, Emmerink, Nijkamp and Rietveld (1996) forthcoming in *Regional Science and Urban Economics*.

second topic: Boyce (1988), Arnott et al. (1991), Mahmassani and Jayakrishnan (1991), Watling and Van Vuren (1993) and Chapter 2]. This chapter aims at investigating the relative efficiency of, and interactions between these two types of instruments, which is in particular important now that both the introduction of high quality information systems and the application of road pricing schemes are likely to become reality in the foreseeable future. Analytical approaches to these questions are rare. The only studies are carried out by de Palma and Lindsey (1994) and El Sanhouri (1994). The current chapter adds to the analysis conducted in Section 7.3 of the present book in that it considers the interaction of various pricing regimes with information provision. In Section 7.3, the analysis was confined to the first-best road-pricing solution, whereas in the present chapter the efficiency impacts of flat (non-fluctuating pricing) are investigated as well. Furthermore, the present analysis is conducted in a two-link network, which enables to study both *mode-* and *route-split* effects. In Chapter 7, the analysis was confined to mode-split effects.

The chapter is structured in the following manner. Section 10.2 discusses the regulatory regimes that will be considered. In Section 10.3, the models to be used are analysed. Next, the efficiency of the available regulatory regimes is investigated in Section 10.4. Finally, Section 10.5 contains the conclusions.

10.2 FIVE REGULATORY REGIMES

In the analysis, five regulatory *regimes* will be considered. The first one is no tolling/imperfect information (**I**), where road users base their behaviour on expected private costs only. The second regime is no tolling/perfect information (**II**), where road users base their behaviour on actual private costs. Thirdly, non-fluctuating (*flat*) tolling/imperfect information (**III**) is considered, where behaviour is based on expected social costs. The fourth regime is a combination of II and III: flat tolling/perfect information (**IV**). Here, the regulator provides perfect information on prevailing levels of congestion, but charges only one single flat toll in all circumstances, the level of which is determined so as to maximize expected efficiency of road use given the fact that drivers are fully informed. Finally, there is the first-best case of fluctuating (*fine*) tolling, implying perfect information (**V**). Although the mere equilibrium tolls as such need not reflect all information, the fact that these tolls may be adapted when necessary secures that they indeed do imply perfect information, at least in the equilibrium approach to be used below. Therefore, the combination of fine tolling and imperfect information will not be analyzed explicitly. Still, this combination is actually described by regime III, since it can be shown that the expected value of the optimal fine fees with imperfect information should be equal to the optimal flat fee with imperfect information, rendering the two regimes identical in terms of user behaviour and welfare effects.

Some important points need to be addressed before proceeding. First, in the analysis *public* information is considered: either no driver or each driver has perfect information; this opposed to Chapter 7. In practical terms, this means that, for information without fine tolling, information given through, for instance, public message signs or radio information is considered (assuming that everybody listens to the radio). The case of *club* information, associated with, for example, on-board information systems, was considered in Chapters 5 to 9.

Second, a SNDUE model with elastic demand and a two-link road network is used. In this respect, this chapter is supplementary to the work of El Sanhouri (1994), in which dynamic bottleneck models with inelastic demand were considered. Also de Palma and Lindsey (1994) mainly consider models with inelastic demand, focusing on route-split and departure time effects.

10.3 OPTIMAL CONGESTION-PRICING UNDER VARIOUS REGULATORY REGIMES

In this section, some basic welfare economic properties of congestion-pricing under the three *pricing regimes* distinguished above are considered. In the analysis, a simple road network, with two alternative routes (denoted by subscripts 1 and 2) between one origin and one destination is assumed. On the basis of this network, one is able to consider both mode-split effects (related to overall demand) and route-split effects. In the problem's purest form, the public regards the two alternative routes as perfect substitutes. One single (inverse) demand function D(N), where N denotes the total number of road users (on both routes), may then be introduced. Furthermore, there are two by two average user cost functions giving the private cost of road use; for either route one representing high capacity (denoted by superscript 0), and one representing low capacity (denoted with superscript 1): $C_1^0(N_1)$, $C_2^0(N_2)$, $C_1^1(N_1)$ and $C_2^1(N_2)$. Naturally, $N=N_1+N_2$ for each of the four possible combinations of cost functions. The probability of low capacity on route i is denoted by p_i, which may run from 0 to 1. Note the equivalence with the formulation of the two-link model in Chapter 6. For notational ease the following probabilities for the four possible states s^{jk} (j=0,1; k=0,1) are introduced: $\rho^{00}=(1-p_1)\cdot(1-p_2)$ to denote the probability of state s^{00} with high capacity on both routes; $\rho^{10}=p_1\cdot(1-p_2)$ and $\rho^{01}=(1-p_1)\cdot p_2$ to denote the probability of low capacity only on route 1 (s^{10}) and route 2 (s^{01}), respectively; and $\rho^{11}=p_1\cdot p_2$ to denote the probability of simultaneous low capacity on both routes (s^{11}). The probabilities for both routes are therefore assumed to be independent (in Section 10.4.5, dependent probabilities are considered). Average user cost and the value of time are assumed to be equal for all road users. The social welfare measure W to be applied below is given by total benefits, represented by the area under the demand curve, minus total costs.

In line with Wardrop's first principle (Wardrop, 1952), at any user equilibrium, depending on the availability of perfect information and on whether tolls are charged, the sum of either the actual or the expected average cost plus the contingent fee should be equal for both routes; otherwise people would shift from the one route to the other. Furthermore, in line with individual optimizing behaviour, this sum should be equal to the benefits D(N) of the marginal road user N.

Next, the derivation of the optimal fees in the three pricing regimes is addressed. First, consider regime V, with fine tolling and perfect information. In this case, the regulator can set first-best fees in each of the four states s^{jk}. For instance, in s^{00}, the optimal fees f_1^{00} and f_2^{00} follow from maximizing social welfare W subject to individual maximizing behaviour based on perfect information:

Problem (10.1) indicates that both demand and cost interdependencies are present between both routes. Consequently, the optimal level of road use on the one route not only depends on the prevailing cost function on that route itself, but also on the

$$\underset{r_1^{00},\, r_2^{00}}{MAX}\ W = \int_0^{N_1+N_2} D(n)dn - N_1'C_1^0(N_1) - N_2'C_2^0(N_2)$$

subject to (10.1)

$$D(N_1+N_2) - C_1^0(N_1) - f_1^{00} = 0$$
$$D(N_1+N_2) - C_2^0(N_2) - f_2^{00} = 0$$

prevailing cost function on the other route. For the expressions for the optimal fees on both routes however, it can be shown that these interdependencies cancel out. The optimal fees for both routes i in state s^{00} are given by:

$$f_i^{00} = N_i'C_i^{0'}(N_i) \quad \text{for} \quad i=1,2 \tag{10.2}$$

For the three other states, the optimal road prices are found by replacing the appropriate superscripts. Therefore, with fine fees and perfect information, the regulator should apply marginal external cost pricing in all circumstances.

In regime III, with flat tolling and imperfect information, the regulator finds the optimal flat tolls f_1 and f_2 by maximizing expected welfare subject to individual maximizing behaviour based on imperfect information:

$$\underset{f_1,\, f_2}{MAX}\ E(W) = \sum_{j=0}^{1}\sum_{k=0}^{1} \rho^{jk}\left(\int_0^{N_1+N_2} D(n)dn - N_1'C_1^j(N_1) - N_2'C_2^k(N_2)\right)$$

subject to (10.3)

$$D(N_1+N_2) - (1-p_1)\,C_1^0(N_1) - p\,C_1^1(N_1) - f_1 = 0$$
$$D(N_1+N_2) - (1-p_2)\,C_2^0(N_2) - p_2\,C_2^1(N_2) - f_2 = 0$$

This leads to the following flat tolls:

$$f_i = (1-p_i)\,N_i\,C_i^{0'}(N_i) + p_i\,N_i\,C_i^{1'}(N_i) \quad \text{for} \quad i=1,2 \tag{10.4}$$

Therefore, with flat fees and imperfect information, the flat tolls are for both routes equal to the expected marginal external costs. It may be stressed here that the total and per route levels of road usage are state independent due to imperfect information.

Finally, in regime IV with flat tolling and perfect information, road usage will of course be different according to the prevailing state, as road users then base their behaviour on actual rather than on expected costs. However, in setting the optimal flat toll, the regulator can do no better than maximizing expected welfare. In order to solve this problem, the variables N_i^{jk}, denoting road usage on route i in state s^{jk}, are introduced. The optimal flat fees f_i then follow from the solution of the following Lagrangian:

$$\mathcal{L} = \sum_{j=0}^{1}\sum_{k=0}^{1} \rho^{jk}\left(\int_{0}^{N_1^{jk}+N_2^{jk}} D(n)dn - N_1^{jk}\, C_1^{j}(N_1^{jk}) - N_2^{jk}\, C_2^{k}(N_2^{jk}) \right)$$
$$+ \sum_{j=0}^{1}\sum_{k=0}^{1} \lambda_1^{jk}\left(D(N_1^{jk}+N_2^{jk}) - C_1^{j}(N_1^{jk}) - f_1\right) \qquad (10.5)$$
$$+ \sum_{j=0}^{1}\sum_{k=0}^{1} \lambda_2^{jk}\left(D(N_2^{jk}+N_2^{jk}) - C_2^{k}(N_2^{jk}) - f_2\right)$$

10.4 THE RELATIVE PERFORMANCE OF THE VARIOUS REGULATORY REGIMES

It is clear that the results derived in Section 10.3 do not lead to straightforward analytical or manageable solutions. Nevertheless, it is important to investigate the properties of the various regulatory regimes. Therefore, the section presents and discusses the outcomes of some experiments that were undertaken in order to gain more insight into the relative performance of the various regulatory regimes. The model was kept as simple as possible by assuming that all demand and cost functions are linear over the relevant ranges (that is, the ranges containing the levels of usage in each of the possible states and in each of the possible regulatory regimes).

10.4.1 The model

The two-link model contains one joint linear demand function, characterized by slope a and intercept d:

$$D(N_1+N_2) = d - a\,(N_1+N_2) \qquad (10.6)$$

Next, for both routes (i=1,2), there is a cost function for both high (j=0, k=0) and low capacity (j=1, k=1). The marginal private cost MPC_1^j (MPC_2^k), which is equal to average social cost ASC_1^j (ASC_2^k), consists of a free-flow cost component k_1^j (k_2^k) and a congestion cost component, which is assumed to be proportional to total usage N_1 (N_2) with a factor b_1^j (b_2^k):

$$MPC_1(N_1) = ASC_1(N_1) = k_1^j + b_1^j N_1; \quad j=0,1$$
$$MPC_2(N_2) = ASC_2(N_2) = k_2^k + b_2^k N_2; \quad k=0,1 \qquad (10.7)$$

All parameters are non-negative, and only *regular* networks are considered, where both routes are at least marginally used under each regime; see also Chapters 6 and 9 for regularity conditions. Apart from the explicit functions, the model is further identical to the one presented in Section 10.3.

For the base case of the experiments, the following parameter values were chosen: a=0.015; d=100; $k_1^0=k_2^0=k_1^1=k_2^1=20$; $b_1^0=b_2^0=0.015$ and $b_1^1=b_2^1=0.04$; and

$p_1=p_2=0.25$.[2] So, both routes are assumed to be identical in the base case. Low capacity is assumed to affect link capacity, leaving free-flow costs unaffected.

By varying the model's respective parameters, it is possible to gain insight into their impact on the relative efficiency of the five regulatory regimes. The results are discussed below. The performance of regimes II-IV will be expressed in the index of relative welfare improvement ω (see also Chapters 5 to 9), which is for instance for regime II defined as:

$$\omega^{II} = \frac{\bar{W}^{II} - \bar{W}^{I}}{\bar{W}^{V} - \bar{W}^{I}} \tag{10.8}$$

where $\bar{W}$ denotes expected welfare. Therefore, ω gives for the regime considered the achievable welfare gains as a proportion of the theoretically possible (optimal) efficiency gains (both compared to welfare in regime I). For comparing regimes I and V, the following index of potential relative welfare improvement ω^V is used:

$$\omega^{V} = \frac{\bar{W}^{V} - \bar{W}^{I}}{\bar{W}^{V}} \tag{10.9}$$

Note that the index ω^V is not directly comparable to the other indices ω. The reason for including ω^V is that it enables a better interpretation of the other ω's. In the base case, $\omega^{II}=0.41$; $\omega^{III}=0.66$; $\omega^{IV}=0.99$ and $\omega^{V}=0.24$.

Obviously, by means of this type of modelling exercises, one may conduct a great variety of experiments. Many of these have been studied; the ones presented below are those that were found to be most interesting.

10.4.2 Varying probabilities

The first parameters studied are the probabilities of cost shocks. Figure 10.1 shows the various ω's when p_1 and p_2 are simultaneously raised from 0 to 1. Clearly, information provision without tolling only makes sense if there is uncertainty: ω^{II} falls to 0 at $p_1=p_2=0$ and $p_1=p_2=1$; that is, when one of the two cost functions applies with certainty. On the other hand, ω^{III} is equal to 1 in these cases: flat tolling is as efficient as fine tolling when the fine fees would always be set at the same levels anyway.

For intermediate values of p_1 and p_2, ω^{II} and ω^{III} show an opposite pattern. This is also found in the experiments reported below, and it indicates that *flat tolling and information provision are highly complementary instruments*. This is underpinned by the fact that ω^{IV} is very close to unity throughout, indicating that the combination of

[2] To give an idea of the quantities and prices generated in this base case, the equilibrium usage under regime I is 1561 on both routes in each state, with expected marginal private cost of 53.17, and an expected total net welfare of 73,099. Expected usage under the optimal regime V is 1203 on both routes, with expected marginal private costs of 41.96 and expected fine fees of 21.96 on both routes, and an expected total net welfare of 96,229.

flat tolling and perfect information provision yields an expected welfare almost as high as does first-best fine tolling. Finally, also the sum of ω^{II} and ω^{III} was included, in particular to examine whether the efficiency gains of flat tolling and information provision are sub-additive or super-additive. In Figure 10.1, sub-additivity appears to hold throughout (except for the two extreme cases).

10.4.3 Varying congestion cost parameter volatility

Next, in Figure 10.2[3], the volatility of the congestion cost parameters, expressed in the difference between the congestion cost parameters in the different states, is shown. On the left-hand side of this figure, the parameters b are all identical, while b_1^1 and b_2^1 simultaneously increase when moving to the right. Therefore, on the extreme left-hand side, there is complete certainty, and in accordance with Figure 10.1, $\omega^{II}=0$ and $\omega^{III}=1$. When volatility increases, however, the relative efficiency of information provision (without tolling) increases, due to the increasing value of information (both from a private and from a social point of view). In contrast, flat tolling becomes less efficient due to the increasingly important shortcoming of fee adaptation. At sufficiently high levels of volatility then, ω^{II} may exceed ω^{III}. As shown by ω^{V}, this happens in situations where potential efficiency gains are relatively large.

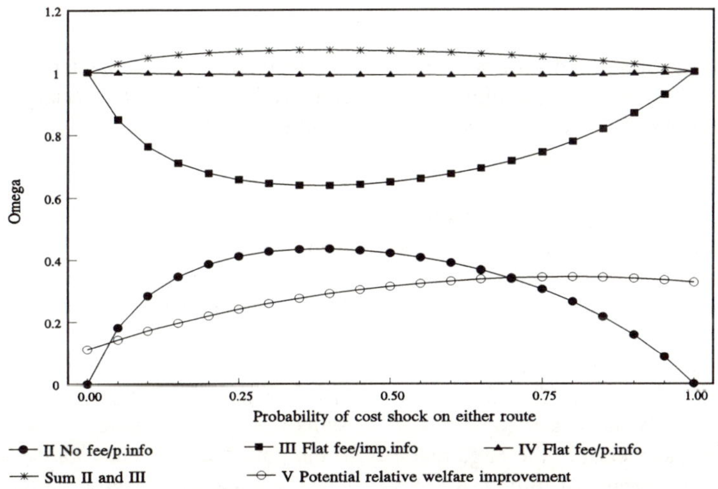

Figure 10.1 Varying probabilities of cost shocks: indices of relative welfare improvement.

[3]Apart from the figures related to probabilities, all figures related to the experiments are logarithmically scaled along the horizontal axis.

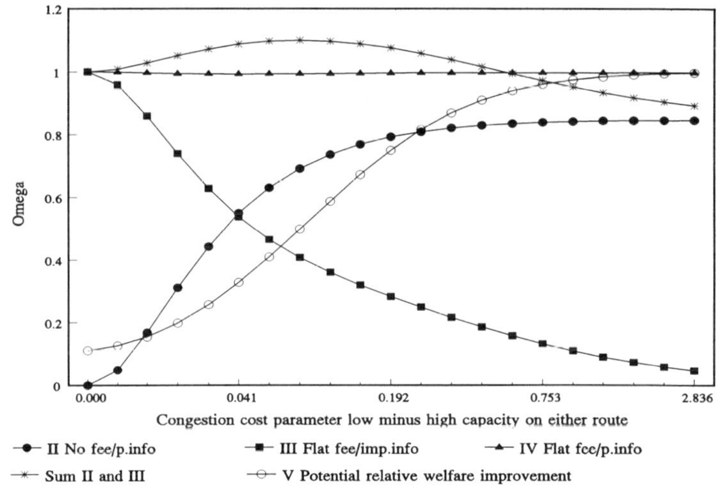

Figure 10.2 Varying congestion cost parameter volatility: indices of relative welfare improvement.

Again, flat tolling and information provision are highly complementary, as reflected by the rather constant and high (close to unity) value of ω^{IV}. Over a large range, the efficiency gains of the instruments are sub-additive; only at extremely high volatility does super-additivity occur.

10.4.4 Varying demand characteristics

Next, demand characteristics are considered in Figures 10.3 and 10.4. In these figures, the demand curve is tilted around the original intersection in regime I, varying from high elasticities on the left-hand side to almost perfect inelasticity on the right-hand side.

In general, the more elastic the demand, the more serious misallocations due to imperfect pricing will be. This explains why in Figure 10.3 ω^V approaches unity in these cases: W^I decreases rapidly. For the same reason, ω^{II} is quite low when demand becomes more elastic: the large majority of the potential efficiency gains can only be obtained by means of tolling. Moreover, when demand is perfectly elastic (implying that marginal benefits are equal to average benefits) and no tolling takes place, expected welfare is zero both in regimes I and II: with imperfect information, marginal and average benefits are equal to expected average costs on both routes; with perfect information, marginal and average benefits are equal to average costs on both routes in all states. In contrast, flat tolling without information provision already yields considerable efficiency gains (see ω^{III}).

When moving to more inelastic demand however, route-split rather than mode-split decisions become increasingly important for overall efficiency, simply because overall demand becomes more sticky. Such route-split decisions are, with otherwise

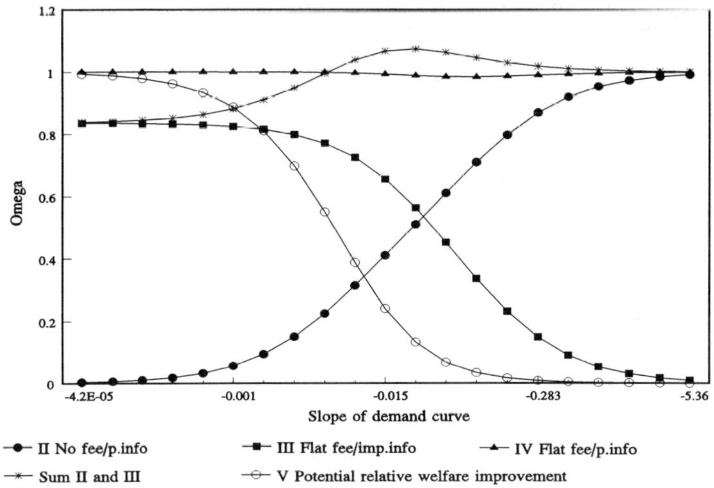

Figure 10.3 Varying demand characteristics with cost-shocks: indices of relative welfare improvement.

equal routes, especially relevant in states s^{10} and s^{01}, when a shock occurs on one of the two routes only. As flat tolling without information provision will then have no effect on route-split decisions, ω^{III} falls to zero when approaching completely inelastic demand. Information provision on the other hand does affect route-split, and ω^{II} increases accordingly. The reason that ω^{II} even approaches unity in Figure 10.3 is that with inelastic demand and *identical* routes, equalization of marginal private costs on both routes (user equilibrium) implies equalization of marginal social costs (optimal route-split) because of the linear cost functions. Hence, individual optimizing behaviour based on perfect information then results in optimal usage of the network.

In Figure 10.4, it is assumed that free-flow costs differ between the two routes (k_2=30 and k_1=10). Then, even with inelastic demand, the user equilibrium without tolling is no longer equal to the user equilibrium with perfect information, and ω^{II} reaches a maximum below unity, whereas ω^{III} remains larger than zero because the (route-specific) flat tolls can to some extent correct for the expected inefficiency in route-split.

Finally, the fact that ω^{V} is practically zero at inelastic demand suggests that information provision without tolling becomes efficient only when potential efficiency gains become very small. These low values for ω^{V} however, are to some extent also caused by the fact that total benefits, measured as the area under the demand curve, are extremely large in all regimes due to the extreme steepness of this curve. It is therefore noteworthy that fine tolling, compared to no tolling/imperfect information, still yields a cost advantage of 6.5% on the right-hand side of Figure 10.3 (at a=5.36).

The almost perfect complementarity of flat tolling and information provision is again clearly demonstrated by the curvature of ω^{IV}. It can finally be noticed that, in

Figure 10.3, the efficiency gains exhibit super-additivity at more elastic demand, whereas sub-additivity prevails at more inelastic demand.

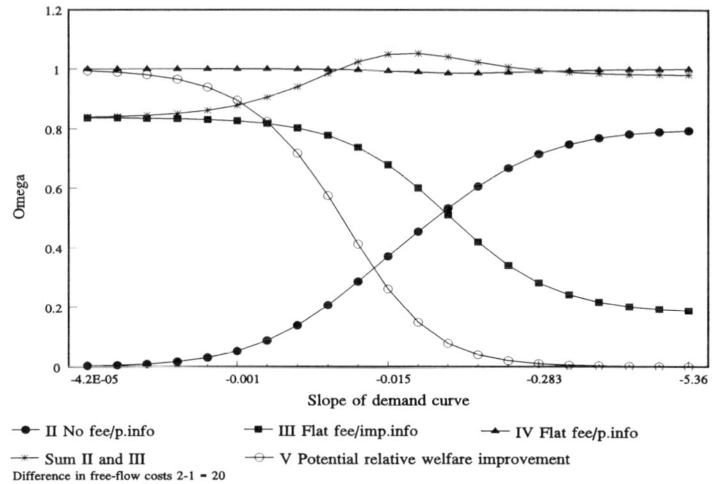

Figure 10.4 Varying demand characteristics with cost shocks and a free-flow cost differential: indices of relative welfare improvement.

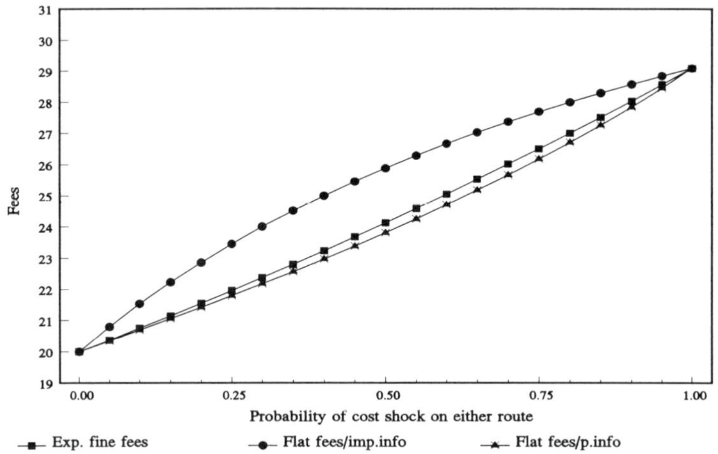

Figure 10.5 Varying probabilities of cost shocks: flat fees and expected fine fees.

10.4.5 Comparing the fees

Finally, the equilibrium values of the (expected) fees in the various pricing regimes are discussed. It is of course hard to give an entirely satisfactory explanation of the equilibrium values of the fees found, as this requires tracing through the entire model. Some interpretation, however, remains possible. The general picture emerging in most of the experiments is as follows.

First, the flat fee with imperfect information was found to exceed both the expected value of the fine fee and the flat fee with perfect information. Without information, more users will use a link when low capacity occurs. The resulting welfare losses are then considerable, and make it apparently worthwhile to set the flat fee relatively high. Consequently, drivers not only benefit from information because of its informational value; a secondary benefit is that tolls will be lower.

Second, perhaps counter-intuitively, the expected value of the fluctuating fee was found to exceed the value of the flat fee with perfect information. This can to some extent be explained by the fact that the latter is a weighted average of marginal external congestion costs, where the weight decreases with a steeper cost curve. Therefore, this fee is biased towards the low congestion value. Figure 10.5 illustrates the findings mentioned above for the simulations related to varying probabilities (the fees for both routes are identical throughout because the routes were assumed to be identical).

10.5 CONCLUSION

This chapter studied the relative efficiency of, and the interaction between various information and pricing instruments for the regulation of stochastic road traffic congestion. Five regulatory regimes were considered: no tolling/imperfect information, no tolling/perfect information, non-fluctuating (*flat*) tolling/imperfect information, flat tolling/perfect information, and fluctuating (*fine*) tolling implying perfect information. Elastic demand and a two-link network were considered in order to be able to consider both mode-split and route-split effects. Many experiments were conducted to gain insight into the influence of some parameters on the relative efficiency of the various regulatory regimes.

Information provision and flat tolling were found to be highly complementary. The combination of these two instruments performed in most situations almost as good as the theoretically first-best option of fine tolling. Given the psychological advantages of flat tolling, related to issues of predictability and transparency, it seems plausible that such a combination of instruments might be more attractive than the use of fine tolls. As long as expected congestion remains the same, the regulator might use the same tolls for certain links at certain times of the day, regardless the actual level of congestion. Moreover, as outlined in the introduction of the chapter, for a smooth usage of fine tolls, one could not do without proper pre-trip and on-route information provision. This means that in both cases the same sort of information technologies would be required anyway.

As far as the performance of flat tolling and information provision in isolation is concerned, the relative efficiency of these two was - in line with the above mentioned complementarity - found to behave in an opposite way. Flat tolling with imperfect information performed relatively well with relatively elastic demand, with modest volatility of congestion and with the more extreme probabilities of shocks (that is, with more certainty). Perfect information provision without tolling was found to behave well in the opposite cases. The efficiency gains of the two instruments were most often found to be sub-additive. Super-additivity was only found with relatively elastic demand and high volatility.

Before proceeding to the next part of the book it is useful to summarise the main findings of the theoretical part. The theoretical models in Chapters 5 to 10 were applied to rather simple road networks with stochastic capacity. The information then provides some of the travellers with knowledge on the *actual* road capacity, while the uninformed travellers base their trip-making decision on the *expected* capacity. In the propositions that were derived the additional assumption of linear demand and link travel cost functions was made. The main results can be summarised as follows:

1. The provision of information leads to a potential Pareto improvement, i.e., expected system welfare increases.
2. The informed travellers are generally better off - except for some situations as described in Chapter 9.
3. The uninformed travellers are - under some network regularity conditions - generally also better off.

Nevertheless, situations where expected system welfare decreases due to information provision have also been detected. In Verhoef, Emmerink, Nijkamp and Rietveld (1996) it is proven that this might be the case with kinked demand functions. In addition, the proofs in the appendices of Chapters 5, 6, 7 and 8 revealed that information provision decreases expected system welfare under the rather unrealistic assumption that the slope of the high capacity travel cost function is steeper than the slope of the low capacity travel cost function.

PART III

SIMULATION MODELS

CHAPTER 11 SIMULATION MODELLING: RECURRENT CONGESTION[1]

11.1 INTRODUCTION

In Chapter 4, a theoretical framework was proposed, in which the impact of information provision and road-pricing could be analysed. Chapters 5 to 10 explored this framework in various directions. In these chapters it was assumed in the static model that some of the road users (the *informed* ones) are perfectly aware of the stochastic traffic conditions, while others (the *uninformed* ones) are unaware of actual traffic conditions, and base their decision-making on expected network conditions. Furthermore, it was assumed that all decision-makers are behaving rationally, that is, maximising some utility measure. In Chapters 11 and 12, a couple of distinct characteristics of decision-making in transport networks is acknowledged that were not taken into account in the models in the previous chapters. First, decision-making is *dynamic* rather than *static*, dynamic in the sense that road users generally take travel decisions from day-to-day. Second, the assumption of perfect rationality of road users might be questioned. In the psychology literature, alternative decision-making models have been proposed that *might* provide a more realistic representation of the actual traveller's choice process.

In order to facilitate some of the above shortcomings of the theoretical model, a simulation model is presented in this chapter. The work follows Mahmassani and co-authors, summarised in Mahmassani and Herman (1990). Attention is confined to one specific mode of travel: the private car. However, rather than analysing either the day-to-day dynamics (cf. Mahmassani and Chang (1985)) or the within-day travel dynamics (cf. Mahmassani and Chen (1991), Mahmassani and Jayakrishnan (1991)), this chapter attempts to integrate these processes.

The simulation model consists of the three components, *driver behaviour*, *control system* and *network model*, suggested in Watling and van Vuren (1993). In this chapter, the control system component reflects the information collecting and information supplying process (Section 11.2). The driver behaviour component is incorporated in the simulation model by a number of simple behavioural models (Section 11.3), while the network model component represents the traffic flow modelling (Section 11.4). The simulation experiments, presented in Section 11.5, focus on the following issues:

1. Performance of boundedly rational model for driver's behaviour in an environment without information.

---

[1]This chapter is based on Emmerink, Axhausen, Nijkamp and Rietveld (1995a) published in *Transportation.*

2. Effectiveness of the provision of after-trip and real-time en route information in relation to the level of market penetration.

The first issue expands the current literature on boundedly rational behaviour. The implications on the overall network performance of different values for the bound are assessed in a combined day-to-day and within-day context, while past research concentrated on modelling and estimating the boundedly rational model seen from an individual perspective. The second issue expands the literature in that it focuses on both after-trip and en route information, and links the day-to-day dynamics and the travel dynamics within a day. Past research concentrated either on after-trip information in a day-to-day travel environment (Iida et al., 1992; Mahmassani and Herman, 1990) or en route information within a day (Mahmassani and Chen, 1991; Mahmassani and Jayakrishnan, 1991; Mahmassani and Peeta, 1993).

The experiments are conducted in a network with recurrent congestion. This is a network congested due to under-capacity or, phrased differently, due to too much demand for mobility. Situations in which congestion is caused by accidents, bad weather or other incidents, that is, non-recurrent congestion, are addressed in the next chapter.

11.2 FOUR DIFFERENT TYPES OF INFORMATION PROVISION

In this section the different kinds of information provision are presented. The four kinds of information provision shown in Table 11.1 are implemented. Throughout the chapter, the code is used to refer to the specific types of information provision.

Code	Information
A	own experience
B	after-trip
C	real-time pre-trip
D	real-time en route

Table 11.1 Information provision types.

Information provision type A assumes that information on different routes is acquired solely through own experience. After a trip has been made, the expected travel time of the chosen route is updated, while the expectations of the other routes remain unchanged.

Information provision type B provides drivers with information on the not chosen routes after their trips have been completed. One could think of radio or TV reports that describe in detail the day's situation on the roads, or of in-home traffic information systems. In the simulation model, the information given is based on the densities realised on the links during the last travel period. These are the input for the travel time calculations of not chosen routes by a driver.

Information provision type C supplies drivers with real-time pre-trip information. This is information based on the actual situation in the network just before the start of a trip, enabling drivers to change route before the trip according to the current situation in the network. As with type B, the information is based on the prevailing densities on the links in the network. No attempt is made to provide predictive information. For research in this direction see, for instance, Ishtiaq and Hounsell (1993), Koutsopoulos and Xu (1993) and Lindveld, Kroes and de Ruiter (1992).

Finally, information provision type D is real-time en route information on the current situation in the network, enabling drivers to switch routes during the trip based on the most recent information. As with information types B and C, the information is based on the prevailing link densities.[2] A prototype of such an advanced information providing system is currently being implemented in Illinois (ADVANCE, 1990). In Europe research in this direction has been carried out within the LISB project in Berlin[3] and within the EU DRIVE I and II programmes.

New information provision types can be specified by combining the types given in Table 11.1. For instance, information provision type A+B implies that drivers base their travel choices on both their own experiences and supplied after-trip information.

11.3 BEHAVIOURAL MODELS

This section presents the models of driver behaviour that were used. These models lie at the heart of the simulation approach, since they generate the drivers' decisions. In addition, these decisions determine the resulting traffic flows in the network completely. Section 11.3.1 addresses the way of updating information, while Section 11.3.2 deals with the decision-making process given the updated information. Throughout the section it is assumed that departure times are fixed. The only choice open to drivers is the route choice.

11.3.1 Updating information

Drivers' travel time expectations are updated after new information is available, and are dependent on the information provided. Throughout this chapter the term expectations rather than predictions is used to indicate that predictive information is not supplied. Subsequently, the updating processes associated with information provision schemes A, A+B, A+B+C, A+B+C+D and A+D are discussed.

The updating mechanisms are based on the following linear equation, known as an exponentially weighted moving average.

$$ET_r^{n+1} = \alpha * NewInformation_r + (1-\alpha) * ET_r^n \tag{11.1}$$

Here, ET_r^{n+1} denotes the expected travel time for route r in period n+1, while the new available information on route r is embedded in $NewInformation_r$. The parameter α lies in the closed interval [0,1] and reflects the weight given to the last travel experience. An α-value close to 1 implies that the driver's expected travel time for the next travel period is largely based on his last experience with a route. On the other hand, a value close to zero means that future expectations strongly rely on past experiences. A

---

[2]As with type C, predictive information is not provided.

[3]The results of a survey of drivers equipped with a route guidance system as part of the LISB trial can be found in Bonsall and Joint (1991).

similar mechanism has been applied by Horowitz (1984) and more recently in a sequential route choice context by Iida et al. (1992) and Vaughn et al. (1992).

Iida et al. (1992) estimated the parameter α and found large discrepancies among different individuals, ranging from 0.3 to 0.7. Vaughn et al. (1992) estimated an equation similar to (11.1), and it appeared that an α-value of 0.2 led to the largest log-likelihood. However, the differences in log-likclihood for varying α-values were rather small.

Information provision type A

Under information provision type A, it is assumed that drivers update the chosen route (r) using equation (11.1), and that the expectations for the other routes (j) remain unchanged. In mathematical terms this can be expressed by equations (11.2) and (11.3)

$$ET_r^{i,n+1} = \alpha * ExperiencedTravelTime_r^{i,n} + (1-\alpha) * ET_r^{i,n} \tag{11.2}$$

$$ET_j^{i,n+1} = ET_j^{i,n} \tag{11.3}$$

in which index i refers to the driver. For simplicity it is assumed that α is equal for all drivers. A more realistic approach is to assign α randomly.

Information provision type A+B

Under information provision type A+B it is assumed that drivers' experience is used to update the expected travel time of the chosen route (the driver thus ignores the information on this route) while the provided information is used to calculate the expected travel times of the unchosen routes for the next travel period. The same updating formula as used in information mechanism A, formula (11.2), is applied if route r was chosen in period n. However, the expected travel times of the unchosen routes are now updated following (11.4),

$$ET_j^{i,n+1} = \alpha * AfterTripInfo_j^{i,n} + (1-\alpha) * ET_j^{i,n} \tag{11.4}$$

in which $AfterTripInfo_j^{i,n}$ denotes driver i's travel time (according to the supplied information) in period n if route j had been chosen.

It is important to note that in the model drivers regard provided information as reliable as own experience, since equal weights are given to both. An extension of the model is to (1) give different weights to own experiences and supplied information, (2) weigh the information according to its (perceived) reliability. A reliability measure could, for instance, be the difference between the experienced travel time of the chosen route and the information provided for this route. The driver, for instance, could perceive the information being reliable if the difference is relatively small. Although investigating drivers' attitudes and reactions towards information is very important it is far beyond the scope of this chapter.

Information provision type A+B+C

This mechanism is based on two sequential phases. The first phase is the application of information mechanism A+B after the trip:

$$ {}^{A+B}ET_r^{i,n+1} = \alpha * ExperiencedTravelTime_r^{i,n} + (1-\alpha) * {}^{A+B+C}ET_r^{i,n} \qquad (11.5) $$

if route r was chosen in period n by driver i and

$$ {}^{A+B}ET_j^{i,n+1} = \alpha * AfterTripInfo_j^{i,n} + (1-\alpha) * {}^{A+B+C}ET_j^{i,n} \qquad (11.6) $$

for all the remaining routes j. Here, ${}^{X}ET_j^{i,n}$ denotes driver i's expected travel time for alternative j in period n after information mechanism X has been applied.

In the second phase the pre-trip information is processed. It is assumed that drivers adjust expected travel time according to formula (11.7).

$$ {}^{A+B+C}ET_j^{i,n+1} = \beta * PreTripInfo_j^{i,n+1} + (1-\beta) * {}^{A+B}ET_j^{i,n+1} \qquad (11.7) $$

This equation is used for all the available routes j. The parameter ß lies in the closed interval [0,1] and reflects how much weight is given to the pre-trip information.

Information provision type A+B+C+D

Pre-trip travel time expectations are made using information mechanism A+B+C. During the trip, en route real-time information is given on the expected remaining travel time for the different routes. The en route decision-making process is discussed in Section 11.3.2.

Information provision type A+D

Pre-trip travel time predictions are made using information mechanism A. During the trip, en route real-time information is given on the expected remaining travel times for the different routes. The en route decision-making process is discussed in the next section.

11.3.2 Decision-making

The previous section dealt with the updating mechanism of information, leading to travel time expectations. The current section explores how these expectations are used in the decision-making process of drivers. The section consists of two parts. First, general model principles are presented, next the models for the information provision types described in Section 11.3.1 are specified.

General model principles

In this section different behavioural models for the driver's route choice problem are discussed. The discussion is restricted to route choice.

In the simulation experiments it is assumed that drivers behave according to the models described in this section. These models consist of two components, a utility maximisation component and a satisficing component.

Utility maximisation component

The models described in this section, have the utility principle as corner stone. The decision-making process of the individual consists of comparing utilities associated with the available route choice alternatives, and choosing the one with highest utility. The utility of a route is calculated using a utility function. A general linear utility function has the following form:

$$U_j^i = \beta_1^i * x_{1,j}^i + \beta_2^i * x_{2,j}^i + \ldots + \beta_n^i * x_{n,j}^i + u_j^i \tag{11.8}$$

in which U_j^i denotes the utility individual i associates to alternative j and $x_{k,j}^i$ the *kth* attribute of alternative j of individual i. Furthermore, u_j^i represents the random error term of alternative j of individual i. In the simulation experiments of Section 11.5, these will be omitted for simplicity. In the simulation experiments the β_k^i are assumed to be equal among the individuals, although it is more realistic to assign the values β_k^i stochastically to the individuals. A similar methodology of modelling drivers' behaviour has been applied by van der Mede and van Berkum (1991).

The utility function used in route choice decision-making is generally assumed to be dependent on the following attributes (Bovy and Stern, 1990): (1) travel time, (2) distance, (3) desired arrival time at destination, (4) schedule delay[4], (5) travel time uncertainty[5], (6) individual's socio economic characteristics. For simplification it is assumed throughout this chapter that the utility function is dependent on travel time only. Under this assumption model (11.8) collapses into model (11.9).

$$U_j^i = -TravelTime_j^i \tag{11.9}$$

Since a decision has to be made every period (day), model (11.9) is expanded to:

$$U_j^{i,n} = -TravelTime_j^{i,n} \tag{11.10}$$

where n denotes the period. Furthermore, since the travel time in period n is unknown at the moment the decision for period n is made, it is estimated. The estimated travel time of driver i in period n for alternative j will be denoted by $ET_j^{i,n}$. If it is assumed that utility is equal to expected travel time (reliability issues are ignored), then model (11.10) turns into (11.11).

$$U_j^{i,n} = -ET_j^{i,n} \tag{11.11}$$

The travel time is estimated using updating mechanism (11.12)

---

[4]Schedule delay is defined as the absolute value of the preferred arrival time minus the actual arrival time. The importance of schedule delay for departure time decisions has been pointed out by Hendrickson and Kocur (1981). A measure, that is to some extent related to schedule delay, was taken into account in the analysis of Chapter 8: the variance of travel time.

[5]A measure related to travel time uncertainty was taken into account in the analysis of Chapter 8: the variance of travel time.

$$ET_j^{i,n} = \alpha * TravelTime_j^{i,n-1} + (1-\alpha) * ET_j^{i,n-1} \quad (11.12)$$

which has been explored in Section 11.3.1.

Satisficing component

The satisficing principle stems from a paper by Simon (1955) and has been introduced in the literature on route and departure time choice by Mahmassani and Chang (1985). Rather than using Wardrop's User Equilibrium (UE) principle (Wardrop, 1952), they introduced a Boundedly Rational User Equilibrium (BRUE) theory. Properties of BRUE have been analysed in Mahmassani and Chang (1987). According to BRUE, individuals are trying to achieve a satisfactory outcome, rather than to maximise their utility. An intuitive argument backing the satisficing principle is based on costs. It is conceivable that an individual would like to avoid the costs (efforts) associated with finding the utility maximising solution, especially if a similar decision has to be made frequently as in the route choice context. Furthermore, if the costs associated with finding the optimal (utility maximising) solution were included in the utility function, it could well be true that the satisficing alternative coincides with the maximising utility solution. In other words, the utility function is not correctly specified if the costs associated with finding the optimal solution are omitted.[6]

The model presented below is an adapted version of the one used by Mahmassani and colleagues, for the first time described in Mahmassani and Chang (1985). In the present context this model needs to be adapted since in the simulation experiments the departure time is assumed to be fixed, so that the preferred arrival time is irrelevant. In Mahmassani and Chang's model, the preferred arrival time and schedule delay are the key variables.

The boundedly rational model of Mahmassani and Chang (1985) assumes that drivers change departure time (and route) only if the schedule delay exceeds a certain threshold value, the so-called bound. The route and departure time decision at day n+1 depends on the schedule delay and bound at day n. In turn, the schedule delay is dependent on the discrepancy between the predicted and actual travel time at day n. Therefore, in their model the decision of the previous day is altered, only if the difference between the predicted and actual travel time on the previous day exceeds the bound. In a similar fashion the boundedly rational theory will be applied in the model here. In this model the bound is an exogenous variable. However, Mahmassani and co-authors (see Mahmassani and Herman, 1990) found some evidence that the bound is in reality dependent on the past experiences. Nevertheless, these considerations are omitted for simplicity. In mathematical terms the decision-making process can be described as (11.13).

---

[6]There is an analogy between search theory - which includes the costs of search (see, for instance, the literature on labour economics) - and models based on satisficing behaviour.

$$\begin{gathered}\textit{Assume route } r \textit{ has been chosen by driver } i \textit{ in period } n.\\ \textit{Driver } i\textit{'s decision in period } n{+}1 \textit{ is route } r \textit{ if the following}\\ \textit{expression holds:}\\ ET_r^{i,n}\cdot(1-bound) \leq ExperiencedTravelTime_r^{i,n} \leq ET_r^{i,n}\cdot(1+bound)\\ \textit{Otherwise the route with the highest utility will be chosen}\\ \textit{in period } n{+}1.\end{gathered} \tag{11.13}$$

The $ExperiencedTravelTime_r^{i,n}$ denotes the experienced travel time of driver i for route r in period n. If the bound is larger than zero, this model does not contain a direct utility maximising incentive. The alternative with highest utility is chosen only if the individual is not satisfied with the previously made decision. Otherwise, the driver sticks to his previous choice. If the bound is zero, however, this model collapses to a utility maximising model. Therefore, a satisficing model can be regarded as a generalisation of a utility maximising model. Model (11.13) will be used in the simulation experiments in Section 11.5.[7]

Decision-making with information

Decision-making with information provision types A, A+B and A+B+C is straightforward because route switching during the trip is not allowed. This implies that the route choice decision is made before the trip. Since it was assumed that the utility function has the form given in (11.10), decision-making consists of applying model (11.13) before departure. The decision-making process becomes more complicated if route switching during the trip is allowed, which is the case for information provision types A+B+C+D, and A+D.

Below the decision-making mechanism corresponding to both information provision type A+B+C+D and A+D are discussed. However, the simulation experiments in Section 11.5 only make use of mechanism A+D for the following reasons:

1. Applying mechanism A+B+C+D requires the estimation of an additional parameter, the ß-parameter in equation (11.7), of which less is known in the literature.
2. Interpreting the results obtained with mechanism A+B+C+D is extremely difficult, since they follow from the application of three (!) information provision types consecutively.

---

[7]Although model (11.13) does not contain an explicit habit component it can be given a habitual interpretation. One could argue that if the last travel experience does not exceed the bounds, the driver is not willing to change alternative because of both satisficing and habit considerations. Furthermore, the longer ago the last change in alternative has been made, the more likely it is that a driver will stick to the same alternative in the future, due to the specification of the updating mechanism of ET. It is likely that the ET component will slowly converge to the experienced travel time thereby making the decision a satisfactory one. Only a large disturbance in the network, due to route changes by many other drivers could cause the driver to change route.

There are different ways to model route switching based on real-time en route information provision of type A+B+C+D. The most logical way is to extend the sequential two-phase mechanism of information provision type A+B+C with a third phase. Then, all the information acquired in the past is used during en route switching. However, from a computational point of view this approach is rather demanding. Since it is not correct to compare pre-trip expected travel time with en route expected remaining travel time, it would require to have, in addition to the expected travel time from the origin to the destination, the expected travel time from every link in the network to the destination. This in turn implies that the ET variable would need to have the form given in (11.14),

$$^{X}ET_{j,k}^{i,n} \tag{11.14}$$

which defines driver i's expected travel time from link k to the destination (after having processed information X) in period n, if route j will be followed. It can be easily seen that the models previously described are encapsulated in this model; they follow by restricting the number of links k to one, the first link.

A, from a computational point of view, less demanding model is described below. Here, it is assumed that a driver regards pre-trip and en route decisions as being independent of each other. More precisely, en route decisions are made without taking into account travel time expectations made before the start of the trip. This approach has to be adopted with care, since in the extreme case it would imply that pre-trip decisions are of no importance at all, since if the driver starts the actual trip, the route switching decisions are completely based on the current situation in the network. Therefore, the following modification is suggested. During the trip, route switching is carried out only if the improvement in expected remaining travel time compared to the expected remaining travel time of the route currently chosen exceeds a certain threshold value, and is in absolute value larger than the parameter τ. As Mahmassani and Jayakrishnan (1991) point out, plausibility is the main justification for this rule. The threshold values bound and τ may reflect perceptual factors, preferential indifference, or persistence and aversion to switching. The value bound is taken to be decreasing in time, dependent on the remaining travel distance, to model a slowly decaying influence of the pre-trip decision during the trip. The specification of the bound used in the simulation experiments is given in Section 11.5.4. In mathematical terms the model is given in (11.15).
The en route switching process in model (11.15) can be found in Mahmassani and Jayakrishnan (1991) and has been proposed by Bonsall (1992a).

The decision-making model used for information mechanism A+D can almost be copied from model (11.15). The only difference being that the driver's trip will be started according to information gathered by own experience rather than pre-trip information.

11.4 SIMULATION FRAMEWORK

This section presents the set-up for the simulation experiments conducted in Section 11.5. Section 11.4.1 discusses methodological issues, while Section 11.4.2 describes the structure of the simulation model.

Start driver i's trip according to the pre-trip decision.
Calculate before entering a new link the expected remaining travel time of all the available routes k based upon the en route information. These travel times will be denoted by RTT_k^i.
Calculate $\min_k RTT_k^i$ *and suppose this occurs for k=m. Assume further that the current route is route r.* (11.15)
If $RTT_m^i < RTT_r^i * (1 - bound)$ *and* $RTT_r^i - RTT_m^i > \tau$
driver i will switch to route m.
Otherwise driver i will continue on route r.

11.4.1 Simulation methodology

Vehicle movement
Following the MPSM model of Chang, Mahmassani and Herman (1985) traffic flows are modelled using a modified form of Greenshield's speed-density model. The speed of a vehicle on a link is calculated just before entering the link using formula (11.16).

$$v = (v_f - v_j) * (1 - \frac{k}{k_j}) + v_j \tag{11.16}$$

v_f and v_j denote the free flow and jam speed; k and k_j denote the current and jam density. Two additional restrictions are imposed:
1. Overtaking does not take place.
2. The jam density cannot be exceeded.

With respect to the latter point, a driver wanting to move to a link that has reached the jam density remains on the current link. A (small) fixed amount of time later, Δ, the driver makes a new attempt to move to the next link.

Network description
The road network depicted in Figure 11.1 is used for the simulation experiments. This network, containing one OD-pair, consists of many routes (25) and decision points (9) for providing an interesting framework for the analysis of the effects of information supplying.

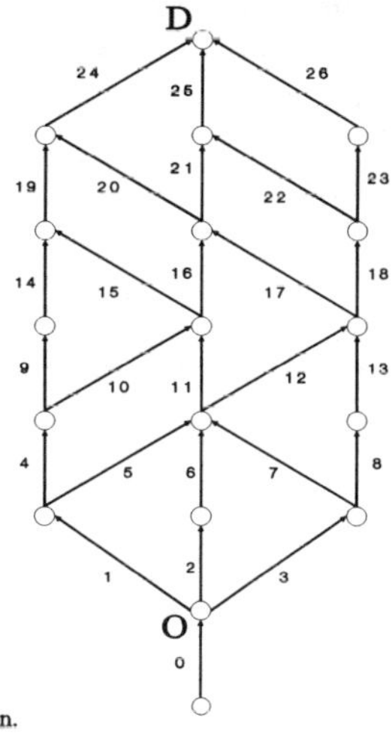

Length of all links is 2 km.

Jam speed is 5 km/h for all links.

Free flow speed is 70 km/h for links 2, 6, 11, 16, 21, 25.

Free flow speed is 50 km/h for other links.

Link 0, having an infinite jam density, is needed to start the simulation.

Figure 11.1 Road network used in simulation experiments.

Number of drivers
The simulation experiments are carried out with 300 drivers. A larger number is, obviously, more realistic, but currently unmanageable with the accessible computing hardware. An implication is that every driver in reality represents a bunch of drivers, or in terms of Chang et al. (1985), every simulated driver is a macroparticle of drivers.

Different levels of congestion
From the speed-density equation (11.16) it is clear that the speed on a link solely depends on the ratio of current and jam density, or more precisely, on the current and jam density in macroparticles. The size of a macroparticle is irrelevant, only the jam density, measured in macroparticles, is relevant. Throughout the chapter the jam density per kilometre in macroparticles will be denoted by the constant K0. This constant then enables to manipulate the level of congestion in the network. A low K0-value, and therefore a low jam density, implies a highly congested network since it assumes that only a few of the 300 drivers can occupy a link at the same time. On the other hand, a high K0-value implies an almost uncongested network. Thus, different K0-values could be interpreted as a network with different capacities. In this chapter three different levels of congestion are analysed. These are shown in Table 11.2.

Departure time structure
In the simulation experiments, departure times are assumed to be fixed. Hence, route choice is the remaining decision. The departure time structure is depicted in Figure 11.2 and assumes that all drivers depart within one hour. The steeper the slope of the curve in Figure 11.2, the higher the departure rate.

level of congestion	K0 (macroparticles)
practically uncongested	12
congested	8
very congested	5

Table 11.2 Three levels of network capacity.

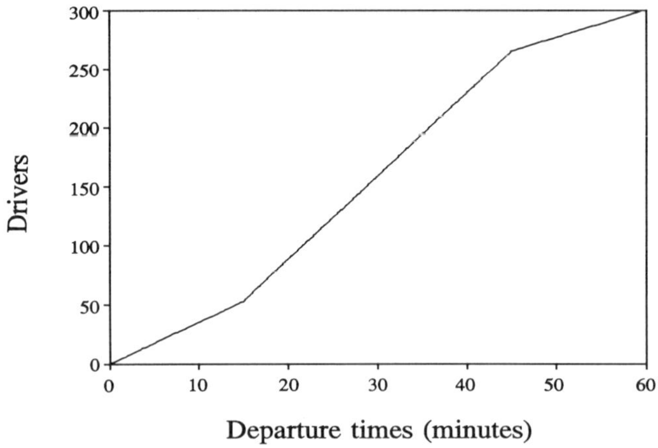

Figure 11.2 Departure time structure.

This structure assumes that:
1. (3/17)*th* of the drivers depart between minutes 0 and 15.
2. (12/17)*th* of the drivers depart between minutes 15 and 45.
3. (2/17)*th* of the drivers depart between minutes 45 and 60.

Thus, the rate of departure is two times as high in the interval [15,45] compared to interval [0,15], and three times as high in [15,45] compared to [45,60]. A similar approach (using different figures) has been applied by Mahmassani and Peeta (1993).

Stochasticity of simulation
The initial ET-values of the drivers are assumed to correspond to a speed of sixty kilometres per hour, thus equalling 12 minutes for every route. To force a route decision in the first travel period, a small random number taken from the interval [0,1] is added to the ET-value of every driver for every route. This divides the drivers randomly over the available routes in the first period. However, as pointed out by Horowitz (1984), different initial distributions can lead to different simulation processes. Hence, the necessity to repeat each experiment a number of times to assess

the significance of the results. According to the statistical theory, the more repetitions the better, but due to computing time limitations, the number of repetitions has been fixed at ten.

Market penetration
In Section 11.5, among other things, an analysis of the impacts of different levels of market penetration will be conducted. The different levels to be analysed are 0, 2, 5, 20, 50, 75, and 100 per cent. In all the cases, the drivers supplied with information will be spread homogeneously throughout the population. For instance, if twenty per cent of the 300 drivers are supplied with information, the drivers with information are respectively, driver 1, driver 6, driver 11, ..., driver 291, and driver 296 of Figure 11.2.

Description of a simulation run
One simulation consists of several subsequent simulated periods. A period represents a day of travel, and the words period and day will be used interchangeably. Every day, the same number of drivers carry out the same trip. The drivers can be considered being commuters. The route choice is the only decision open to them. The route choices lead to a certain traffic situation. The overall network performance will be measured in terms of travel time averaged over all drivers. During the next period, drivers will incorporate the experience gained during their last travel period in the next day's route choice decision which in turn will lead to a changed overall network performance etc. The overall network performance is taken as the sum of the drivers' travel times. This process will be repeated until either the system has reached a steady state[8] for at least ten subsequent periods, or the number of simulated periods exceeds 400. In the former case, the network performance at the end of the simulation process equals the steady state travel time, while in the latter case the network performance is taken to be equal to the average network travel time over the last twenty simulated travel periods, that is, the average over the periods 381 to 400. Every simulation will be repeated ten times, which will give an average network performance over the ten conducted runs. These averages will, unless stated otherwise, be used to draw inference from the simulation experiments.

Steady state
The term steady state is used to indicate a situation (during a simulation run) characterised by the fact that none of the drivers has an incentive to change routes in future periods. The term steady state, rather than equilibrium, is employed to underline that a steady state in the simulation model does not necessarily coincide with Wardrop's user equilibrium (Wardrop, 1952). It is important to note that if drivers did not change routes in the current period (compared to the last one), it does not necessarily imply that a steady state is reached. A steady state is reached only if the expected travel time equals the experienced travel time. This does not need to be the case if drivers did not change routes in the last period, and this is the reason for adopting the approach described above, which assumes that a steady state has been

[8]The term steady state will be explained below.

reached after the drivers did not change routes for ten consecutive periods. After ten periods without route choice changes, it is highly unlikely that drivers will change routes in future periods, since the difference between expected and experienced travel time will have become very small.[9] The steady state that will be reached is not unique but depends on the initial knowledge of the drivers, see Horowitz (1984) and Mahmassani and Jayakrishnan (1991).

11.4.2 Structure of simulation model

The simulation model is implemented in the language C and adopts an event based simulation approach. In such an approach, the simulation is conducted using an ordered list of events, the so-called event list. An event is an occurrence that alters the state of the simulated system. In the road network, an event corresponds to the potential movement of a driver (macroparticle) to another link.

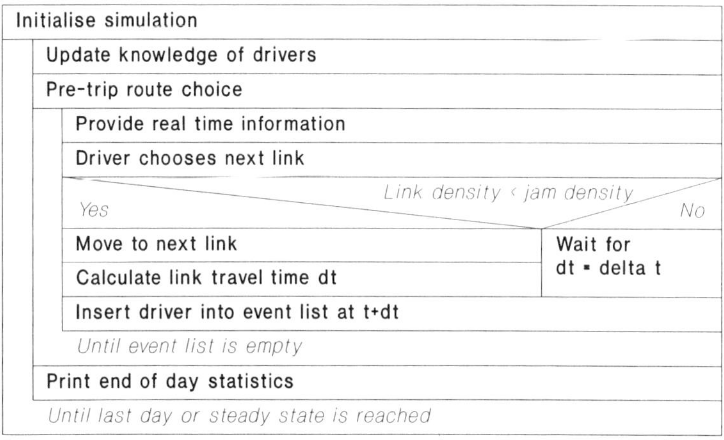

Figure 11.3 Flow of control in simulation model.

An event based simulation model proceeds by taking the first event from the event list, executing this event, and inserting newly generated events in their appropriate position in the event list. An empty list characterises the end of the simulation process. The flow of control in the simulation model is depicted in Figure 11.3.

The model deals with information provision at two different levels. First, the box *Update knowledge of drivers* updates drivers' travel expectations, and, in addition, gives information to drivers prior to their trip. One could think of either historic after-trip information, or pre-trip information on the current conditions in the network. Second, box *Provide real-time information* supplies the driver with real-time en route

[9]Given the model specification in this chapter, a steady state is equivalent with a situation in which the *ExperiencedTravelTime* is equal to *ET* for all drivers.

information before entering a new link, thereby enabling him to change his route dependent on the prevailing network conditions.

11.5 RESULTS OF SIMULATION EXPERIMENTS

The results of the simulation experiments will be explained in four sections. Section 11.5.1 describes some general characteristics prevailing in all the conducted experiments. Section 11.5.2 analyses the performance of the boundedly rational model using information based on own-experience (information provision type A). Section 11.5.3 investigates the implications of providing after-trip information (information provision type A+B), while Section 11.5.4 focuses on real-time en route information (information provision type A+D).

Before analysing the results, it should be stressed that these are obtained in a hypothetical setting: (1) Drivers behave according to a simple decision-making model; (2) there is only one OD-pair; (3) traffic is simulated using a linear speed-density relationship, and (4) provided information is never ignored. Nevertheless, this is an interesting environment for investigating the potential of information provision.

11.5.1 General characteristics of simulation experiments

Throughout this chapter, the parameter α (see equation (11.1)) is set equal to 0.4. Given the empirical evidence in Iida et al. (1992) and Vaughn et al. (1992) this seems a reasonable estimate.[10] The decision model used in the simulation experiments is model (11.13).

The different congestion levels lead to large discrepancies in steady state travel time. Under congestion level K0=5, the average travel time is just under 40 minutes, implying an average speed of 18 kilometres per hour. Congestion level K0=8 leads to a travel time slightly over 20 minutes (average speed of 36 kilometres per hour), while congestion level K0=12 causes a travel time of approximately 15 minutes, reflecting an average speed of 48 kilometres per hour. The congestion level K0=5 leads to a heavily congested network, while congestion level K0=12 leaves the network relatively uncongested.

Different runs of the same experiment (differing only in starting values) lead to different steady state values. These findings underline the theoretical arguments given in Horowitz (1984).

A typical average travel time pattern (travel time averaged over all the drivers) during one simulation run is depicted in Figure 11.4. Two conclusions can be drawn:

1. The average travel time is a highly sensitive performance measure. A relatively small number of drivers taking the same route in a small time interval can cause a highly congested link, while at the same time the other links will be relatively uncongested, but nevertheless leaving the average travel time in the network excessively large.
2. Figure 11.4 indicates that the particular simulation run depicted reaches a steady state. This is the case for most of the experiments. However, the number of days

---

[10] However, it is acknowledged that a stochastically assigned α-parameter (randomly differing among the drivers) would have been more realistic.

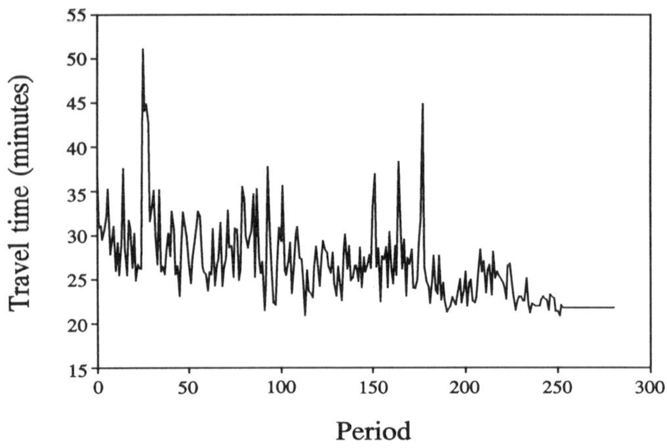

Figure 11.4 Evolution of daily average travel time during a simulation run. K0=8, bound=0.

needed to reach a steady state varies strongly, depending on the parameters chosen in the behavioural model and the kind of information provided.

11.5.2 Comparison of models with different bounds

In this section the influence of the bound on the network performance is analysed under information provision type A (information based on own experience). Model (11.13) is used and the bound varies between 0 and 0.4 in steps of 0.1. Combined with the three congestion levels of Table 11.2 this gives 15 different experiments.

Figure 11.5 depicts the travel time as a percentage of the travel time under the model with bound equal to zero, which is used as base case throughout this section. It can be seen that the models with bound generally outperform the base case, especially if the level of congestion is equal to 8. These results suggest a steady state travel time reduction of 7 per cent compared to the base case.

The number of routes used[11] decreases as the bound increases. This intuitively appealing result is shown in Figure 11.6. The number of routes chosen decreases dramatically if the bound increases to 0.4. Furthermore, Figure 11.6 shows that a higher level of congestion induces a larger number of routes used. This could be explained by noting that it takes longer to reach a steady state if the recurrent

[11]Throughout this chapter, the performance indicator *number of routes used* refers to the number of different routes used by a driver during a simulation run averaged over all drivers.

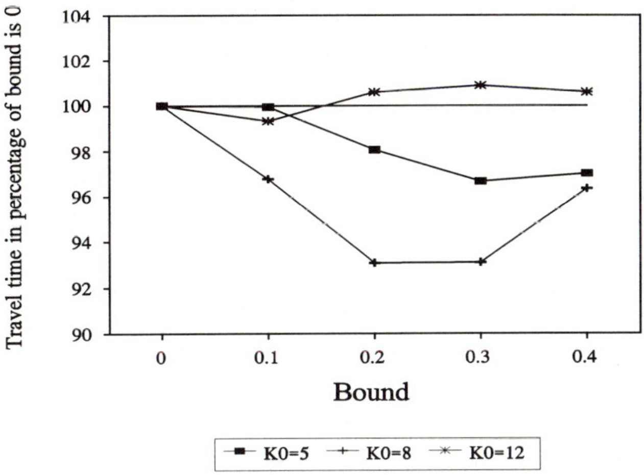

Figure 11.5 Travel time as a percentage of the travel time under the model with bound=0 for three levels of congestion.

congestion is more severe, which is underlined by Figure 11.7.[12]

As shown in Figure 11.8, the switching propensity (percentage of drivers switching routes from day-to-day) is obviously larger the smaller the bound. In addition, it can be seen that the curve gets smoother the larger the bound.

Combining the results in Figure 11.6, Figure 11.7 and Figure 11.8, Figure 11.5 could be explained by arguing that drivers use the different route alternatives more efficiently in a model with non zero bound. Furthermore, they can better rely on their own expected travel time, since the switching propensity of the other drivers is smaller. However, if the bound gets too large, the overall network performance deteriorates due to too many missed good route opportunities. The optimal value for the bound seems to lie, dependent on the level of congestion, between 0.2 and 0.3.

---

[12]In Figure 11.7 one point is missing. In this case a steady state was not reached after 400 days.

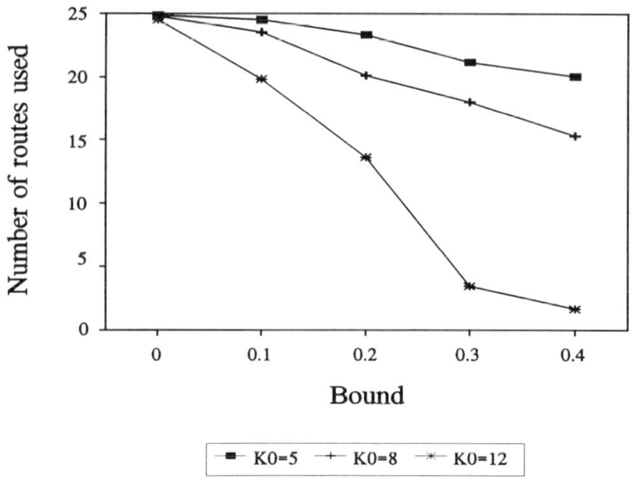

Figure 11.6 Number of routes used for models with different bounds and three levels of congestion.

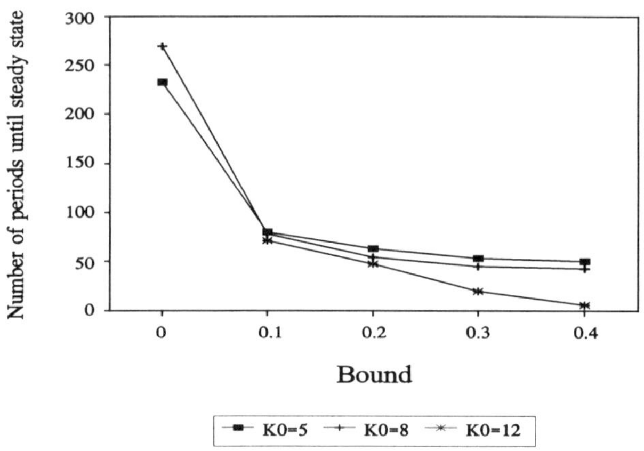

Figure 11.7 Number of periods till steady state is reached for models with different bounds and three levels of congestion.

These results are in agreement with the insights gained by the work of Mahmassani and co-authors. They argued that, in general, better system wide performance is attained when path switching behaviour is dampened by an indifference band (the bound in the model). Potential negative effects of extreme behaviour could occur if drivers' behaviour is modelled with a *myopic* switching rule, that is, drivers will always select the best path in terms of travel time (Mahmassani and Chen, 1991). In

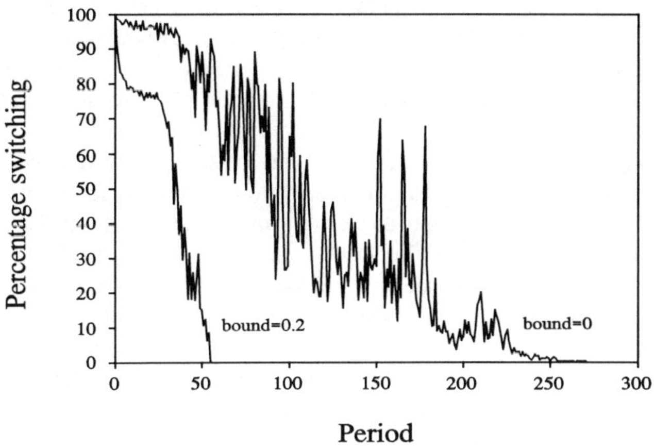

Figure 11.8 Evolution of drivers' switching propensity under model with different bounds and congestion level K0=8.

the model this corresponds to a value of the bound equal to zero.

11.5.3 After-trip information

In this section, the implications of after-trip information provision are investigated. Throughout the section it is assumed that drivers behave as explained in Section 11.3.1. Furthermore, it is assumed that drivers make decisions according to the boundedly rational model explained in (11.13), with the bound being equal to 0.2. All combinations of three levels of congestion and five levels of market penetration (0, 2, 5, 20, 50%) were investigated. The results of the experiments are compared with the no information case.

Figure 11.9, Figure 11.10 and Figure 11.11 show the results for drivers with and without information for different levels of congestion.[13] It should be noted that the y-axis is not scaled equally across these figures. Different scales were needed to preserve the information in each figure. The following points can be made:

1. Drivers with information benefit compared to the base case as long as the level of market penetration is under 20 per cent. In addition, they are best off if the level of congestion is not too severe, that is, if K0=8.[14]

[13]The *with information* curve is not defined for zero per cent market penetration.

[14]The 20% mentioned in this section is indicative, but of course, to some extent dependent on the other parameters used in the simulation experiments.

2. As the level of market penetration exceeds 20 per cent, drivers without information perform better than drivers with information.[15] In a different context, the same phenomenon can be recognized in some of the experiments of Mahmassani and Chen (1991) and Mahmassani and Jayakrishnan (1991).[16] This could be explained by referring to the phenomenon of overreaction and the informativeness of the information. The information given to the drivers is based on densities realised during the previous day. If too many drivers respond to this historic information, it will not reflect the current situation in the network. Thus too many drivers responding to *old* information decreases the informativeness and accuracy of the information and causes *overreaction*. A similar result was obtained by Koutsopoulos and Xu (1993). In these situations, Mahmassani and Jayakrishnan (1991) argued that coordinated information is necessary.
3. After-trip information is most beneficial to everyone if the level of market penetration is below 20 per cent. This contradicts the results in Mahmassani and Tong (1986). In experiments concerning real commuters they found that after-trip information improved the system performance, even at a market penetration level of 100 per cent. However, experiments in the next chapter and research by van Vuren and Watling (1991) suggests that the optimal level of market penetration is highly dependent on the kind of information provided.

The figures presented so far are related to the steady state average travel time. However, if after-trip information is provided it is also interesting to investigate the process leading to the steady state for both the drivers with and without information. Figure 11.12, shows a typical pattern if the level of market penetration is low. It can be seen that drivers with information outperform the ones without every period. It seems that if drivers with information benefit in the steady state situation, they have been better off during the process leading to the steady state as well.

During some simulation experiments a steady state was not reached after 400 days. In these cases a cyclical pattern arose. These findings underline theoretical work by Horowitz (1984). He argued that even in a two-link network under an information mechanism as given by equations (11.2) and (11.3) the situation in the network could oscillate perpetually.

---

[15] One might argue that it is counter intuitive for drivers with information to be worse off compared to the situation without information. The argument supporting this case could, for instance, be based upon the fact that if the drivers would be worse off by using the information, they would ignore it. However, in the model presented here it is assumed that drivers supplied with information will always use it. Because of these unrealistic results, no experiments with a level of market penetration higher than 50 per cent have been conducted.

[16] Compare in their paper Figure 2 (p. 299) and Figure 4 (p. 300) for the model without bound and a high level of market penetration.

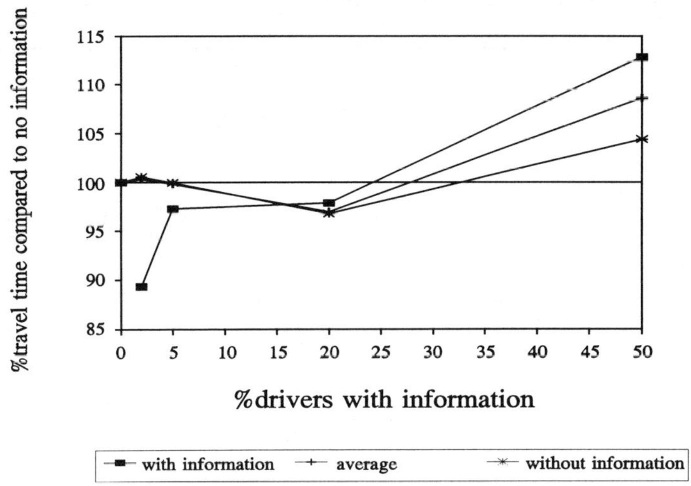

Figure 11.9 Travel times compared to no information case. K0–5.

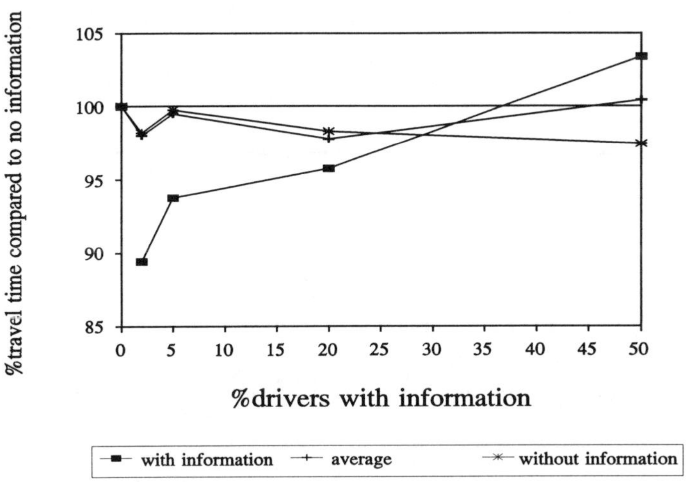

Figure 11.10 Travel times compared to no information case. K0=8.

Finally, it was observed that drivers with information try a considerable smaller number of routes than drivers without information. Summarising the results in this section, drivers with information experience shorter travel times if the level of market penetration is under 20 per cent. They achieve this using a significantly smaller number of routes. If the level of market penetration exceeds 20 per cent, overreaction takes place and the network performance deteriorates.

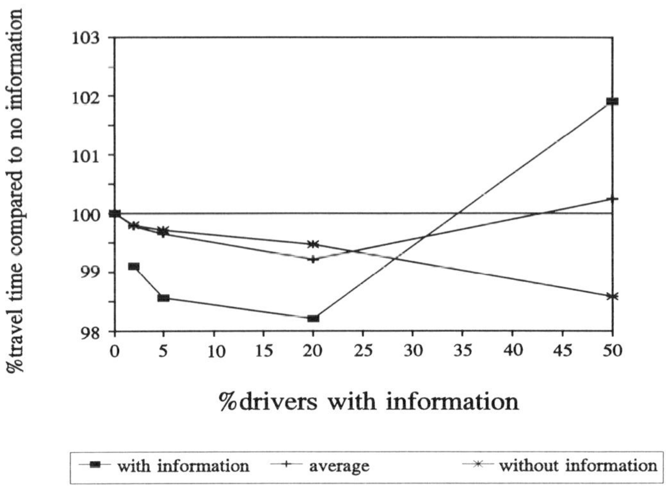

Figure 11.11 Travel times compared to no information case. K0=12.

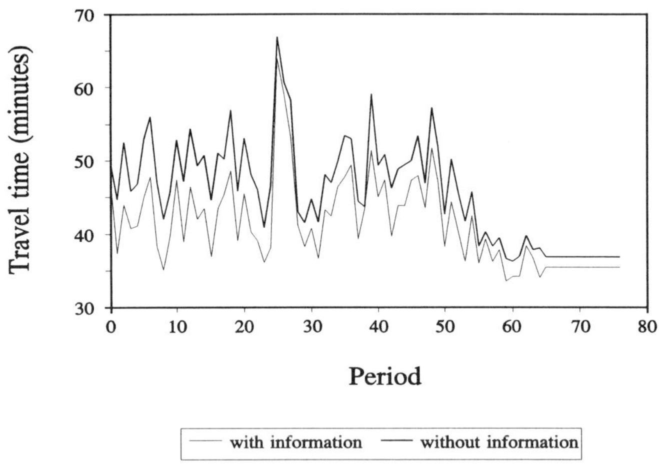

Figure 11.12 Average daily travel times for drivers with and without information. K0=5, level of market penetration is 5%.

11.5.4 Real-time en route information

This section analyses the effects of real-time en route information provision. It is assumed that all the drivers update their travel time predictions following Section 11.3.1. The parameter τ in the decision-making model is set at 1 minute, and the bound is specified as in (11.17).

$$\text{bound} = (\#\text{ of links remaining}) * 0.05 \quad [\,] \qquad (11.17)$$

where the parameter 0.05 is referred to as the *en_route_bound*. Figure 11.1, shows that each route consists of 7 links. Therefore, the bound in (11.17) decays from 0.30 (6 links remaining) to 0.05 (1 link remaining). Unequipped drivers make their decisions following model (11.13), in which the bound is set equal to 0.2.

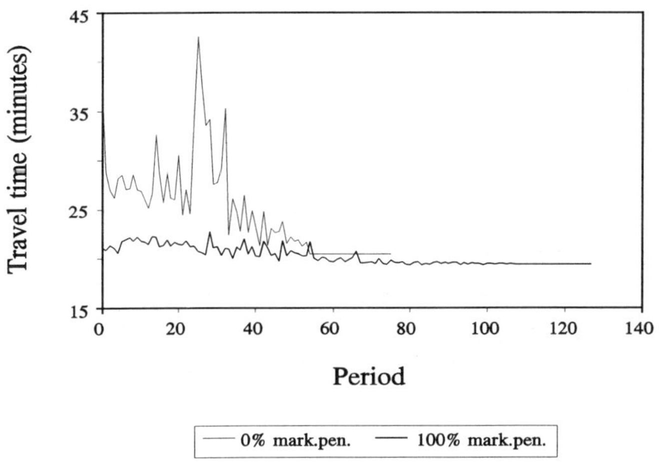

Figure 11.13 Daily travel time pattern for no and full market penetration. K0=8.

The pattern in Figure 11.13 prevailed in all the simulations experiments. Figure 11.13 shows that a high level of market penetration has a strongly decreasing effect on the variance in travel time, but an increasing effect on the number of days to steady state. Furthermore the steady state travel time under information provision does not differ largely from the situation without information provision. The gains of information are reached in the process leading to a steady state, the gains in steady state travel time itself are marginal. This is a result that one could have expected in a network with only recurrent congestion.

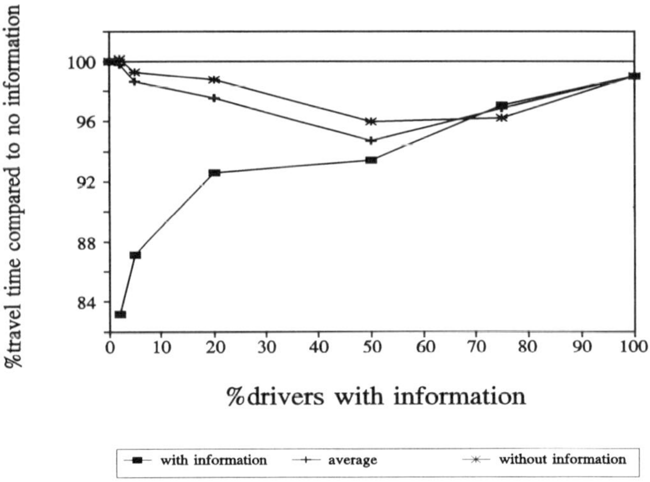

Figure 11.14 Travel times compared to no information case. K0=5.

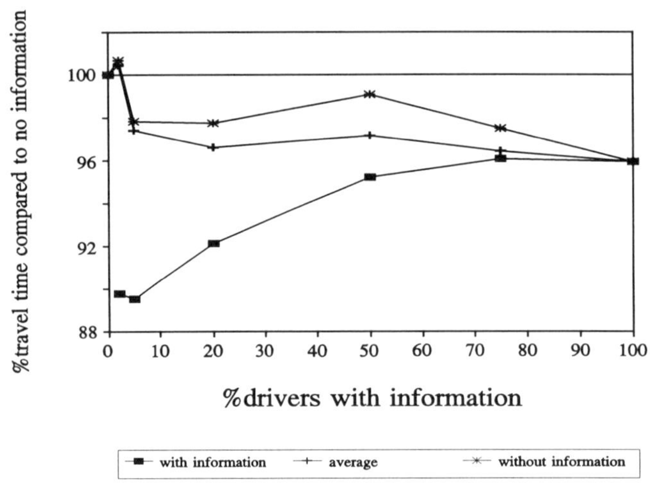

Figure 11.15 Travel times compared to no information case. K0=8.

The results of the simulation experiments are summarised in Figure 11.14,

Figure 11.15, and Figure 11.16.[17] The numbers plotted are averages over ten simulated runs. The following conclusions can be drawn:

1. If K0=12, the gains of providing real-time en route information are very small, due to the fact that the network is practically uncongested.
2. For K0=5 and K0=8, there are considerable gains for the drivers with information, up to a level of market penetration of 75 per cent. If K0=8, a level of market penetration equal to 100 per cent still provides substantial gains for the drivers.
3. There are savings between 3 and 5 per cent in overall travel time for a level of market penetration between 10 and 75 per cent.
4. Drivers without information benefit as well. Their travel time savings are between 1 and 4 per cent.[18]
5. The difference between the *with* and *without* information curve converges as the level of market penetration increases and is already relatively small at a 75 percentage level.

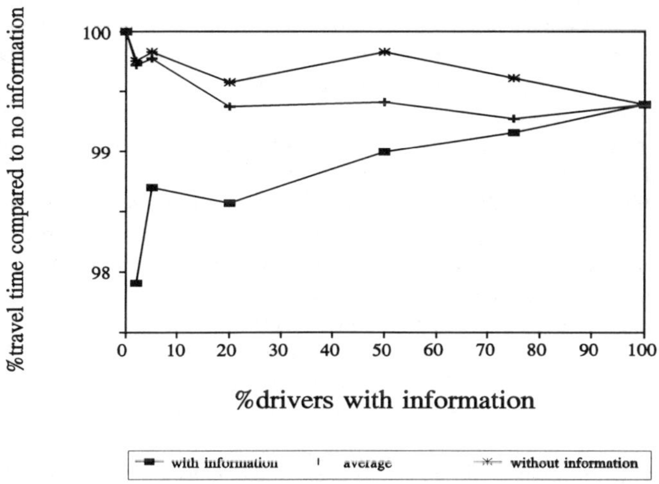

Figure 11.16 Travel times compared to no information case. K0=12.

The difference between the *with* and *without* information curve has an interesting interpretation. It reflects the benefits to the population of drivers equipped with the motorist information system. It can therefore, in combination with the costs of the

[17]To preserve informativeness, the y-axis's of these Figures are scaled differently.

[18]In one case (K0=5, market penetration=75%) drivers without information outperform the ones with information. The same phenomenon prevailed in some simulation experiments in Section 11.5.3.

system, be viewed as an indicator of the market potential of these new technologies. The difference between these two curves is called the *information benefits to equipped drivers*, see Chapter 2. In Figure 11.14, Figure 11.15 and Figure 11.16 it can be seen that the *information benefits to equipped drivers* are high at low market penetration levels, but relatively small at a market penetration level of 75 per cent.

11.6 CONCLUSION

The simulation model presented in this chapter integrates the within-day route choice dynamics with day-to-day dynamics. The computer model allows the investigation of the implications of these two levels of decision-making on the travel time patterns. Four different types of information have been described (own experience, after-trip, pre-trip, en route), and information mechanisms based on combinations of these have been specified using boundedly rational models. Before drawing conclusions, it should be stressed that these results were obtained in a hypothetical setting and could be network dependent. Nevertheless, they provide some useful insights into the potential effects of information provision in a network with recurrent congestion under different information schemes.

The results of the simulation experiments concerning the boundedly rational model (in an environment with information based on own experience) are in agreement with the work summarised in Mahmassani and Herman (1990) and Mahmassani and Chen (1991). A model with a bound between 0.2 and 0.3 performs best (in terms of total network travel time) in the recurrent congested environment.

The experiments with information provision were restricted to after-trip information and real-time en route information. The results suggested that after-trip information is beneficial to all drivers if the level of market penetration does not exceed 20 per cent. In this case, the drivers with information benefit most; in some cases travel time savings of up to 10 per cent were achieved under low levels of market penetration. If real-time en route information is provided, there are benefits to all drivers for most levels of market penetration. Depending on the level of congestion, the benefits might be as large as 15 per cent to equipped drivers under low levels of market penetration. With high levels of market penetration the average network travel time savings are between 3 and 5 per cent compared to the situation without information. It should be emphasised that the percentages mentioned are related to steady state travel times. In addition, the process leading to a steady state shows, in particular with real-time en route information, a considerable smaller day-to-day travel time variance. Real-time en route information seems to stabilise the traffic flows, and these are probably the most relevant gains of information provision in a network with recurrent congestion.

Comparing the results obtained with after-trip and real-time en route information it emerges that with the latter high levels of market penetration can be achieved without the occurrence of significant overreaction. It can be concluded that high quality (actual, accurate and informative) information allows a relatively high level of market penetration, possibly close to 100 per cent, while information of low quality (uninformative and inaccurate) causes overreaction taking place already at low levels of market penetration. However, given the shape of the *information benefits to equipped drivers* curve, it is unrealistic to expect a 100 per cent level of market penetration when these technologies were commercially marketed.

The next chapter is devoted to studying the case of non-recurrent congestion. The impact of information is analysed with the occurrence of unpredictable, random incidents in the transport network.

12 SIMULATION MODELLING: NON-RECURRENT CONGESTION[1]

12.1 INTRODUCTION

Due to large mobility demand (road users) and scarce supply of road infrastructure, road users are facing large, undesired, travel time delays, particularly throughout the morning and evening peak-hours. This kind of congestion is referred to in the literature as *recurrent* congestion, that is, congestion taking place regularly. The impact of traffic information under these conditions was analysed in the previous chapter. However, travel time delays are even larger if additional, random incidents occur on the roads, thereby diminishing the available road capacity during a certain period of time. Congestion of this type is referred to as *non-recurrent* congestion and is caused by bad weather (fog, heavy rain, snow etc.), traffic accidents, lost cargo, sudden lane closures etc. In the literature, it has been claimed that non-recurrent congestion accounts for up to 60 per cent of total congestion delays (Lindley, 1986, 1987, 1989). Consequently, it is more realistic to assume that the capacity of a transport network is stochastic rather than deterministic, see also Chapter 4.

Al-Deek and Kanafani (1993) modelled the benefits of motorist information systems in corridors with incidents. They considered a network, consisting of two routes, with random incidents occurring on the more attractive one (under normal conditions). Their model is based on *user equilibrium* assumptions, that is, with motorist information systems, drivers will be diverted, so to re-establish the user equilibrium traffic flows. As a consequence, in equilibrium there are no additional benefits to equipped drivers; only in the phase leading to equilibrium, equipped drivers outperform the non-equipped ones. Depending upon their model parameters, the network wide travel time savings are quite substantial; up to 30 per cent compared to the situation without information.

In another study, Hall (1993) presents a different picture of the benefits owing to motorist information systems in networks with non-recurrent congestion. He analysed non-recurrent congestion using the definition of *effective capacity*, which is the expected roadway capacity after accounting for the random occurrence of incidents. He concluded that in reality, based on Lindley's (1986) data, the average effective loss is surprisingly small, ranging between 2 and 9 per cent, depending on the length of the bottleneck under investigation. Hence, he argued that the application of motorist information systems might be justified as giving a reasonable assurance of arriving on time to the road users (for instance, diminishing stress and anxiety), but should not be viewed as a substitute for building new roads.

---

[1]This chapter is based on Emmerink, Axhausen, Nijkamp and Rietveld (1995b) published in *Transportation Research C.*

In this chapter the potential of motorist information systems in the case of non-recurrent congestion is analysed from another perspective, and it is therefore complementary to the work by Al-Deek and Kanafani (1993) and Hall (1993). The analysis will centre around the behavioural responses of the road users. These responses, in fact, determine the traffic flows in the road network. In this approach, these flows are not determined by *user equilibrium* considerations as in Al-Deek and Kanafani (1993). Neither is the present analysis led by the *supply-side*, as Hall's (1993) study. Here, it is argued - as did Bonsall (1992a), Bonsall et al. (1991), Dehoux and Toint (1991), Iida et al. (1992), Mahmassani and Herman (1990), Watling and van Vuren (1993) - that the effects of motorist information systems on the network wide performance should be analysed where the information systems technology penetrates into the transportation system, at the driver's level. However, doing so reduces the available research tools quickly; analytical models become either unrealistic or too complex to solve; only a simulation approach remains.

The methodology developed in the previous chapter is extended to deal with stochastic incidents as well. Mahmassani and Jayakrishnan (1988) used a similar approach to analyse a transportation network during periods of perturbations, but their analysis only dealt with historical information, that is, the road users' own experience in previous periods. In this book, real-time en route information is also addressed.

As in the previous chapter, it is assumed that drivers behave following boundedly rational rules. It can be questioned whether these behavioural models are able to capture drivers' decision making behaviour. However, when analysing the *potential* of information technologies in road networks, it seems justified to assume a kind of rational behaviour. It is a further step to *predict* the impact of these technologies on network wide performance. In order to do so, the behavioural models used in this and other studies need to be validated. Research addressing this issue is still in its infancy, mainly due to the unavailability of rich data. A modest attempt in that direction is made in the next chapter.

The simulation experiments in the network with non-recurrent congestion focus on the following issues:

1. The influence of the level of market penetration on the network wide performance.
2. The additional benefits to equipped drivers.
3. The quality of the information in relation to drivers' behaviour and network wide performance.

The first point has been mentioned repeatedly in the literature. It has been envisaged that the level of market penetration is an important parameter in determining the success of these advanced technologies (Chapter 11; Mahmassani and Jayakrishnan, 1991; Watling and van Vuren, 1993). The second issue is particularly important from a commercial point of view. These systems are economically viable only if there are substantial additional benefits to the equipped drivers, as discussed in Chapter 2. And finally, the third issue has not been given much attention in the literature. However, it seems intuitively clear that low quality information could have an adverse effect on network wide performance. In the simulation experiments in this chapter, the analysis is confined to the updating frequency of the information. It is acknowledged that this is only one aspect affecting the quality of the information. Some other issues involved are briefly discussed in Section 12.3.

The analysis builds on the simulation model developed in the previous chapter. Section 12.2 discusses the differences between recurrent and non-recurrent congestion; Section 12.3 focuses on the quality aspect of information. Section 12.4 adds the new component to the model of the previous chapter. In Section 12.5, the parameters used in the chapter are discussed. Section 12.6 presents the results of the simulation experiments, and finally, Section 12.7 summarises the main arguments.

12.2 RECURRENT VERSUS NON-RECURRENT CONGESTION

In the transportation literature, the distinction between *recurrent* and *non-recurrent* congestion is frequently made. Although these terminologies are widely accepted and understood, a formal definition is hard to trace. In this section, an attempt to provide such a definition will not be made; the discussion will be confined to clarifying these notions.

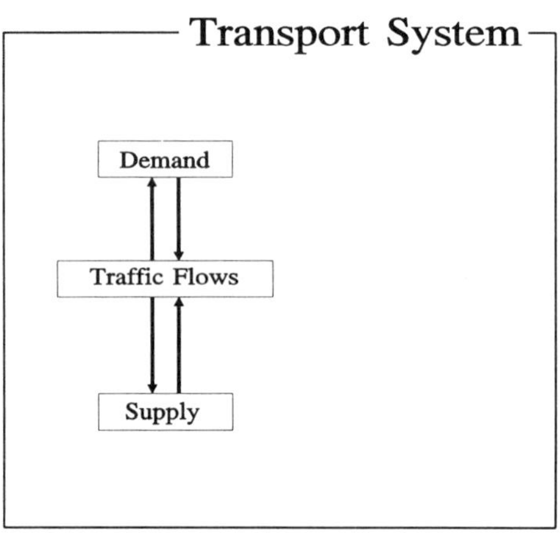

Figure 12.1 Transport system without shocks.

Figure 12.1 depicts a schematical illustration of a transport system. As with other economic commodities, a distinction can be made between the demand and supply side. On the one hand, the demand side may be thought of as the demand for mobility by road users. On the other hand, the infrastructure may be regarded as the supply side of the system. The evolution of the system depicted in Figure 12.1 might be referred to as *recurrent* in the sense that unexpected or unpredicted incidents do not take place. The *market* on which demand for mobility and supply of infrastructure meet results in traffic flows in the network, where the level of congestion can be viewed as a surrogate for the price. Given these flows, demand might evolve over time, since road users learn by experience about the traffic situation. If congestion takes place, it occurs at predictable recurrent periods of time and is caused by a relatively large demand, or

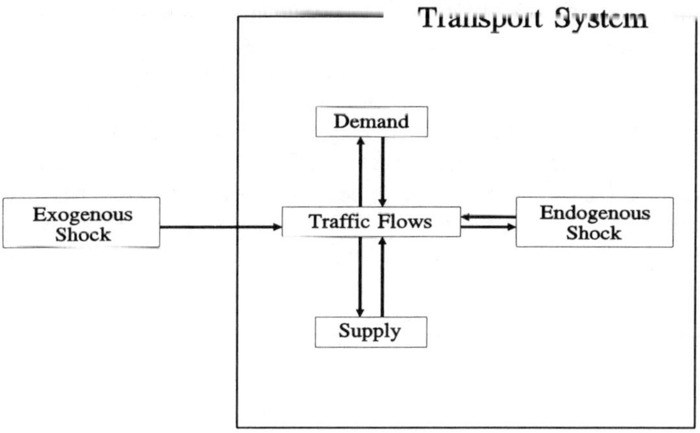

Figure 12.2 Transport system with shocks.

a relatively low supply during a certain period of time.

Clearly, Figure 12.1 is a stylised representation of the real-world situation. In reality, transport networks suffer both from *exogenous* and *endogenous* shocks as depicted in Figure 12.2. The endogenous shocks reflect unpredictable incidents occurring from within the system; particularly traffic accidents come to mind. These shocks are stochastic, but might also be a function of supply and demand. For instance, in relatively congested periods, accidents are more likely to take place. These stochastic incidents may severely affect the traffic flows in the network.

In reality, the system is from time to time affected by exogenous shocks as well, that is, shocks from outside the transport system. Particularly, weather conditions, traffic accidents and sudden lane closures might be thought of. These may have a significant adverse effect on network capacity.

In the chapter, the effects of these endogenous and exogenous shocks will be analysed in a network with a motorist information system. The details of the implementation are given in Section 12.4; the next section addresses the quality of the information in further detail.

12.3 INFORMATION ACCURACY

The information provided to the road users will ideally reflect the actual traffic flows in the network, or even better, the traffic flows in the future. However, as has been pointed out by Watling and van Vuren (1993), this is still a hypothetical case. Some sources affecting the quality of the information are listed below:

1. level of precision and reliability of the traffic measurement technique
2. reliability of broadcasting channel
3. delay in transmitting the information
4. updating frequency of the information
5. format in which information is provided

In this section, only the inaccuracy of the information caused by (1) a discrete updating frequency and (2) a delay in transmitting the information will be dealt with. No attention will be paid to unreliability due to measurement errors or other biases.

First, a discrete updating frequency implies that traffic data is not continuously collected, but only at discrete points in time. The reason being, for example, technical limitations of the information gathering system, or psychological arguments that continuously fluctuating information might confuse the road users. Second, the delay in transmitting the data reflects the time needed to transform *raw* data into a format suitable for transmission to the drivers. The process from detection to information provision is schematically depicted in Figure 12.3, where time is shown on the horizontal axis.

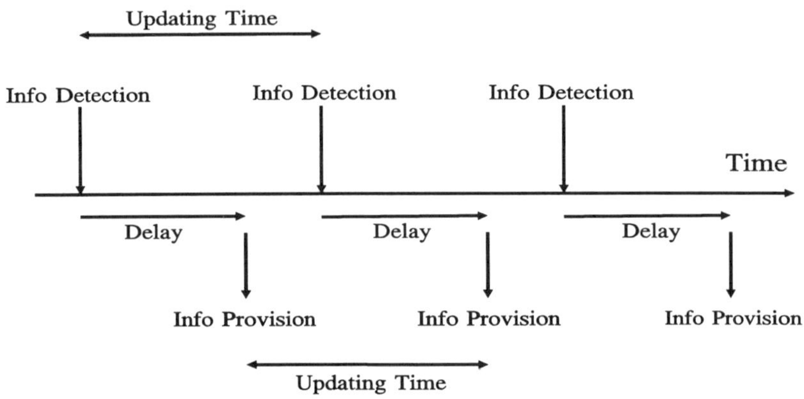

Figure 12.3 From information detection to information provision.

The age of the information as a function of time can now easily be derived. Figure 12.4 shows three different cases. In Figure 12.4A, there is both a delay and a discrete updating interval, Figure 12.4B illustrates a system with a delay and continuous updating, while Figure 12.4C shows discrete updating and no delay.

For simplicity, the simulation experiments in Section 12.6 only deal with a discrete updating time (Figure 12.4C); it is assumed that there is no delay in transmitting the information. Three updating times will be considered, 1, 5 and 10 minutes, respectively. These should give a greater understanding of the impact of the age of the information, as a surrogate for information quality, on network wide performance.

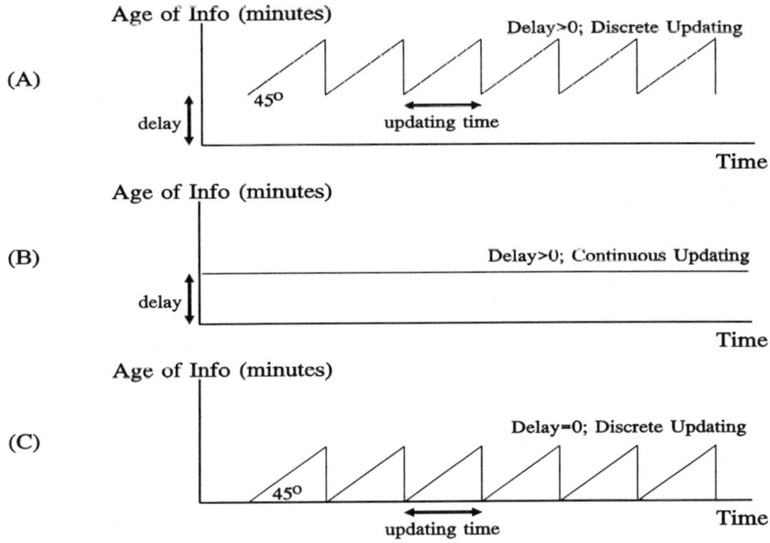

Figure 12.4 Updating time and delay. Three cases.

12.4 THE MODEL

The model that is used is an adapted version of the one described in the previous chapter. For the details, see Chapter 11. In this section, the model description is confined to highlighting the main characteristics and introducing non-recurrent congestion.

In the simulation model, 1000 drivers travel daily from the same origin to the same destination, using their past trip experience in making route choice decisions. It is assumed that departure times are fixed and that all drivers depart during one hour, see Figure 12.5. Clearly, the model would gain much in realism when the departure time choice would have been added to the drivers' choice set. Unfortunately, incorporating departure time choice in the behavioural model would substantially increase the computational and behavioural complexity. Therefore, it was decided to confine ourselves in this research project to route choice only.

The road network being used is depicted in Figure 11.1. It consists of one origin and one destination (one OD-pair), connected by 25 different routes and 9 decision points.

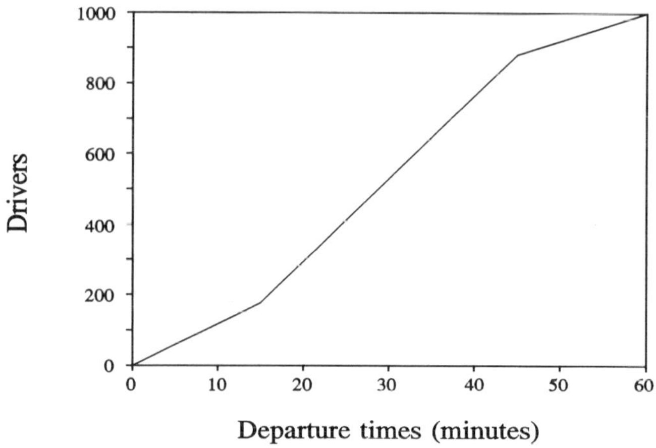

Figure 12.5 Departure time structure.

Each simulation run consists of 200 consecutive days, and is repeated five times, to account for the stochastic start of a run.[2]

It is assumed that non-recurrent congestion can take place on links 1 to 26 in Figure 11.1. An incident can only occur once a day at a specific link, and in such a situation the capacity of that link is always halved (the maximum density decreases 50 per cent). The duration of any incident is equal to 15 minutes; the starting time of an incident is a random number, taken from the interval [0,45].

Every day and for each link a Bernoulli trial is performed to decide whether or not an incident will take place. In Section 12.5, the *success* probability of the Bernoulli trial is discussed, together with the other parameters used in the model.[3]

12.5 MODEL PARAMETERS

The parameters used in the simulation experiments are listed in Table 12.1. Table 12.2 shows the parameters that have been varied throughout the simulation experiments, leading to 96 different cases.[4] The updating time, ranging from 1 to 10 minutes, was conceived as being a reasonable estimate for the available technologies. The *en_route_bound* (see Section 11.5.4) has been varied to test the interaction between the network wide performance, the en route *willingness-to-switch* and the quality of

[2]Test runs showed that a length of 200 days captures most of the dynamics in the model. Also from a practical perspective, 200 days - almost being equal to 10 months - seems a reasonable period.

[3]The term *success* refers to the occurrence of an incident.

[4]The updating frequency is irrelevant when there is no market penetration.

the information (the updating frequency). Clearly, the lower the en_route_bound, the higher the switching propensity, and visa versa.

The decision for the Bernoulli(p)-value was a difficult one. The literature does not agree on certain p-values, and, in addition, this value seems to be heavily dependent upon the kind of road network under consideration. Hall (1993) briefly reviewed some results in the literature; these suggested a very low value for p, and an incident duration ranging between 30 and 70 minutes. However, Keller et al. (1983) reported significantly different figures; highly varying p-values, and significantly lower incident durations (between 10 and 30 minutes). Since the focus of this chapter is on the *potential* of information provision during incident periods, it was decided to generate a relatively large number of incidents in the network and therefore relatively high p-values were chosen. To be precise, the p-values have been chosen to have on average 2, 5 and 10 daily incidents in the network. This corresponds to p-values of 0.08, 0.19 and 0.38. However, to compensate for these high probabilities, relatively short incident durations (15 minutes) were chosen. By doing so, a volatile network is created, which leaves ample room for studying the potential of information technologies on network performance.

Parameter	Value
Length of run	200 days
Number of runs	5 runs
Number of drivers	1000 drivers
Departure times of drivers	see Figure 12.5
Link capacity per km in macroparticles	32 macroparticles
Link capacity per km under an incident	16 macroparticles
α in formula, see (11.1)	0.4 []*
Bound in model, see (11.13)	0.2 []
τ in model, see (11.15)	1 minute
delta t, see Figure 11.3	0.2 minutes
incident duration	15 minutes
starting time of incident	random number from [0,45]

*[] indicates a parameter with no dimension

Table 12.1 Model parameters.

Parameter	Values
Market penetration	0, 5, 20, 50, 75, 100 (%)
Information updating time (upd)	1, 5, 10 (minutes)
en_route_bound, see (11.17)	0.02, 0.05 []
Bernoulli(p)	0.08, 0.19, 0.38 []

Table 12.2 Experimental design of simulation experiments.

12.6 RESULTS OF THE SIMULATION EXPERIMENTS

As discussed in the previous section, 96 different cases have been simulated, and each case has been repeated 5 times, to account for the stochasticity of a run. Unless stated otherwise, the performance indicators used to analyse these cases are the average travel time for drivers *with* information, *without* information, and the *network wide* travel time. Formula (12.1) shows the calculation of these performance indicators.[5]

[5]On purpose, a *run-in* period has not been used; the performance indicator is averaged over all periods. The reason is that to exclude the first days from the performance indicator calculation would imply that the different speed of knowledge acquisition for drivers with and without information is ignored.

$$X^m = \sum_{s=1}^{5} \sum_{p=0}^{199} X^m_{p,s} \qquad m = \textit{average, with, without.} \tag{12.1}$$

X^m : *average travel time for group m, averaged over periods* [0...199] *and runs* [1...5].

$X^m_{p,s}$: *average travel time for group m in period p and run s.*

In studies concerning recurrent congestion, attention is generally focused on the network performance at *steady state*. In these circumstances, steady state is defined as a situation in which no driver has an incentive to switch alternative (Mahmassani and Herman, 1990). In a road network with non-recurrent congestion, however, drivers equipped with an information device might switch alternatives when faced with an incident, even after the system has been simulated for a long period of time. As a consequence, a steady state might not be reached. Due to the relatively large number of incidents generated in the road network, this is the case with the simulation experiments conducted in this chapter. Therefore, the travel time performance indicator is based on a run average as specified in (12.1). In the next sections, the results of the simulation experiments are discussed.[6]

12.6.1 General observations

The uninformed drivers base their behaviour on information type A (own experience); the informed ones use information type A+D (own experience and real-time en route; see Table 11.1). Figure 12.6 depicts a trend prevailing in all conducted experiments. The variance in daily network wide travel time decreases as the level of market penetration increases. Furthermore, the network wide travel time is significantly lower as a large part of the drivers has access to the information. This leads to an important conclusion regarding network wide performance when drivers behave boundedly rationally.

> *If drivers behave according to a boundedly rational model without being provided with information in a network with non-recurrent congestion, they are unable to choose routes leading to an efficient usage of the road network in terms of total travel time.*

Figure 12.6 shows that a significant efficiency improvement is possible through information provision. Comparing these results with simulation experiments in a network with recurrent congestion (Chapter 11; Mahmassani and Herman, 1990), it becomes clear that information provision is more useful in a network with non-recurrent congestion. In Chapter 11 it was found that information provision in a network with recurrent congestion is particularly useful during the process leading to a steady state; at steady state, the reduction in travel time did not exceed 5 per cent.

---

[6]The variance of the five different runs of the conducted experiments was very small; the results reported in this chapter are statistically significant.

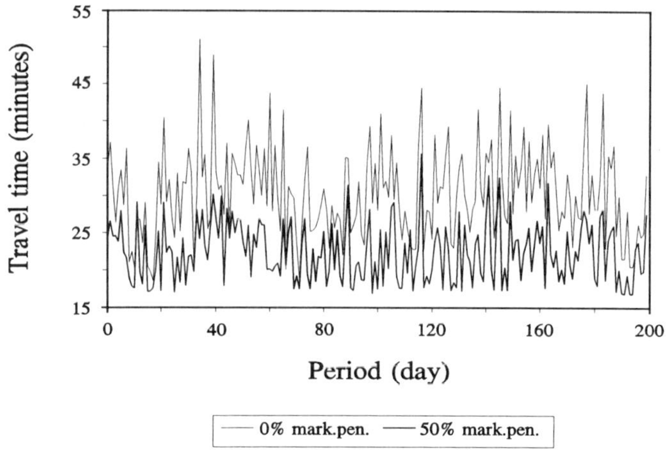

Figure 12.6 Daily network wide travel time for 1 run. upd=1, p=0.19, en_route_bound=0.05.

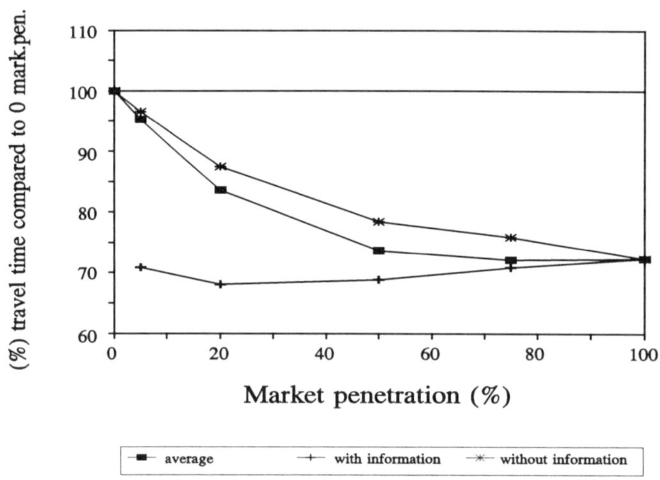

Figure 12.7 Benefits for different groups of drivers. upd=1, p=0.19, en_route_bound=0.05.

Figure 12.7 shows a typical graph of both the network wide benefits and the benefits to the different groups of drivers as a function of the level of market penetration.[7] As discussed in Chapter 2, a motorist information system is an economic good for which the benefits to the buyers depend upon the level of market penetration. This is underlined by the experiments in this chapter, as shown in Figure 12.7. There it can be seen that:

the additional benefits to equipped drivers decrease as the level of market penetration increases.

Moreover, non-equipped drivers are also affected by the information device, and their benefits depend upon the level of market penetration as well.

12.6.2 Updating frequency and network wide performance

A decrease in the updating frequency has an adverse effect on network wide performance, as depicted in Figure 12.8. The size of this negative effect is dependent on the level of market penetration. As the level of market penetration increases, more drivers respond to the *old* (not frequently updated) information, and hence the information does not accurately reflect the situation in the road network. As a consequence, overreaction as defined in Ben-Akiva et al. (1991) and discussed in Chapter 2 takes place. At full market penetration, the size of this negative effect exceeds 12 per cent. However, the network wide situation at full market penetration is still considerably better than without information. The negative effect associated with overreaction does not completely offset the benefits owing to information provision.

The relatively large discrepancies between the curves shown in Figure 12.8 might have been affected by the structure of the test network, depicted in Figure 11.1. Each driver has to make a maximum of six (!) consecutive decisions. Clearly, the larger the number of decisions, the more likely it is that one of these was based on relatively *old* information. Conversely, one might also argue that many decision points leave ample room to the drivers to correct past route choices.

Furthermore, Figure 12.8 shows that the *optimal* network wide performance is (1) not necessarily reached at full market penetration of the information technology, and (2) is dependent on the quality (updating frequency) of the information. The first point has been mentioned repeatedly in the literature, see for instance Mahmassani and Jayakrishnan (1991), Watling and van Vuren (1993). Concerning the second point, in Chapter 11 similar results regarding the dependency between the optimal level of market penetration and the quality of the information were found. Mahmassani and Jayakrishnan (1991) argued that in such a situation *coordinated guidance* becomes necessary. Coordinated guidance could lead the traffic flows - in theory - towards a system optimum. To achieve this, some drivers have to be diverted from their user optimal routes to system optimal ones. However, as discussed by Bonsall et al. (1991),

---

[7]The *with information* curve is not defined for zero per cent market penetration. Furthermore, since there are clearly benefits to the first equipped driver, the *with information* curve should not start at 100 per cent travel time at zero market penetration. In addition, the *without information* curve might be discontinuous at full market penetration in a network with non-recurrent congestion.

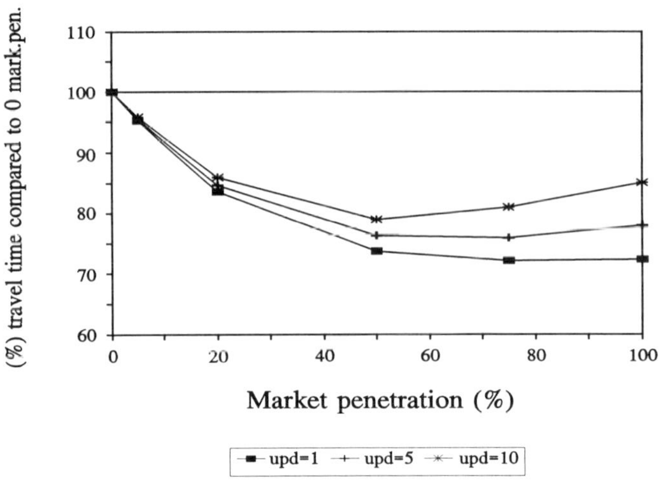

Figure 12.8 Network wide performance as a function of market penetration for three updating times. p=0.19, en_route_bound=0.05.

it is highly unlikely that these drivers will comply with the information. An incentive to *force* them to comply is needed; a kind of pricing is the most obvious one, see Chapter 3. The theoretical models that were presented in Chapters 5 to 10 also showed that the user equilibrium is generally not equal to the system optimum without some kind of pricing.

12.6.3 Number of incidents and network wide performance

The Bernoulli success parameter was set equal to 0.08, 0.19 and 0.38, to have on average 2, 5 and 10 incidents in the road network. Figure 12.9 shows the effectiveness of information provision for these three incident rates with an updating time of 1 minute.

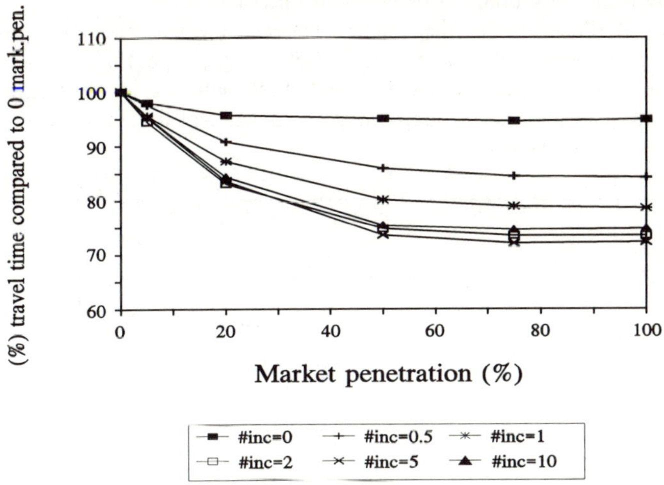

Figure 12.9 Effects of information for six incident rates. upd=1, en_route_bound=0.05.

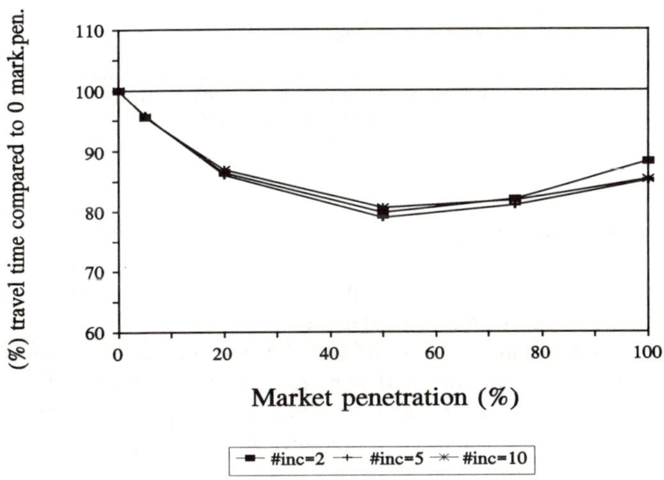

Figure 12.10 Effects of information for three incident rates. upd=10, en_route_bound=0.05.

Surprisingly, the percentage of travel time improvement for the three different scenarios (2, 5 and 10 incidents) is practically the same.[8] With an updating time of 10 minutes, the result is similar, as depicted in Figure 12.10.

Figure 12.11 illustrates the dependency between the network wide travel time and the updating interval for the three incident rates at 50 per cent market penetration. Clearly, the network wide travel time increases both with the incident rate and the updating interval. The increase is approximately linear, and the curve seems to shift parallel for different incident rates.

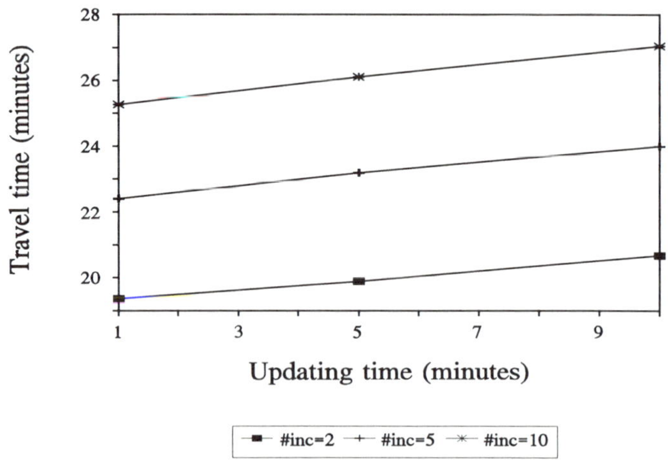

Figure 12.11 Dependency between updating frequency and incident rates. 50% market penetration, en_route_bound=0.05.

12.6.4 En route switching propensity and network wide performance

The en_route_bound determines the willingness-to-switch routes during the trip. If the en_route_bound is relatively large, then switching during the trip will only take place if the gain in expected travel time is substantial.

Figure 12.12 and Figure 12.13 depict the effects of the en_route_bound on the network wide performance for two different updating times. Firstly, Figure 12.12 shows that if the updating time is small - rendering high quality information - the model with the lowest bound outperforms the other one. However, at higher levels of

[8]This result was so surprising, that it was decided to carry out some experiments with smaller incident probabilities, namely 0, 0.02 and 0.04, respectively. For these incident probabilities, the efficiency improvement owing to information provision was clearly smaller, see Figure 12.9. Obviously, the curve with zero incidents reflects the network with recurrent congestion that was studied in Chapter 11.

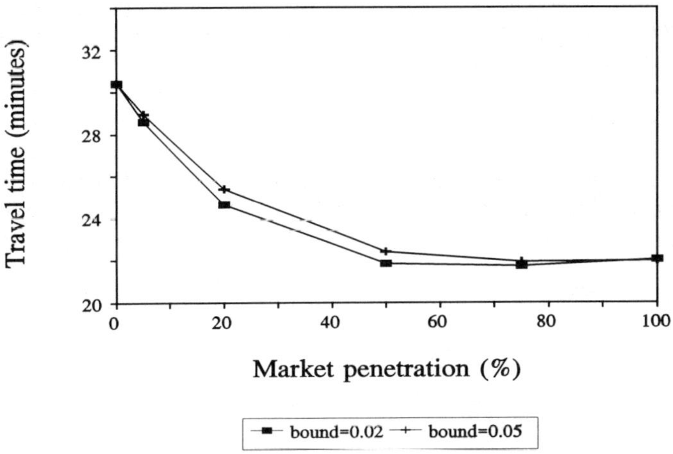

Figure 12.12 Effects of en_route_bound on network wide performance. upd=1, p=0.19.

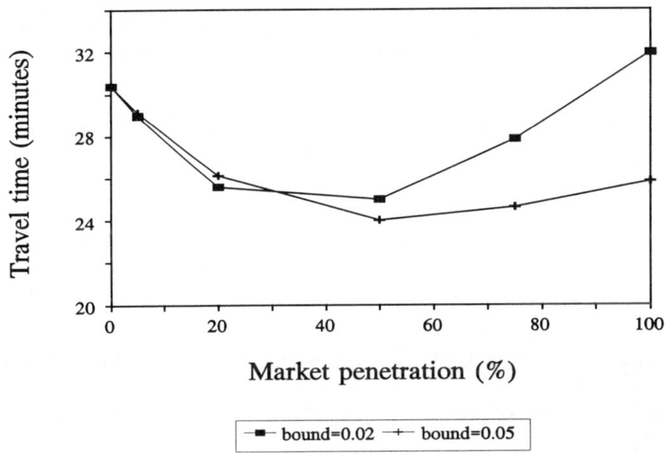

Figure 12.13 Effects of en_route_bound on network wide performance. upd=10, p=0.19.

market penetration the gains are almost identical. This situation has changed in Figure 12.13, where the updating time is 10 minutes. Here, at low levels of market penetration, the model with the higher en route switching propensity performs best.

However, at higher levels of market penetration, this is reversed, and at full market penetration, the model having the largest bound performs significantly better. This result corresponds to intuition:

With relatively unreliable information, drivers should switch only if the gain in expected travel time is large.

In connection with this:

At high levels of market penetration, the network wide performance may get worse, even with a small updating interval.

The relationship between these three parameters (switching propensity, market penetration, quality of information) and the network wide performance is schematically illustrated in Figure 12.14.

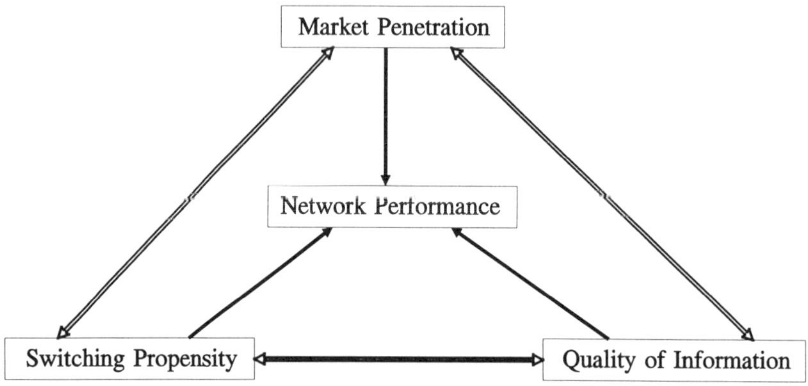

Figure 12.14 Network wide performance in relation to switching propensity, market penetration and information quality.

Figure 12.14 should be interpreted in the following manner. To preserve a relatively *efficient* road network performance the *rules* listed below should be taken into account:

1. If the level of market penetration is relatively low, then the quality of the information is not of crucial importance.
2. If the level of market penetration is relatively high, then the switching propensity should be relatively low.
3. If the quality of the information is relatively low, then the switching propensity should be relatively low.

12.7 CONCLUSION

Recently, the potential of motorist information systems to resolve part of the non-recurrent congestion problem has been attracting more attention. This chapter follows that trend through a simulation of traffic flows in a road network in which incidents are generated in a random fashion. A behavioural approach has been chosen to acknowledge that the effects of motorist information systems on the network wide performance have to be analysed where the new technology penetrates into the transportation system, i.e., at the drivers level. Hence, traffic flows in this chapter were not determined by user equilibrium considerations, but by the aggregation of driver's decisions.

As in Chapter 11, driver's behaviour was modelled using boundedly rational principles. These models have not yet been rigorously validated in a transportation context. As a consequence, the results derived in this chapter cannot simply be generalised to real road networks. However, they provide a clear indication of what to expect in a *boundedly rational world.*

The simulation experiments focused on the relationship between the network wide performance in the road network with randomly generated incidents and:

1. the level of market penetration,
2. the quality of the information as reflected by the updating frequency,
3. the switching propensity during a trip.

The following results were obtained:

1. If drivers behave according to a boundedly rational model without being provided with information in a network with non-recurrent congestion, they are unable to use the road network efficiently in terms of total travel time.
2. The additional benefits to equipped drivers decrease quickly as the level of market penetration increases.
3. Ceteris paribus (see Table 12.1), the network performance is dependent on the level of market penetration, the quality of the information and the en route switching propensity.

Comparing the results in this chapter with the ones obtained in Chapter 11, there is clearly more scope for motorist information systems in road networks with non-recurrent congestion. Motorist information systems have a potential to improve the efficiency of road networks. However, particularly due to the complexity of human behaviour with respect to these new technologies, the beneficial effects remain uncertain. Further research should focus on a greater understanding of the behavioural responses of users of motorist information systems. Without such understanding, estimates of the real-world effects of these technologies will always be unreliable.

In the next chapter a very modest attempt is made to empirically investigate driver's route choice behaviour under the availability of information provision.

PART IV

EMPIRICAL MODELS

13 RADIO TRAFFIC AND VARIABLE MESSAGE SIGN INFORMATION; AN EMPIRICAL ANALYSIS[1]

13.1 INTRODUCTION

Obtaining a greater understanding of the impact of motorist information systems on driver's behaviour is extremely important for judging the performance and market success of telematics. Since traffic flows are simply the aggregation of individual decisions, road user's behaviour will completely determine whether these new technologies are able to meet their objective of using scarce road space in a more efficient manner. Even the most sophisticated information system might be unsuccessful if the behavioural consequences are not fully understood in advance.

Besides being one of the most important areas of the motorist information systems research, the analysis of driver's behaviour is also very complex. It does not only involve modelling driver's behaviour (and therefore the psychology of decision-making), but in addition it involves the interaction of driver's behaviour with these new (and often to the public yet unknown) technologies, while finally it involves modelling the interaction among the drivers themselves. De Palma and Lefèvre (1983) addressed the difficulties of modelling these interactions mathematically. Chapters 11 and 12 and Mahmassani and Jayakrishnan (1991) (amongst others) used a simulation technique to analyse these complex interactions. Chapters 5 to 10 used a static economic equilibrium approach to model the above mentioned interactions, and to infer conclusions on the impact of information provision (and congestion pricing) on the welfare economic principles of efficiency and equity.

In this chapter, an empirical analysis of driver's route choice behaviour is conducted. The impacts of two types of traffic information on road user's behaviour are considered: radio traffic information and variable message sign information. It is particularly important to gain more insight into the impacts of variable message sign information, since these might provide a relatively cheap alternative for the more sophisticated and expensive on-board navigation systems. In the next few years, the Dutch government is planning to substantially increase the number of variable message signs that provide dynamic traffic information (Dutch Ministry of Transport, 1992). It is expected that these will play an important role in improving both the efficiency and control of traffic flows.

---

[1]This chapter is based on Emmerink, Nijkamp, Rietveld and van Ommeren (1996) published in *Transportation Research A*. I am grateful to the participants of the EU DRIVE II project BATT for allowing me to use their data. In particular, I would like to thank Gerard Pepping, Free University, Amsterdam, for designing and conducting the survey, and Cathelijne van der Burg for data assistance.

The empirical analysis will consist of univariate and bivariate discrete choice models. The bivariate discrete choice models acknowledge the interaction between the different types of information: radio traffic information and RIA traffic information. The data, used for estimating the models that will be presented in this chapter, are stemming from a survey that has been conducted by the participants of the EU DRIVE II project BATT (Behavioural and Advanced Transport Telematics) in the month of July of 1994 (BATT Consortium, 1994). Drivers on the Amsterdam ringroad were asked to fill in a questionnaire about the influence of radio traffic information and variable message sign information on their route choice behaviour. This chapter aims to analyse the behavioural issues in this survey.

The chapter is organised in the following manner. Section 13.2 reviews relevant studies in the area. Section 13.3 briefly describes the design of the questionnaire and the data that has been collected, while the analysis (using univariate and bivariate discrete choice modelling techniques) is performed in Section 13.4. Section 13.5 contains the conclusions.

13.2 THE LITERATURE

Research addressing empirically the behavioural issues of motorist information systems has been sparse. This is mainly due to the still very limited implementation of motorist information systems around the world, so that rich data is not available. In this section the studies that have attempted to analyse (part of) the behavioural impacts of some aspects of motorist information systems are briefly discussed.

In the eighties, Mahmassani and associates have analysed the route and departure time choice in an information context, using laboratory experiments with real commuters (see Mahmassani and Herman (1990) for an overview). These laboratory experiments tended to give some justification for the boundedly rational decision-making theory, originally developed by Simon (1955) and applied to the transportation field by Mahmassani and Chang (1987).

Using data from a mail survey, Caplice and Mahmassani (1992) applied a binary logit model to explain the propensity to listen to radio traffic reports. They found that older drivers were more likely to listen to radio reports and so were female drivers. They also used the same survey to analyse route and departure time switching decisions, and reported that a commuter will tend to switch route and departure time rather than route alone, and female commuters tend to switch departure time more often than males. Jou and Mahmassani (1994) used a sophisticated econometric model for the day-to-day dynamics of departure time and route choice decisions but did not (yet) include traffic information in their analysis. Yang et al. (1993) used data from interactive simulation experiments and applied neural network concepts to model route choice behaviour in a motorist information systems environment. Their major finding was that route choices were mainly based on recent experiences. Khattak et al. (1993) used a multinomial logit model to analyse driver's diversion and return choices, and identified some of the issues relevant for implementing motorist information systems. They, for example, argued that: (1) information can encourage drivers to divert from their regular routes; (2) information on travel times should be supplied on numerous alternative routes; (3) drivers are more likely to divert the more precise the information on the location of congested road sections.

Lotan and Koutsopoulos (1993) suggested the use of concepts from fuzzy set theory and approximate reasoning to model route choice behaviour, although they did not actually develop specific models. Adler et al. (1993) proposed a structural model for predicting en route behaviour under motorist information systems, based on latent factors of conflict arousal and motivation. Due to lack of data availability, these models were not estimated.

Spyridakis et al. (1991) reported on a survey conducted in 1988 with responses of 3,893 Seattle commuters. The survey focused on issues of trip-planning flexibility, quality of the commuting trip (such as driving stress, the importance to save commuting time and commuting distance etc.), the likelihood of changing route, the helpfulness of traffic information from radio and TV, and the anticipated use of phone hot lines for traffic information. In this survey commuters were also asked about the importance of variable message signs on one particular motorway. They found that motorists: (1) value saving commuting time; (2) are more likely to change their routes from work than from home; (3) are more likely to divert to known routes than to unknown routes; (4) are to be influenced by traffic information. Conquest et al. (1993) applied cluster analysis to the same survey to define four commuter groups: (1) route changers; (2) nonchangers; (3) route and departure time changers; (4) pre-trip changers. Mannering et al. (1994) used the Seattle survey to apply ordered logit models to analyse commuter's flexibility regarding route, mode and departure time.

Van Berkum and van der Mede (1993) estimated an individual route choice model and included as a new element a habit component. Panel data consisting of four waves allowed them to estimate the importance of this component through time. Their results indicated that route choice is mainly based on habitual considerations, although traffic information might play an important role in breaking this habitual pattern.

13.3 THE AMSTERDAM SURVEY

In this section some necessary background information on the Dutch survey itself and the context in which it was held is provided.

13.3.1 The Amsterdam ringroad

With the completion of the *Zeeburgertunnel* in September 1990, the Amsterdam ringroad was completed. The road has a length of 32 kilometres and connects the northern suburbs (above the river IJ) with the centre and south of Amsterdam. See Figure 13.1 for more details on the relevant area.

The crossing of the river IJ has always been Amsterdam's major road transport problem. Many commuters living north of the river and working south of the river have to cross it twice a day. In the past, this led to large daily traffic pile-ups before the *Coentunnel* and other tunnels. With the completion of the ringroad, the situation has improved, but during the peak-hours (part of) the ringroad is still very congested (Dutch Ministry of Transport, 1991a, 1991b).

The Amsterdam highway network (see Figure 13.1) consists of the ringroad (numbered A10), and four highways that provide access to the ringroad (numbered A1, A2, A4, and A8). Additional to these highways, there are a number of surface alternatives available to drivers in the Amsterdam area. However, most of these routes are not a practical alternative for motorway users, as they have both relatively low

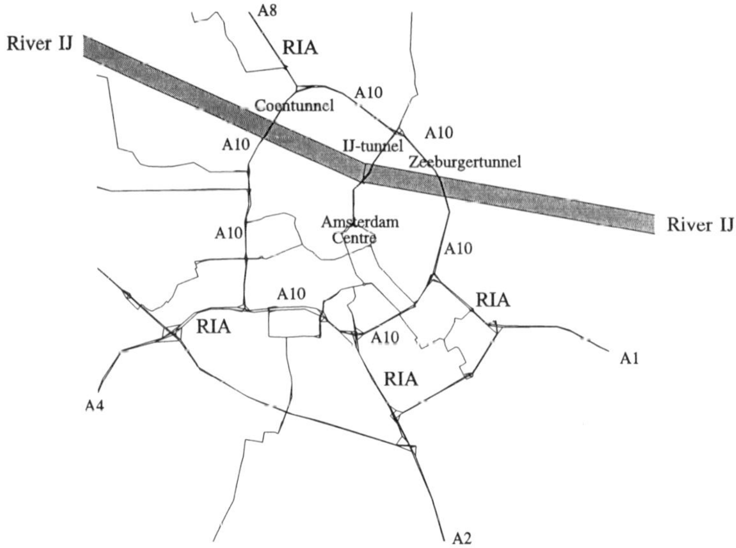

Figure 13.1 Major road network of the Amsterdam region.

speed limits and relatively many traffic lights.

Depending on the origin and the destination of a trip of an arbitrary driver in the Amsterdam area, the following route choice options are available:

(a) one side of the ringroad is so superior that the other side of the ringroad will only under exceptional circumstances be considered as a real alternative;

(b) the attractiveness of both sides of the ringroad is rather similar: the information might influence the driver's decision which side to take.

Furthermore, in both cases the information might influence the driver's decision on where to leave the ringroad. Then, the congestion on the ringroad will be traded off against the trip costs on the surface route.

An information systems application, consisting of variable message signs that provide dynamic traffic information before entering the ringroad, has been implemented to provide up-to-date information to drivers of the ringroad. Throughout this chapter these variable message signs are referred to with the acronym RIA (Route Information Amsterdam). The first RIA sign was installed on the motorway A8 since November 1991, just before entering the Amsterdam ringroad. In April 1994, an additional three RIA signs were installed on the ringroad's access motorways A1, A2 and A4.

The RIA signs provide dynamic information to road users approaching the ringroad on the length of traffic queues (in kilometres), and on the closure of driving lanes and tunnels. The information is collected by the Motorway Control and

Signalling System Amsterdam, and is updated every four seconds. The information is based on prevailing traffic intensities as collected by many loop detectors.

It is interesting to compare the impact of radio traffic information and RIA traffic information since both have the same objective: to facilitate the road users to make better route choice decisions, and hence, to improve the efficiency of the road network. Radio traffic information is broadcasted by both the national and regional radio channels. The information is generally given in terms of queue lengths on heavily congested parts of the road network, and broadcasted approximately every twenty minutes. The RIA signs provide a similar kind of information but there are two main differences. First, the information is updated very frequently, and second, the information reaches the motorway users at the right moment, just before they have to decide which side of the ringroad to use.

13.3.2 The survey

The participants of the EU DRIVE II project BATT have designed a survey with the objective to analyse the behavioural impacts of traffic information (BATT Consortium, 1994). In this chapter, the data from this survey are used to investigate the role that radio traffic information and RIA signs play in the driver's route choice decision-making process.

The survey consisted of four parts:

(a) **Travel characteristics:** travel purpose, usage of ringroad Amsterdam, trip distance, travel time, availability of alternative routes etc.;

(b) **Use of radio traffic information:** listening propensity, influence on route choice;

(c) **Use of RIA traffic information:** passing frequency, influence on route choice, type of alternative route if diverted owing to RIA information, being better or worse off owing to RIA, willingness-to-pay for having RIA information in the vehicle;

(d) **Socio-economic characteristics:** age, gender.

2,145 questionnaires were distributed in the third week of July 1994 between 7 AM and 8 PM. 826 questionnaires were returned by post, that is, a response rate of 38.5%. The questionnaires were handed out to drivers at four petrol stations on the motorway, each of these stations located nearby one of the four RIA signs.[2]

The survey characteristics of interest are given in Table 13.1. For an extensive descriptive report of the survey see BATT Consortium (1994). In summary, the sample consists mainly of 25 to 60 year old male full time workers. More than 80% uses the ringroad frequently and over 50% makes a trip with a length exceeding 50 km. More than half of the drivers has the possibility to arrive late, while 57% of the sample indicates the availability of alternative route choice options. A relatively high

---

[2]Clearly, the people reached in this way may not be representative of the population of motorway users. However, it is unlikely that this will bias the econometric results, because there is no a priori conditioning on the endogenous variables of interest, such as the use of radio traffic information and RIA traffic information, or the willingness-to-pay for in-vehicle RIA traffic information.

	per cent *(1)*		per cent *(1)*
sex		age	
male	83.8	<24	4.4
female	16.2	25-34	44.1
		35-44	26.9
		45-59	22.8
		>60	2.8
trip-purpose		frequency of using ringroad	
commute-to-work	46.4	≥ 5 days a week	47.8
business	22.2	3, 4 days a week	17.4
otherwise *(2)*	31.4	1, 2 days a week	18.0
		< once a week	16.8
flexibility of arrival time		average trip distance	
impossible to arrive late	43.9	< 10 km	1.0
possible to arrive late	56.1	10 - 25 km	10.5
		25 - 50 km	35.0
		> 50 km	53.5
alternative routes available information		listening propensity to radio traffic	
yes	57.1	never	23.9
no	42.9	regularly	38.3
		frequently	37.9
route choice influence of radio traffic information		passing frequency of RIA sign	
never	26.1	> once a week	77.9
a few times	62.2	once a week	8.4
frequently	11.6	< once a week	13.6
route choice influence of RIA sign information		willingness-to-pay for in-vehicle RIA	
never	27.5	nothing	50.1
a few times	40.5	fl. 0-25 *(3)*	18.9
frequently	31.0	fl. 25-50	14.4
		fl. 50-100	9.7
		> fl. 100	6.9

(1) Missing values were omitted.
(2) Otherwise consists of people that either indicated another trip purpose, or a multiple trip purpose.
(3) At the time the survey was conducted, the exchange rate of the Dutch guilder and the American dollar was approximately 1.8 fl./$.

Sample size : 826 observations

Table 13.1 Survey characteristics.

percentage of 72 of the motorway users has at least a few times been influenced by either radio traffic information or RIA traffic information. Just over half of the road users is not prepared to pay anything at all for having in-vehicle RIA information.

13.4 THE ANALYSIS

The outline of the empirical analysis of the survey is shown in Figure 13.2.

Section 13.4.1.1 investigates the listening propensity to radio traffic information, while Section 13.4.1.2 analyses the impact of radio traffic information on route choice behaviour. Section 13.4.2 then turns to RIA traffic information. Section 13.4.2.1 analyses the effects of RIA traffic information on route choice behaviour, while Section 13.4.2.2 analyses the satisfaction derived from using alternative routes. In Section 13.4.3, the dependency between the use of RIA traffic information and radio

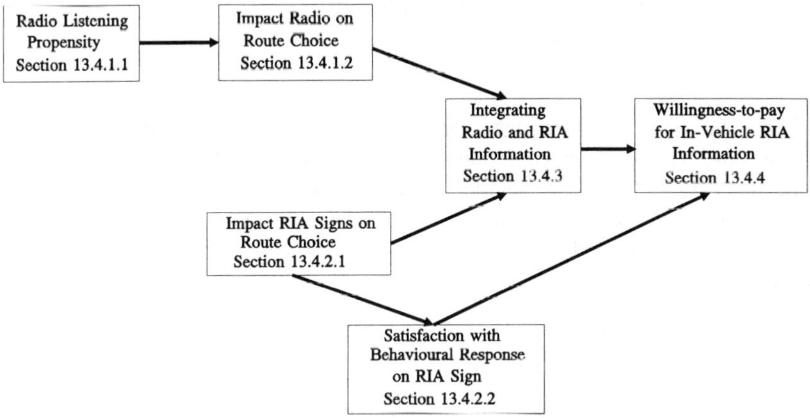

Figure 13.2 Outline of the empirical analysis.

traffic information is acknowledged by estimating a bivariate ordered probit model for the impact of these types of information on route choice behaviour. The analysis is concluded in Section 13.4.4 by a simultaneous analysis of the willingness-to-pay for having the RIA traffic information in the vehicle and the use of radio traffic information.

13.4.1 Radio traffic information

The effects of radio traffic information on driver's behaviour are analysed in the current section. Section 13.4.1.1 identifies the factors affecting the listening propensity, while Section 13.4.1.2 discusses the influence of radio traffic information on route choice.

13.4.1.1 Explanation of listening propensity

In this section an attempt is made to identify the group of drivers that is likely to listen to radio traffic information. As dependent variable, the listening propensity to radio traffic information will be used. This variable is defined using the following ordering:

1: never listening to radio traffic information[3];

[3]The survey did not contain a question addressing ownership of a vehicle equipped with a radio. Therefore, the category of drivers who are never listening to radio traffic information also includes the group of drivers that do not have an on-board radio (although this is likely

2: regularly listening to radio traffic information;
3: frequently listening to radio traffic information.

In order to explain the listening propensity an ordered probit model was estimated. The underlying response model can be described as:

$$Y = \beta' x + u \tag{13.1}$$

where Y is the underlying response variable, ß is a vector of parameters, x is a vector of explanatory variables, and u is the standard normally distributed error term. Next, the constants $\alpha_0<\alpha_1<...<\alpha_m$ are defined, where m denotes the number of categories of the dependent variable, and where α_0 and α_m are defined as $\alpha_0=-\infty$, $\alpha_m=\infty$. Y is not observed, but it is known to which of the three categories it belongs. It belongs to the *jth* category if:

$$\alpha_{j-1} \leq Y < \alpha_j \qquad (j=1,...,m) \tag{13.2}$$

The model can now be estimated using the maximum likelihood criterion, see e.g. Maddala (1983). Among the variables available in the questionnaire, all potentially explanatory variables, which were exogenous, were included.

a small portion in the sample, which consists mainly of long-distance drivers).

variables	reference group
* age	$\geq$ 45
* gender	female
* trip purpose	otherwise
* trip distance	$\geq$ 50 km
* arrival time flexibility	no flexibility
* route alternative	no alternative available
* listening propensity	never and regularly
* additional length of alternative	$\geq$ 5 km longer
* RIA appreciated	not appreciated
* perceived correctness RIA	sometimes or never correct
* route choice influence RIA	never
* satisfaction alternative route	don't know
* passing frequency RIA sign	< once a week
* type of alternative	off the motorway

Table 13.2 Reference group of dummy variables.

The parameter estimates, based on 712 valid observations, are shown in the first column of Table 13.3. The reference groups of the dummy variables are given in Table 13.2. The difference between the α_1 and α_2 estimates and corresponding standard errors indicate that the ordered probit model is able to make a clear discrimination between the three different groups of the dependent variable. For a more precise statistical interpretation of the threshold parameters α, see Emmerink, Nijkamp, Rietveld and van Ommeren (1996).

The parameter estimates show that a driver is more likely to listen to radio traffic information if:

(a) the length of the trip is long;
(b) there is more than one route alternative available;
(c) the driver is on a business trip;
(d) the driver is under 45 years old;
(e) the driver is male.

The first three factors make intuitive and economic sense. The effect of the first factor is a statistical artifact since the probability of switching the radio on increases with the length of the trip. The second finding is consistent with economic theory as it indicates that there is a demand for information only if the driver has an opportunity to actually use the information. Similar results have also been reported by Caplice and Mahmassani (1992). The reason why drivers on business trips make more often use of radio traffic information is not totally clear. It is hypothesized here that their higher value of time (or of their employer's) is the main reason. There are a priori no reasons to expect age or sex differences, hence no reason to predict the signs of the coefficients associated with these variables. The findings here are in disagreement with Caplice and Mahmassani (1992). Notice that the flexibility of arrival time has no significant effect on the propensity to listen to radio traffic information. However, the estimates of the bivariate model (see Section 13.4.4) reveal a significant negative effect of flexibility of arrival time on listening propensity. The difference between the

variables influence	Radio Traffic Information: listening propensity		Radio Traffic Information: route choice	
* constant				
α_1	-0.38**	(-2.24)	-0.64**	(-2.21)
α_2	0.66***	(3.88)	1.64***	(5.47)
* age				
< 35	0.29***	(2.77)	0.21	(1.12)
35 < ... < 45	0.26**	(2.25)	0.21	(1.11)
* gender				
male	0.20*	(1.68)	0.36*	(1.78)
* trip purpose				
commute	0.17	(1.64)	-0.16	(-0.97)
business	0.21*	(1.68)	0.28	(1.36)
* trip distance				
< 25 km	-0.62***	(-4.48)	0.20	(0.70)
25 km < ... < 50 km	-0.40***	(-4.21)	-0.35**	(-2.23)
* arrival time flexibility				
flexible	-0.07	(-0.76)	-0.16	(-1.08)
* route alternative				
alternative available	0.16*	(1.85)		
* listening propensity				
frequently			0.50***	(3.43)
number of observations	712		319	
-2 Log Likelihood for full model	1487.4		479.2	
-2 Log Likelihood for restricted model	1537.7		513.9	
Likelihood Ratio Chi-square	50.3	(9 d.f.)	34.7	(9 d.f.)

*: significant at 10%; **: significant at 5%; ***: significant at 1%

Table 13.3 Estimation results of listening propensity to radio traffic information and route choice influence due to radio traffic information: Two ordered probit models (t-values in parentheses).

estimates presented in this section and the ones in the bivariate model are likely due to the fact that the bivariate model uses additional information on the correlation between the willingness-to-pay and radio listening propensity.[4]

13.4.1.2 Influence of radio traffic information on route choice

A second model was estimated to analyse the influence of radio traffic information on route choice, or to be more precise, to analyse whether radio traffic information plays a role in the route choice decision making process. In order to answer this question,

[4]In the univariate model, the willingness-to-pay may not be used as an explanatory variable, since it is clearly endogenous with respect to radio listening propensity.

an ordered probit model was estimated. The discrete dependent variable used in this model can take on the following three values:

1: route choice has never been influenced by radio traffic information;
2: route choice has only a few times been influenced by radio traffic information;
3: route choice has frequently been influenced by radio traffic information.

To build a useful model, the data was conditioned on listening to radio traffic information (regularly or frequently), and on having more than one route choice alternative available. This reduced the number of relevant observations from 826 to 319. The estimates of the parameters of the ordered probit model are shown in the second column of Table 13.3. The reference groups of the dummy variables are shown in Table 13.2.

All signs of the parameters point into the a priori expected direction, although the significance is low due to the limited number of valid observations.

The significant positive sign of the dummy for gender clearly shows that the female part of the population is more reluctant to changes in route choice due to radio traffic information. This phenomenon has frequently been reported in the literature, see Conquest et al. (1993), Khattak et al. (1993), Mannering et al. (1994), Spyridakis et al. (1991). Caplice and Mahmassani (1992) however, reported the reverse effect.

As in the model in Section 13.4.1.1, the flexibility of arrival time has no significant influence on en route route choice behaviour. Caplice and Mahmassani (1992) found a negative effect of arrival time flexibility on switching behaviour. Mannering et al. (1994) however, found a positive effect of departure time flexibility.

In general, one might assume that the longer the trip, the more route alternatives are available. However, the results found here do not strongly support this assumption. Finally, the significant positive sign of the listening propensity to radio traffic information indicates the not surprising effect that the more a driver listens to radio traffic information, the more likely he is to change his route choice due to radio traffic information. This result was also found by Caplice and Mahmassani (1992).[5]

Another interesting independent variable might have been the driver's familiarity with the alternative route choice options available. In Mannering et al. (1994) this variable had a significant positive effect on the frequency of route changes. Although, ninety per cent of the drivers in the survey uses the ringroad regularly, this does not necessarily imply that they are familiar with the alternative routes (off the motorway).

13.4.2 RIA traffic information

In this section the impact of the RIA signs on motorway users will be analysed. Section 13.4.2.1 investigates the influence on route choice, while Section 13.4.2.2 analyses the satisfaction with alternative routes.

---

[5]The variable listening propensity might in fact be endogenous. However, the estimates of an ordered probit model in which the listening propensity variable was omitted showed that the inclusion of this variable does not affect the other estimated parameters.

13.4.2.1 Influence of RIA traffic information on route choice

The influence of the RIA traffic information on route choice behaviour has been analysed using an ordered probit model. For comparison purposes, the dependent variable was divided into similar categories as the dependent variable in Section 13.4.1.2:

1: route choice has never been influenced by RIA traffic information;
2: route choice has only a few times been influenced by RIA traffic information;
3: route choice has frequently been influenced by RIA traffic information.

To let the dependent variable make intuitive sense, only those observations were used that satisfied the following three conditions:

(a) the people should have seen the RIA signs;
(b) they should have understood the messages given by the RIA sign;
(c) they should have a route choice alternative available.

This reduced the number of valid observations from 826 to 397. The parameter estimates of the ordered probit model are given in the first column in Table 13.4. For the reference groups of the dummy variables see Table 13.2.

The significance of the passing frequency parameters imply that road users who have more frequently passed the RIA sign are more willing to change their route. Two effects might play a role here: (1) The probability of route switching increases with the number of times the RIA sign is passed; (2) The probability of route switching increases with experience.

It is informative to compare the influence of the RIA traffic information with the influence of radio traffic information, hence to compare the second column of Table 13.3 with the first column of Table 13.4. Then it follows that all estimates point into the same direction. Apparently, these two types of information have similar effects on road users. This leads to the important conclusion that drivers use radio traffic information and RIA traffic information in a similar way.

It is interesting to note that the flexibility of the arrival time is significantly different from zero and has the expected sign in Table 13.4. The significant negative sign implies that drivers with flexible arrival times are less likely to be influenced by the RIA traffic information than those with fixed arrival times. Or in other words, this result indicates that motorway users with a fixed arrival time might be prepared to change their route based on the RIA information more frequently than those with flexible arrival times; they might be prepared to take more risk in order to arrive on time. This result has also been reported by Caplice and Mahmassani (1992).

The results in Table 13.4 also indicate that commuters are significantly less likely to be influenced by the RIA traffic information than business travellers; familiarity with the area, the need for drivers with a business trip purpose to arrive in time, and habitual behaviour by commuters (see van Berkum and van der Mede (1993)) might play an important role here.

As in the previous sections, it seems that the female part of the population is less influenced by the RIA traffic information than their male counterparts. This phenomenon was discussed in Section 13.4.1.2.

	RIA Traffic Information			
	route choice influence	satisfaction alternative route		
variables		yes vs. no	yes vs. don't know	no vs. don't know
constant				
α_1	-0.44* (-1.76)	-1.04 (-1.45)	-0.66 (-1.23)	0.38 (0.57)
α_2	0.86*** (3.44)			
age				
< 35	0.04 (0.30)	0.72* (1.92)	0.37 (1.39)	-0.35 (-0.99)
35 < age < 45	0.08 (0.53)	0.19 (0.49)	-0.01 (-0.03)	-0.20 (-0.56)
gender				
male	0.33** (2.13)	-0.12 (-0.28)	-0.25 (-0.84)	-0.13 (-0.31)
trip purpose				
commute	-0.34** (-2.31)	-0.40 (-1.09)	-0.35 (-1.40)	0.05 (0.14)
business	0.21 (1.24)	-0.15 (-0.36)	-0.20 (-0.72)	-0.05 (-0.13)
trip distance				
< 25 km	0.10 (0.50)	0.02 (0.04)	-0.21 (-0.64)	-0.23 (-0.47)
25 km < distance < 50 km	-0.18 (-1.38)	-0.12 (-0.37)	-0.29 (-1.26)	-0.16 (-0.52)
arrival time flexibility				
flexible	-0.23* (-1.90)	0.12 (0.39)	-0.18 (-0.86)	-0.29 (-1.04)
passing frequency RIA sign				
> once a week	0.69*** (3.92)	0.92* (1.95)	0.40 (1.13)	-0.53 (-1.21)
once a week	0.53** (2.11)	0.64 (1.00)	0.21 (0.45)	-0.43 (-0.73)
type of alternative				
motorway		1.19*** (3.86)	0.41* (1.94)	-0.78*** (-2.68)
additional length of alternative				
< 2 km longer		1.24*** (2.99)	0.63** (2.29)	-0.61 (-1.54)
2 km < length < 5 km		0.35 (0.99)	0.07 (0.28)	-0.28 (-0.84)
number of observations	397		486	
-2 Log Likelihood for full model	773.6		926.6	
-2 Log Likelihood for restricted model	812.0		965.6	
Likelihood Ratio Chi-square	38.4 (10 d.f)		39.0 (26 d.f.)	

*: significant at 10%; **: significant at 5%; ***: significant at 1%

Table 13.4 Estimation results of route choice influence due to RIA traffic information: Ordered probit model. Estimation results of satisfaction with alternative route: Multiple logit model (t-values in parentheses).

13.4.2.2 Level of satisfaction of route choice adjustment through RIA traffic information

Seventy-two per cent of the drivers in the data set indicated to sometimes change route due to the provision of RIA traffic information. The satisfaction that is derived from the alternative route based on the RIA traffic information will be further analysed in the present section. A successful introduction of route guidance systems will strongly depend on the level of perceived satisfaction with alternative routes. This, in turn, strongly influences the level of compliance with the route guidance system.

In the survey, the question was asked whether people felt they were generally better off with the alternative route that was chosen based on the RIA traffic information. Initially 37% answered this question with *yes*, 14% with *no*, while 49% gave the answer *don't know*. In fact, the answer *don't know* seems to make most sense, since in the Amsterdam RIA system road users are *ex post* not provided with information on the alternative route, and are therefore unable to give an objective estimate.

A multiple logit model with the answer to the question "Are you generally better off with the alternative route?" as dependent variable was used.[6] Following Maddala (1983, Section 2.10) the multiple logit model can be described by:

$$\ln\left(\frac{P_j}{P_m}\right) = \beta_j' x \tag{13.3}$$

where m denotes the number of categories, and P_j the probability associated with category j. A positive β_j parameter indicates that the probability of choosing category j increases.

The three categories of the dependent variable, being the answer to the above question, were defined according to the following categorisation:

1: *yes*;
2: *no*;
3: *don't know*.

The parameter estimates of the model are given in the last three columns of Table 13.4. The reference groups of the dummy variables are shown in Table 13.2. The parameters of the independent variables age, passing frequency, type of alternative route and length of alternative route are significant. The passing frequency parameter indicates that experience with the RIA sign seems to have a positive effect on the level of satisfaction with a diversion.

It is found that the estimates of the parameters corresponding with type of alternative route and length of alternative route are strongly significant. First, the results indicate that drivers whose alternative route is still on the motorway are relatively satisfied with their alternative, whereas drivers whose alternative is off the motorway are relatively dissatisfied. Second, the results reveal that a motorway user whose alternative route is not much longer than his usual one (measured in kilometres) is relatively satisfied with his alternative, while drivers with a long alternative route are relatively dissatisfied.

Of course, the above results relate to a road network in a particular region. For this region, these results suggest that a route guidance system should (preferably) not send drivers off the motorway, and in addition should not direct them via alternatives that are much longer than the initial route, measured in kilometres. However, it is difficult to generalise these results to other regions. As pointed out in Section 13.3.1,

---

[6]See for a discussion on the advantages of the multiple logit model (ß-parameters are alternative specific) compared to the multinomial logit model (ß-parameters are identical across alternatives) Judge et al. (1985, Section 18.3.1). (Also note that the multiple logit model is often also referred to as the multinomial logit model.)

many of the available surface routes are not as attractive as the ringroad to most road users in the Amsterdam region.

13.4.3 Integrating radio and RIA traffic information

Until now, univariate models have been applied to investigate the impact of radio traffic information and RIA traffic information on route choice behaviour. Empirical work in this area has generally been based on univariate econometric models. These models, however, implicitly assume that different types of information have no impact on each other. Clearly, this assumption does not hold true in reality. Radio traffic information and RIA traffic information both give information on the same traffic situation, and are therefore highly dependent. From a theoretical point of view, any model that investigates the impact of these kinds of information on route choice behaviour should at least allow for dependency. In this chapter, the current models are extended by using a bivariate ordered probit model. This model allows for a positive or negative correlation in the respective error terms of the underlying response models. If a positive (negative) correlation is found, it implies that an unobserved variable is influencing both endogenous variables in the same (reverse) direction.[7]

The endogenous variables will be similar to the ones in Sections 13.4.1.2 and 13.4.2.1:

Y_1:
- 1: route choice has never been influenced by radio traffic information;
- 2: route choice has only a few times been influenced by radio traffic information;
- 3: route choice has frequently been influenced by radio traffic information.

and

Y_2:
- 1: route choice has never been influenced by RIA traffic information;
- 2: route choice has only a few times been influenced by RIA traffic information;
- 3: route choice has frequently been influenced by RIA traffic information.

The underlying response model is defined as:

$$\begin{aligned} Y_1^* &= \beta_1 x_1 + u_1 \\ Y_2^* &= \beta_2 x_2 + u_2 \end{aligned} \tag{13.4}$$

where Y_1^* and Y_2^* are unobserved variables. The error terms u_1 and u_2 are assumed to follow a standard normal distribution, and the correlation coefficient between u_1 and u_2 is given by ρ. Although Y_1^* and Y_2^* are unobserved, it is known to which category they belong:

$$Y_i = j \quad if \quad \alpha_{j-1}^{(i)} \leq Y_i^* < \alpha_j^{(i)} \quad i=1,2 \tag{13.5}$$

Therefore, Y_1=i and Y_2=j if and only if:

---

[7]In addition, a bivariate model uses information on two endogenous variables, which generally improves the parameter estimates and decreases the size of the standard errors.

$$\alpha_{i-1}^{(1)} \leq Y_1^* < \alpha_i^{(1)} \quad \text{and} \quad \alpha_{j-1}^{(2)} \leq Y_2^* < \alpha_j^{(2)} \qquad (13.6)$$

This bivariate ordered probit model was estimated using the maximum likelihood criterion. The results are given in Table 13.5. The univariate models estimated before were re-estimated as the number of observations has declined. The reference groups of the dummy variables are shown in Table 13.2.

	Bivariate				Univariate (correlation=0)			
	radio		RIA		radio		RIA	
variables								
constant								
α_1	-0.71**	(-2.37)	-0.53*	(-1.71)	-0.70**	(-2.33)	-0.42	(-1.31)
α_2	1.60***	(5.16)	0.73**	(2.28)	1.62***	(5.23)	0.85***	(2.66)
age								
< 35	0.15	(0.85)	0.03	(0.19)	0.15	(0.79)	0.03	(0.18)
35 < age < 45	0.18	(0.95)	0.15	(0.82)	0.18	(0.93)	0.14	(0.75)
gender								
male	0.37**	(1.68)	0.31*	(1.44)	0.39**	(1.74)	0.30*	(1.41)
trip purpose								
commute	-0.20	(-1.14)	-0.30**	(-1.83)	-0.21	(-1.22)	-0.31**	(-1.87)
business	0.24	(1.20)	0.27*	(1.37)	0.23	(1.13)	0.29*	(1.51)
trip distance								
< 25 km	0.32	(1.24)	0.14	(0.59)	0.32	(1.27)	0.14	(0.61)
25 km < dist. < 50 km	-0.38***	(-2.45)	-0.22*	(-1.49)	-0.37***	(-2.41)	-0.21*	(-1.48)
arrival time flexibility								
flexible	-0.17	(-1.18)	-0.48***	(-3.50)	-0.18	(-1.24)	-0.47***	(-3.44)
listening propensity								
frequently	0.55***	(4.08)			0.57***	(3.95)		
passing frequency RIA sign								
> once a week			0.68***	(3.20)			0.81***	(3.60)
once a week			0.73***	(2.47)			0.90***	(2.83)
correlation								
$\rho(u_1,u_2)$	0.50***	(7.67)			---			

309 observations
-2 Log Likelihood for full model 1006.5
-2 Log Likelihood for restricted model (estimate α's and ρ) 1079.0
Likelihood Ratio Chi-square 72.5 (19 d.f.)

309 observations
-2 Log Likelihood for univariate model (sum of univariate models) 1048.2
Likelihood Ratio Chi-square 41.7 (1 d.f.)

*: significant at 10%; **: significant at 5%; ***: significant at 1%

Table 13.5 Bivariate ordered probit model of route choice adaptations due to radio traffic information and RIA traffic information (t-values in parentheses).

Comparing these results with the estimates of the univariate ordered probit models, it can be concluded that all the estimates of the explanatory variables are very similar. However, the correlation coefficient is very significant and equal to 0.5, which is extremely high taking into account the number of exogenous variables. Also, the increase in the loglikelihood using the bivariate model is very substantial. It is therefore concluded that there exists an unobserved variable that has a positive effect on both the route choice influence due to radio traffic information and the route choice influence due to RIA traffic information. This unobserved variable might be the *propensity to use information*. If an individual has a larger *propensity to use information*, he or she will be more influenced by both RIA and radio traffic information.

13.4.4 Willingness-to-pay for having in-vehicle RIA traffic information

To investigate the driver's willingness-to-pay for motorist information systems technologies, the survey contained a question addressing this issue. Previous research has not been paying much attention to this issue empirically. A theoretical exposition of the willingness-to-pay and market potential of motorist information systems can be found in Chapter 2. Before proceeding, it is important to underline that the analysis below is based on *stated* preference willingness-to-pay figures, rather than on *revealed* preference willingness-to-pay.

The question in the Amsterdam survey was concerned with the driver's willingness-to-pay for having in-vehicle RIA traffic information for one year. The respondents answered according to the following five categories:[8]

1: nothing;
2: between fl.0 - fl.25;
3: between fl.25 - fl.50;
4: between fl.50 - fl.100;
5: more than fl.100.

The results in Section 13.4.3 clearly indicated the existence of a very significant unobserved variable reflecting the *propensity to use information*. In the model in Section 13.4.3, the impact of this variable was captured in the correlation coefficient of the bivariate disturbance term. In the current section, the same methodology will be used. As before, a bivariate ordered probit model is estimated with the listening propensity to radio traffic information and the willingness-to-pay for having in-vehicle RIA traffic information as the endogenous variables. This model (see Section 13.4.3 for the exact specification) allows for correlation between the radio traffic information and the RIA traffic information. Hence, it provides the opportunity to test whether the unobserved variable is still a significant factor.

To be able to give a useful interpretation to the willingness-to-pay question, the data was conditioned on whether:

(a) the people have seen the RIA signs;
(b) the people have understood the RIA traffic information.

The endogenous variable representing the listening propensity to radio traffic information was categorised in the same manner as before:

1: never listening to radio traffic information;
2: regularly listening to radio traffic information;
3: frequently listening to radio traffic information.

The endogenous variable representing the willingness-to-pay for having in-vehicle RIA traffic information is defined using the five categories of the question in the survey.

---

[8]At the time the survey was conducted, the exchange rate of the Dutch guilder and the American dollar was approximately 1.8 fl./$.

The results of the maximum likelihood estimation procedure of the bivariate ordered probit model, based on 450 valid observations, are shown in Table 13.6. The reference groups of the dummy variables are shown in Table 13.2.

The correlation coefficient of the bivariate normally distributed error term is significantly positive, although the value is substantially smaller than in the bivariate ordered probit model of Section 13.4.3. This result implies that there is a small, positive correlation between listening to radio traffic information and the willingness-to-pay for on-board RIA traffic information. It is concluded that the unobserved variable *propensity to use information* is still significantly influencing both endogenous variables.

The parameter estimates of the listening propensity equation of the bivariate model are in agreement with the results found in Section 13.4.1.1,[9] and will not be discussed further. Interesting interpretations follow from the significant parameter values from the willingness-to-pay equation of the bivariate model. The significant sign of the parameter related to the level of satisfaction with an alternative route implies that drivers who have had good experiences with the RIA traffic information in the past (drivers that were better off by using an alternative route due to the RIA traffic information) tend to have a positive willingness-to-pay. The significant positive sign of the business trip purpose parameter makes intuitive and economic sense; it implies that during working hours, the costs of travelling (the value-of-time) are larger. The parameters reflecting age suggest that drivers under 45 years old are prepared to pay significantly more for these technologies than the elderly. Finally, the significant positive sign of the variable related to gender indicates that men tend to be willing to pay more for the RIA technology than women. The latter result follows logically from the tendency found throughout the chapter that women are more reluctant to route changing during the trip due to radio or RIA traffic information. The gender results might, for example, be due to the different position in the labour market (less wage, more part-time jobs) which were not included in the survey. In addition, the variable gender might also be a proxy for the risk attitude. It might be the case that women are more risk-averse in their route choice behaviour than men.

In the analysis no attention has yet been paid to the fact that fifty per cent of the drivers in the Amsterdam survey indicated to have a zero willingness-to-pay for having in-vehicle RIA traffic information. This relatively large percentage might be partly due to strategic behaviour, and might therefore bias the estimates obtained above.[10] To investigate this potential bias, it was decided to repeat the above analysis for the group of drivers who indicated a positive willingness-to-pay. Conditioning the data in this way reduced the number of valid observations to 255. The results are given in the third and fourth column of Table 13.6. The parameter estimates are obtained from a bivariate ordered probit model. The insignificant estimate of the

[9]In the bivariate model the flexibility of arrival time parameter is significantly different from zero.

[10]The drivers in the questionnaire might hypothesize that the results will be used for government's policy purposes. This might lead some respondents to answer the willingness-to-pay question with nothing while their actual willingness-to-pay is positive.

variables	All Observations: listening propensity	All Observations: willingness-to-pay	Conditioned on Positive WTP: listening propensity	Conditioned on Positive WTP: willingness-to-pay
constant				
α_1	-0.43* (1.87)	1.49*** (2.66)	-0.95*** (-2.97)	1.69 (1.24)
α_2	0.68*** (2.96)	2.06*** (3.68)	0.28 (0.90)	2.48 (1.81)
α_3		2.59*** (4.63)		3.16** (2.31)
α_4		3.12*** (5.47)		
age				
< 35	0.20 (1.38)	0.19 (1.38)	0.23 (1.25)	0.33 (1.61)
35 < age < 45	0.09 (0.54)	0.35** (2.34)	0.25 (1.32)	0.40** (2.01)
gender				
male	0.38** (2.43)	0.37** (2.27)	0.24 (1.09)	0.70** (2.54)
trip purpose				
commute	-0.04 (-0.26)	0.05 (0.35)	-0.21 (-1.18)	-0.01 (-0.03)
business	0.14 (0.86)	0.45*** (2.91)	-0.12 (-0.58)	0.44** (2.22)
trip distance				
< 25 km	-0.41** (-2.40)	-0.05 (-0.27)	-0.50** (-2.24)	-0.46** (-1.98)
25 km < distance < 50 km	-0.39*** (-3.25)	-0.10 (-0.88)	-0.48*** (-2.88)	-0.34** (-1.96)
arrival time flexibility				
flexible	-0.21* (-1.92)	0.15 (1.39)	-0.38** (-2.50)	0.10 (0.71)
route alternative				
alternative available	0.23** (2.03)	0.18 (1.55)	0.28** (1.81)	0.04 (0.24)
passing frequency RIA sign				
> once a week		0.12 (0.64)		0.29 (1.19)
once a week		0.36 (1.45)		0.10 (0.33)
RIA appreciated				
appreciated		0.02 (0.15)		-0.01 (-0.05)
perceived correctness RIA				
always		0.37 (1.48)		0.38 (0.94)
frequently		0.22 (0.93)		-0.10 (-0.27)
route choice influence RIA				
frequently		0.40 (0.91)		0.91 (0.73)
a few times		0.38 (0.88)		0.94 (0.76)
satisfaction alternative route				
satisfied		0.28** (2.31)		-0.08 (-0.46)
dissatisfied		-0.07 (-0.32)		-0.09 (-0.37)
correlation				
$\rho(u_1,u_2)$	0.14** (2.28)		0.016 (0.2)	
number of observations	450		255	
-2 Log Likelihood for full model	2174.1		1139.3	
-2 Log Likelihood for restricted model (estimate α's and ρ)	2250.7		1206.9	
Likelihood Ratio Chi-square	76.6 (27 d.f.)		67.6 (27 d.f.)	

*: significant at 10%; **: significant at 5%; ***: significant at 1%

Table 13.6 Bivariate ordered probit models of willingness-to-pay for in-vehicle RIA traffic information and listening propensity to radio traffic information (t-values in parentheses).

correlation coefficient indicates that these data provide no evidence for correlation between the listening propensity to radio traffic information and the willingness-to-pay for on-board RIA traffic information. Hence, eliminating the drivers who indicated to have a zero willingness-to-pay reduces the impact of the unobserved variable substantially. Thus, the robustness of the outcome for the inclusion of the unobserved variable *propensity to use information* is something which deserves attention in future research.

Furthermore, it is interesting to compare the second and fourth column of Table 13.6. There it can be seen that the trip distance parameters are significant in the latter and not in the former. The other parameters in the two models do not lead to significantly different conclusions. To conclude, the data indicate that the main market segment for the RIA technology under consideration is the male driver on a business trip who is under 45 years old.

13.5 CONCLUSION

In this chapter, data from the EU DRIVE II project BATT were used for analysing the impact of both radio traffic information and variable message signs displaying dynamic traffic information on route choice behaviour. These variable message signs are the so-called RIA (Route Information Amsterdam) traffic signs. Descriptive statistics of these data revealed that over 70% of the motorway users is sometimes influenced by these types of information.

The analysis showed that:

(a) Women are more reluctant to be influenced by information during the trip than men;

(b) Commuters tend to be less influenced by information than motorway users with other trip purposes;

(c) Flexibility of arrival time is of no importance in the models concerning radio traffic information, but it is of importance in the models explaining route choice influence due to RIA traffic information. It was found that drivers who have a possibility to arrive late are less inclined to change route due to RIA traffic information.

(d) Radio traffic information and RIA traffic information are used in a similar way by road users.

Furthermore, it was found that when a motorway user is influenced by the RIA traffic information and is opting for an alternative route, then the associated level of satisfaction is likely to be large if: (1) the alternative was still on the motorway; (2) the alternative route was not much longer in distance than the original route.

The bivariate ordered probit models indicated the existence of a significant positive correlation between radio traffic information and RIA traffic information. It was found that drivers who are influenced by radio traffic information are more likely to be also influenced by RIA traffic information. This result suggests the existence of an unobserved variable reflecting the *propensity to use information*. This variable is significantly influencing both the route choice influence due to radio and RIA traffic information.

The willingness-to-pay for having in-vehicle RIA traffic information is relatively large for male drivers on a business trip under 45 years old. In addition, there appears to be a small correlation between listening to radio traffic information and the willingness-to-pay for having in-vehicle RIA traffic information.

Future research should focus on obtaining more understanding of the interaction between information and the day-to-day dynamics of driver's decision-making. Clearly, this kind of research can only be carried out if a large, rich panel data set on individual choice behaviour is available. In addition, future research should pay attention to the interaction between different types of information. The research in this chapter clearly showed the existence of an unobserved variable that is influencing both types of information. Even though this variable suggests a strong interaction between the two types of information, the survey did not provide the data to explicitly measure the level of interaction.

14 EMPIRICAL ANALYSIS OF THE WORK SCHEDULE FLEXIBILITY[1]

14.1 INTRODUCTION

In this chapter, the flexibility of work schedules as a key factor of possible success or failure to road-pricing and motorist information systems is examined. Of course, the relationship between departure time and the flexibility of work schedules is also of great interest to many other important policy issues in transportation research. To name a few, one might, for instance, think of the implementation and analysis of variable work time schedules, flexible work hours, staggered work hours, telecommuting and compressed work weeks, etc.

The analysis of all types of technologies within the telematics field is becoming increasingly popular in transportation research. These instruments - such as variable message signs, route guidance systems, dynamic traffic management systems, etc. - are expected to relieve part of the road congestion problems in urban and suburban areas by improving the drivers' decision-making process. Better individual decision-making is assumed to ameliorate the efficiency on road networks.[2] The importance of these new technologies has further been strengthened by the impossibility to relieve congestion with the more traditional tools of building more roads; this statement holds in particular for urban and suburban areas. From both an environmental and financial point of view, expanding the existing road infrastructure is often regarded undesirable.

As a consequence, the controversial policy instrument of road-pricing is gaining public and political momentum. While on theoretical grounds it can easily be shown that road-pricing is the *first-best* (optimal) solution to traffic congestion, the instrument could not gain much support in the past. This situation is slowly changing as an increasing number of rather successful road-pricing pilot studies have recently been carried out; for an overview, see Chapter 3.

As the behavioural adjustments to road-pricing and motorist information systems fully determine the impact of these advanced technologies as congestion relieving instruments, the shift towards more empirically oriented research is a logical one. Unfortunately, the empirical literature on both the impact of motorist information systems and road-pricing is still in its infancy. Due to the fact that most of the technologies have not yet been implemented there is a lack of (rich) data sources. Empirical papers related to more advanced information systems are generally based on laboratory-like experiments (see Mahmassani and Herman (1990)), while empirical work using revealed preference data is concerned with the less sophisticated telematics technologies such as variable message signs (Chapter 13). Furthermore, much of the

---

[1]This chapter is based on Emmerink and van Beek (1997) published in *Urban Studies*.

[2]This is not necessarily the case as was discussed in Chapter 2.

empirical work in this area is related to investigating the factors that determine route and departure time choice without the availability of an advanced traveller information system and without a road-pricing scheme being in place (Jou and Mahmassani, 1994; Small, 1982).

In this chapter, one important empirical issue related to the likely effectiveness of road-pricing and motorist information systems as congestion relieving policy instruments is addressed. It will be analysed how much scope there is for departure time adjustments given the population's present work constraints. If a major part of the population is not constrained by a single departure time option, then both road-pricing schemes and motorist information systems might provide a more effective and efficient policy instrument for dealing with congestion.[3] However, if the opposite holds true, then the effectiveness of both instruments is rather limited. For example, in those circumstances the impact of motorist information systems is restricted to rerouting drivers in road networks when non-recurrent (unpredictable) incidents take place.

The chapter is organised as follows. First, Section 14.2 discusses the survey on which the empirical analysis is based. Next, evidence from this survey is analysed in Section 14.3. The implications for road-pricing and motorist information systems of the empirical evidence found is discussed in Section 14.4. Finally, Section 14.5 concludes.

14.2 THE DATA

The data to be used in the present analysis was collected in the spring of 1992 by means of telephone interviews. As The Netherlands is rather densely populated, and as most inhabitants are familiar with some form of congestion, the present data might be indicative for the impact of transport policies in urban and suburban areas in other Western countries. Telephone numbers were randomly selected from complete telephone lists for the Netherlands. A possible problem with this sampling method is the exclusion of either households without telephone or households with unlisted telephone numbers. However, research from the Dutch Central Agency of Statistics (CBS) showed that mobility research is likely to be unaffected by this deficiency, as households without telephone have an on average lower mobility profile, while households with unlisted telephone numbers have an on average higher mobility pattern. CBS concluded that the *net-effect* of excluding these households is expected to be rather small (CBS, 1993).

The size of the sample is 1,983 respondents, out of which 965 respondents are employed. After deleting all missing and inconsistent data values, a sample size of 457 valid cases of workers remained. However, as the analysis in the present chapter focuses on the peak-hours, only respondents travelling during these hours were

---

[3]For a theoretical exposition of the benefits of simultaneously implementing road-pricing and driver information systems, see Chapter 10.

selected.[4] In the remainder of the chapter, the focus is on these remaining 352 respondents.

	per cent[1,2]		per cent[1,2]
gender		type of household	
male	59.9	single person	18.2
female	40.1	couple, no children	34.9
		couple, children	41.5
age		otherwise	5.4
<26	9.1		
26-35	40.6	#hours working per week	
36-45	30.4	<31	15.3
46-55	15.1	31-40	70.5
>55	4.8	>40	14.2
net household income[3]		flexible work start time	
<2000	13.4	yes	33.0
2000-3000	27.8	no	67.0
3000-4000	28.4		
4000-5000	14.5	firm size (#people)	
>5000	15.9	<6	8.8
		6-10	5.4
educational level		11-100	41.5
high school	25.4	>100	44.3
vocational	36.4		
polytechnic and university	35.0		
otherwise	3.2		

[1]Missing values were omitted.
[2]Due to rounding errors, the columns do not necessarily sum to 100.
[3]Net household income measured in Dutch guilders per month. The exchange rate between the Dutch guilder and the American dollar is approximately 1.7 fl./$.

Sample size 352 respondents.

Table 14.1 Survey characteristics.

The survey characteristics of interest are shown in Table 14.1. For a more detailed descriptive analysis of the data, the reader is referred to NHTV/BGC (1992). The sample consists mainly of 26 to 55 year old people of which approximately 60% is male. 33% of the respondents have the option to (within certain bounds) freely choose their work start time.

14.3 ANALYSIS OF THE WORK START TIME

The present section is organised in the following manner. In Section 14.3.1 the factors that influence the flexibility of the work start time during the peak-hours from both the employers' and employees' point of view are identified, while the associated level of satisfaction derived from the actual work start time is analysed in Section 14.3.2. Finally, in Section 14.3.3 the focus is on those workers with flexible work start times. It is investigated: (1) how large the work start time flexibility during the peak-hours actually is (both from the employer's and employee's point of view), and (2) which are the factors that constrain the flexibility from being larger.

---

[4]In order to obtain as many observations as possible, the morning peak-hours were defined as the hours between 5 am and 11 am.

14.3.1 Explanation of flexibility in work start time

In order to study the flexibility of the work start time, the formulation of the Tobit model as given by Maddala (1983, Section 6.3, p. 151) is followed. The model, sometimes referred to as a censored regression model, is defined as:

$$\begin{aligned} y_i &= \beta' x_i + u_i \quad \text{if RHS} > 0 \\ y_i &= 0 \quad \text{otherwise} \end{aligned} \tag{14.1}$$

where ß is a vector of unknown parameters, x_i is a vector of known exogenous variables, u_i is the normally distributed error term with mean zero and common variance σ^2, and finally, y_i is the endogenous variable under consideration. Given the features of the data, the tobit model is the most appropriate one as the size of a work start time interval has a natural lower limit equal to zero; the size of an interval cannot be negative.

The maximum likelihood estimates of the tobit model, i.e. the parameters that maximise the likelihood of the data, can be found by maximising the loglikelihood function, which is given by:

$$\log L = \sum_0 \log\left(1 - \phi\left(\frac{\beta' x_i}{\sigma}\right)\right) + \sum_1 \log\left(\frac{1}{\sqrt{2\pi\sigma^2}}\right) - \sum_1 \frac{1}{2\sigma^2}(y_i - \beta' x_i)^2 \tag{14.2}$$

Here, $\phi(x)$ denotes the distribution function of the standard normal evaluated at x, the summation range 0 represents the observations of the endogenous variable with $y_i=0$, and the summation range 1 reflects the observations with $y_i>0$. It can be shown that the likelihood function of the tobit model has a single maximum (Maddala, 1983, Section 6.4, p. 157), hence independent of the starting values an iterative maximisation procedure will converge to a global maximum. In addition, Amemiya (1973) proved that the estimators of the parameters ß thus obtained are consistent and asymptotically normal.

14.3.1.1 Work schedule interval allowed for by employer

In order to analyse the work schedule flexibility during the peak-hours, only those workers travelling within these hours were selected. Of the 352 remaining valid observations, 67% of the respondents indicated that their employer does not allow for any flexibility at all. The work schedule flexibility is defined as the number of minutes between the latest and earliest work start time as allowed for by the employer. The endogenous variable is taken to be the logarithm of the size of this interval if this is larger than zero. Otherwise, the dependent variable is defined as zero. For two reasons, the endogenous variable is transformed in terms of logarithms: (1) to facilitate that an increase of the work start time interval from 5 to 10 minutes has a greater impact than an increase from, for example, 100 to 105 minutes; (2) to diminish the impact of outliers, i.e. to scale outliers. The estimates of the tobit model are shown in the first column of Table 14.2.

Among the variables available in the questionnaire, all potentially explanatory variables, which were exogenous, were included. Statistically significant factors are:

Variables	Employer's	Employee's indif.	Employee's intol.	Satisfaction
constant	-5.42**(-2.43)	3.28***(4.11)	4.14***(9.52)	-0.34 (-0.37)
age				
<26 (reference)				
26-35	1.55 (1.04)	0.53 (1.31)	0.21 (0.89)	0.59 (1.15)
36-45	2.27 (1.50)	-0.05 (-0.12)	0.22 (0.86)	-0.35 (-0.66)
46-55	1.40 (0.85)	-0.83* (-1.73)	0.28 (1.05)	-0.59 (-1.04)
>55	2.70 (1.38)	-1.71***(-2.72)	0.11 (0.31)	-1.05 (-1.49)
gender				
male	0.99 (1.32)	-0.65**(-2.33)	0.05 (0.35)	-0.43 (-1.25)
female (reference)				
net household income				
<2000	-2.11 (-1.58)	-0.65 (-1.32)	0.37 (1.33)	-0.16 (-0.27)
2000-3000	-1.71* (-1.66)	-0.68* (-1.73)	0.16 (0.74)	-0.31 (-0.67)
3000-4000	-1.09 (-1.15)	-0.18 (-0.52)	0.21 (1.08)	-0.22 (-0.51)
4000-5000	-0.40 (-0.38)	-0.22 (-0.55)	0.28 (1.28)	-0.31 (-0.64)
>5000 (reference)				
educational level				
high school (reference)				
vocational	0.75 (0.88)	0.35 (1.22)	0.09 (0.49)	0.75** (2.20)
polytechnic and univ.	1.50* (1.72)	-0.12 (-0.40)	0.00 (0.01)	0.25 (0.69)
otherwise	-0.97 (-0.45)	0.64 (0.99)	0.55 (1.55)	-0.39 (-0.55)
type of household				
single person		0.65* (1.91)	0.04 (0.19)	0.08 (0.18)
couple, no children (reference)				
couple, children		0.09 (0.31)	-0.08 (-0.48)	-0.03 (-0.09)
otherwise		-0.07 (-0.13)	0.12 (0.45)	-0.47 (-0.78)
#hours working per week				
<31 (reference)				
31-40	2.14* (1.90)	0.33 (0.94)	-0.04 (-0.23)	-0.03 (-0.06)
>40	1.01 (0.73)	0.36 (0.81)	0.10 (0.40)	0.32 (0.58)
ln(employer's flexibility interval)		0.11* (1.94)	0.04 (0.11)	0.11* (1.65)
ln(commuting time)		0.06 (0.45)	-0.06 (-0.80)	0.13 (0.82)
firm size				
<6 (reference)				
6-10	-1.31 (-0.69)	0.26 (0.43)	-0.24 (-0.72)	1.13 (1.56)
11-100	-1.58 (-1.26)	0.50 (1.16)	0.29 (1.26)	0.60 (1.33)
>100	0.87 (0.70)	0.36 (0.79)	0.50** (2.11)	0.91* (1.90)
employee's work start time				
<7:30 am	-2.54 (-1.35)			0.14 (0.21)
7:30 am - 8:00 am	1.19 (1.12)			0.62 (1.15)
8:00 am - 8:30 am	0.80 (1.07)			0.22 (0.56)
8:30 am - 9:00 am	0.01 (0.04)			0.33 (0.80)
>9:00 am (reference)				
employee's desired work start time				
<7:30 am		0.54 (1.19)	0.37 (1.54)	
7:30 am - 8:00 am		-0.03 (-0.06)	-0.33 (-1.44)	
8:00 am - 8:30 am		-0.16 (-0.46)	-0.24 (-1.33)	
8:30 am - 9:00 am		-0.59* (-1.66)	-0.22 (-1.17)	
>9:00 am (reference)				
σ	4.67***(12.78)	1.94***(23.34)	1.04***(24.89)	
number of observations	352	352	330	352
-2 Log Likelihood (full)	917.2	1382.5	974.8	381.7
-2 Log Likel. (const and σ)	974.2	1435.3	1006.2	422.0
Likelihood Ratio Chi-square	57.0 (21 d.f.)	52.8 (26 d.f.)	31.4 (26 d.f.)	40.3

*: significant at 10% (|t|>1.645);**: significant at 5% (|t|>1.960);***: significant at 1% (|t|>2.576)

Table 14.2 Tobit estimation results of work flexibility intervals, first, second and third column. Logit estimation of satisfaction, fourth column. (t-values in parentheses)

1. net household income;
2. educational level;
3. number of hours working per week.

With respect to net household income it is found that lower incomes are more likely to have a fixed (by the employer determined) work start time. This exogenous variable is probably a surrogate for the type of job. Apparently, high income groups have jobs that provide a greater choice of freedom related to the work start time.

The coefficient of the variable educational level indicates that the type of education is strongly influencing the flexibility of the work start time. Employees with a relatively high level of education have an on average more flexible work start time than those with a relatively low level of education.

Another very significant factor reflects the number of hours working per week. It is found that part-time workers have a much less flexible work start time than full-time workers. This indicates that part-time workers are much more likely to work according to by the employer determined fixed schedules.

The likelihood ratio χ^2-test statistic shows that the exogenous variables included are (as a group) significantly explaining the dependent variable under consideration.

The empirical results presented in this section have some interesting implications for the debate on the regressiveness/progressiveness of road-pricing: an issue that has been subject of controversy from the time road-pricing was proposed as a congestion relieving policy (Foster, 1974; Kulash, 1974; Richardson, 1974; Vickrey, 1955). Clearly, if the redistribution of the revenues is not considered, then those people with a high marginal utility of time and a low marginal utility of income benefit, while travellers with a low marginal utility of time and a high marginal utility of income are worse off (Cohen, 1987; Else, 1986; Giuliano, 1992; May, 1992; Nowlan, 1993). To overcome this equity issue, Small (1992) proposed a redistributional scheme in which the revenues would be partly used to replace existing regressive taxes, and to partly subsidise alternative transport modes. In this way, he argued, road-pricing might be turned into a progressive policy.

The empirical evidence from the data however, supports the regressiveness of road-pricing. As it is found that particularly the high income groups have a greater flexibility with respect to work start time, these travellers are more easily able to avoid paying the toll. On the other hand, the low income groups generally do not have the option to change their work start time, and therefore either turn to another transport mode or pay the toll. In both cases these income groups will be worse off as "congestion theory *requires* that peak period commuters on average be made worse off as a result of the tolls. If this were not the case, insufficient numbers of commuters would be tolled off ..." (Giuliano, 1992, p. 346).

One additional remark is worth making. The regressiveness/progressiveness of a road-pricing scheme is also dependent upon the commuting distance. If, for example, high income groups have a relatively larger commuting distance, then a road-pricing scheme is relatively more disadvantageous for these workers, thereby rendering the scheme regressive. The data however, did not provide any evidence for the existence of such relationships. A regression with the logarithm of the commuting distance as the dependent variable and all potentially explanatory variables as exogenous variables did not give significant results with respect to income or level of education. The only significant parameter reflects gender, and indicates that males are more likely to have a larger commuting distance than females. This result has already been reported in the literature (Madden, 1981; Rouwendal and Rietveld, 1994; Wheeler, 1969; White, 1986).

In order to take the above discussed undesirable redistributional consequences of road-pricing into account, Daganzo (1995) proposed a combination of road-pricing and rationing and showed under which conditions such a policy would lead to strict Pareto improvements. With simultaneously applying rationing and road-pricing, he is able to construct strict Pareto improvements, as this policy "recognizes people's personal needs and differences by allowing them to choose the form of the penalty they must pay for using the bottleneck" (p. 153).

14.3.1.2 Employees' work schedule indifference and intolerance intervals
The questionnaire contained questions related to the actual and preferred work start time. Moreover, a work start time interval was asked for in which the respondent would be almost as well off as at the preferred work start time, an interval which will be labelled [indifference early , indifference late]. In addition, these employees were asked to provide the bounds of an interval outside which they would be definitely worse off: [intolerably early , intolerably late]. The situation for an individual respondent is schematically depicted in Figure 14.1.

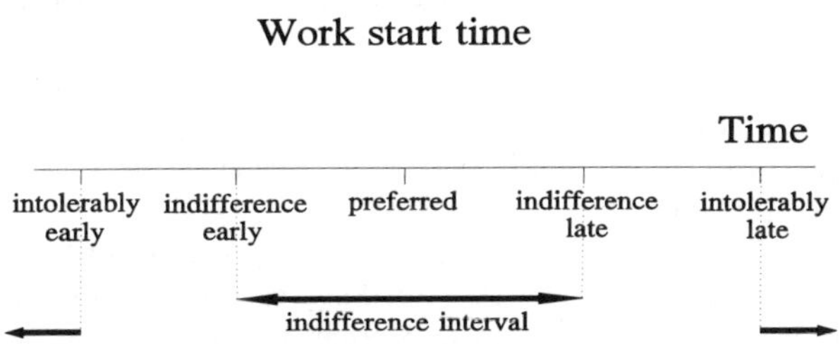

Figure 14.1 Work start time intervals.

Again, for the reasons given in Section 14.3.1.1, the logarithm of the size of the indifference interval is taken as the endogenous variable. As the data is censored at 0, a tobit model was estimated in order to determine the key factors that influence the size of this indifference interval of the employee. The parameter estimates are shown in the second column of Table 14.2. Statistically significant factors are (1) age; (2) gender; (3) net household income; (4) type of household; (5) employer's flexibility, and (6) employee's desired work start time.

With respect to age, older people are more likely to have a small indifference interval: Flexibility decreases as age increases. Next, males are relatively less flexible than females; the explanation for this finding is unclear. A relatively low net household income seems to decrease the indifference flexibility, while single person households are relatively flexible: A finding that corresponds with intuition. Furthermore, if the employer allows for more flexibility, then the estimates show that the employee is also more likely to have a larger indifference interval. Finally, the employee's desired work start time seems to play a significant role; unfortunately, the explanation of this phenomenon is rather unclear.

Next, the estimation results of the logarithm of the size of the intolerable intervals are presented in the third column of Table 14.2. The significance of the parameters estimates of this tobit model is rather low. Hence, the employees' intolerable work start time interval is not significantly influenced by the exogenous variables that were included in the model.

14.3.2 Explanation of the level of satisfaction with the actual work start time

As the questionnaire contained information on the actual work start time and the size of the indifference work start time intervals, some kind of satisfaction measure can be computed. Clearly, if the actual work start time falls within the indifference interval, then the respondent under consideration is relatively satisfied with the present work start time. Otherwise, the respondent is relatively dissatisfied; here, restrictions imposed by others prevent the respondent from choosing a satisfactory work start time.[5]

Descriptive statistics reveal that out of 352 valid observations, 71% of the respondents' work start time falls within this indifference interval, while 29% of the respondents is relatively dissatisfied with the actual work start time.

To further analyse the factors that determine the level of satisfaction with the actual work start time a binary logit model is estimated. The dependent variable is defined as:

0: unsatisfied with actual work start time: actual work start time is outside indifference interval;

1: satisfied with actual work start time: actual work start time is within indifference interval.

Before analysing the estimates of the model a remark is in order. For simplicity, dissatisfied workers who prefer to start working earlier than they actually do are treated in a similar fashion as workers who prefer to start working later.

The fourth column of Table 14.2 shows the estimates of this binary logit model. The educational level, the size of the employer's flexibility interval and the size of the firm have a significant effect on satisfaction with actual work start time.

With respect to educational level, it is found that those with a vocational type of education are more likely to have a satisfactory work start time. Apparently, vocational education serves as a surrogate for particular types of jobs.

Next, it is found that employees whose employers allow for a relatively large flexibility are more likely to be satisfied with their actual work schedule. This finding corresponds with intuition and seems to suggest that employees to some extent actually make use of the by the employer provided flexibility.

Finally, the estimates indicate that the firm size has a significant influence on the level of satisfaction with the actual work start time. It is found that workers of larger firms are more likely to be satisfied with the actual work start time than those working with smaller firms. This might indicate that agreements concerning working conditions of large firms, bargained upon by trade unions, have a positive effect on the workers' position.

---

[5]These restrictions are further explored in Section 14.3.3.

The relatively low level of satisfaction with present work start times (as shown above almost 29% of the respondents is dissatisfied), was also underlined by empirical evidence related to the completion of the Amsterdam ringroad in 1990. Results of a *before* and *after* survey indicated that one of the main effects of adding a new tunnel (the *Zeeburgertunnel*) to the existing road network was the so-called *return to the peak* phenomenon. It was found that the decrease in traffic congestion was partly offset by travellers returning to the peak (Dutch Ministry of Transport, 1991a, 1991b). This phenomenon can be explained by travellers returning to - from an individual point of view - more desirable departure times.

14.3.3 The amount of flexibility in work start times during the peak-hours

Table 14.3 depicts statistics on the size of the respective intervals during the peak-hours. The figures are based on 105 respondents, namely those who indicated to have a by the employer allowed flexible work schedule. As the number of observations is relatively small, the figures should be interpreted with great care. Nevertheless, they provide some useful insights into the main factors that determine the amount of work schedule flexibility.

interval	(a) indif. late - indif. early		(b) intol. late - intol. early		(c) employ. late - employ. early	
minutes	%[1]	(cum%)[2]	%[1]	(cum%)[2]	%[1]	(cum%)[2]
[0 , 15]	13.3	(13.3)	1.9	(1.9)	17.1	(17.1)
<16 , 30]	21.0	(34.3)	1.0	(2.9)	9.6	(26.7)
<31 , 60]	46.7	(81.0)	6.6	(9.5)	18.1	(44.8)
<61 , 120]	17.1	(98.1)	44.8	(54.3)	50.4	(95.2)
<120 , 180]	1.0	(99.0)	21.9	(76.2)	2.9	(98.1)
<180 , 240]	0.0	(99.0)	12.4	(88.6)	1.0	(99.0)
<240 , ∞ >	1.0	(100.0)	11.4	(100.0)	1.0	(100.0)

[1]Percentage of workers in peak-hours with flexible departure time.
[2]Cumulative percentage of workers in peak-hours with flexible departure time.

Sample size 105 respondents. The sample is conditioned on employer's flexibility.

Table 14.3 Results of indifferent and intolerable work start time intervals during the morning peak-hours. (a)+(b): employees; (c): employers.

It is interesting to notice that the flexibility in work start time provided by employers who allow for flexibility is relatively large. As it is a well-known phenomenon that the level of congestion in road networks is rather sensitive to small shifts in departure time patterns (see, for example, Ortúzar and Willumsen (1994)), the work start time restrictions imposed by employers seem to be sufficiently flexible to diminish levels of congestion considerably. In fact, Table 14.3 shows that the employees themselves are less flexible than the employers; the employees do not make full use of the flexibility that is allowed for. For example, while 17% of the employees have an indifference band interval of <61,120] minutes, 50% of the employers allows for such a flexibility.

The findings above lead to the preliminary conclusion that the work start time restrictions imposed by the employers do not necessarily form the bottleneck for the successful implementation of a road-pricing scheme and a motorist information system as often suggested. The employers leave some room to render these instruments successful. Rather, the employees do not make full use of the possibilities provided by the employers. Restrictions from the side of the employees have to be removed to yield these instruments effective.

Consequently, the restrictions preventing employees from being more flexible in terms of work start times have to be better understood. The survey contained a question related to this issue. Respondents were asked for the reason why the intolerably early and late bounds were leaving them with difficulties. Respondents were able to choose between various categories, and were allowed to opt for more than one category. In order to interpret the empirical results, the following classification was made:

1. reasons related to the family: one might think of restrictions related to the children, work start time of the partner, having breakfast together, etc.;
2. reasons related to work: one might think of work start times of colleagues, fixed appointments, being busy at work, etc.;
3. reasons related to transport to work: one might think of congestion on the roads, timetable public transport, difficulties with parking, etc.;
4. strong dislike to wake up early or late;
5. other reasons: one might think of the opening hours of shops, other activities outside work, or a mix of the reasons listed above.

The results are shown in Table 14.4. With respect to the intolerably early bound, both family, work and wake up reasons are quite important to the respondents. Also the category other plays a very substantial role. It is likely that a substantial share of the category other is a combination of one of the previously listed categories. The category transport is relatively unimportant. An explanation lies in the fact that implementation of the intolerably early bound would imply that most of the workers in the sample would face less rather than more congestion. Hence, in terms of reducing the travel time to work, it is for most respondents actually beneficial to depart much earlier or much later.

reason	% intolerably early[1,2]	% intolerably late[2,3]
family	19.0	23.8
work	22.9	42.9
transport	10.5	4.8
wake up	30.5	1.9
other	36.2	33.3

[1]Percentage of the respondents who indicated this reason as a relevant restriction to start working earlier than the intolerably early bound.
[2]The columns do not necessarily sum to 100 as respondents were allowed to fill in more than one category.
[3]Percentage of the respondents who indicated this reason as a relevant restriction to start working later than the intolerably late bound.

Sample size 105 respondents.

Table 14.4 Workers' restrictions preventing them from having more flexible work start times during the peak-hours.

Furthermore, it is important to notice that the work restraint is rather important. Although it was previously found that employers are relatively flexible in terms of imposing work start time constraints onto their employees, it is detected here that the employees impose at least as severe restrictions onto themselves and onto other employees. These, which will be called *employee-imposed* constraints, appear to play a major role in curbing the employee's work start time flexibility. The impact of these constraints is particularly strong with respect to the intolerably late bound; being unable to attend fixed appointments when arriving relatively late at work is the underlying reason.

Finally, the strong dislike to wake up relatively early is a dominant factor in the explanation of the intolerably early bound. Clearly, this factor plays a very modest role in the intolerably late bound.

To conclude, many different factors play an important role in determining the flexibility bounds of employees' work start times. Particularly interesting is the work restraint imposed by the employees themselves. From a government policy point of view, it is extremely difficult to relax this constraint by means of regulation as it is not imposed by the employer, but rather by the employees themselves.

14.3.4 Economic interpretation of empirical results

In the empirical analysis in Section 14.3.2 it was found that a relatively large share (almost 29%) of the employees is dissatisfied with their actual work start time, i.e. the actual work start time does not fall within their own indifference interval. Assuming that employees behave *rational* with respect to choosing their work start time, Section 14.3.3 identified the existence of particular restrictions and cost structures subject to which the employees choose the *optimal* work start time.

The restrictions and cost structures affecting the employees' *optimal* (and hence *actual*) work start time, can be classified into three categories. First, restrictions and cost structures related to work; second, restrictions and cost structures related to the home situation, and third, restrictions and cost structures related to the employees' environment.

First, concerning restrictions and cost structures related to work, one might think of *formal* (well-defined) restrictions such as work start times as allowed for by the employer (see the analysis of the employers' flexibility in Section 14.3.1), but also of the in Section 14.3.3 defined employee-imposed restrictions. Work schedules of colleagues clearly *informally* influence the optimal work start time, and might even be of greater importance than the formally imposed restrictions. In addition, it is hypothesized that psychological aspects such as "always arriving relatively late at work negatively affects my career prospects" might play a role.

Second, the choice process is influenced by restrictions and cost structures directly imposed by the employees' household situation, see also Section 14.3.3. Here, the most important factors are the partner's work schedule and the presence of children.

Third, one might identify cost structures of the environment that play a role in determining the *optimal* work schedule. The most important considerations are related to the travel time and costs of the commute-to-work trip. One might, for example, think of the level of congestion on the roads, the road-price (if any) that has to be paid for using particular road segments, price, comfort and frequency of public transport, etc.

The three factors identified above influence the employee's *optimal* work start time decision. In fact, the influence of these factors might be so strong that the employee is relatively dissatisfied with the *actual* work start time, even though the *actual* work start time is the *optimal* one.

14.4 IMPLICATIONS FOR ROAD-PRICING AND MOTORIST INFORMATION SYSTEMS

The empirical analysis has shown that a sufficient fraction of employers provides their employees with freedom to adapt their work start time - given the sensitivity of congestion levels to small changes in traffic flows -, and hence to render a road-pricing scheme aimed at spreading departure times successful. However, the analysis has also revealed that the flexibility from the employees' point of view is more restricted. Various factors were found to play a key role here, such as family reasons but also the so-called employee-imposed work related reasons. Even though the employer might provide sufficient space for more flexible working hours, this does not necessarily imply that the employee is willing to make full use of the options provided.

In the literature it is shown that there is considerable scope for synergy effects - both from a theoretical and a practical perspective - by simultaneously implementing road-pricing and motorist information systems (Brett and Estlea, 1989; Emmerink, Nijkamp and Rietveld, 1995b; Chapter 10). The empirical evidence reveals that there is some scope for modest shifts in departure time patterns. The financial incentive provided by a time-dependent road-pricing scheme, i.e. the road-price is dependent on the time of the day, might accomplish the desired shift in departure time patterns to better match socially optimal levels of congestion. On the one hand, workers with a sufficiently flexible work start time will try to avoid the road-price by departing relatively early or late. On the other hand, workers with a relatively small work start time interval will not change their departure time much. However, these employees will incur a reduction in travel time as levels of congestion will decrease due to the road-price. In addition, the road-price might be able to break some of the workers' habitual departure time patterns.

The empirical material presented gave some support for the hypothesis that road-pricing is a regressive policy instrument. It was found that particularly high income workers are relatively flexible with respect to their work start time. These travellers are better able to avoid paying the road-price, while the relatively low income workers have the option to either pay the road-price or turn to an alternative transport mode. Consequently, when implementing a road-pricing scheme, much attention should be paid to the issue of redistributing the revenues. Investment projects in public transport and the replacement of some existing regressive taxes are two means by which the fairness of road-pricing (and therefore the public and political acceptability) could be raised.

As the road-pricing scheme shifts departure time patterns to more socially desirable ones, it is the role of the motorist information system to further fine-tune the traffic flows with respect to drivers' route choice. Particularly when non-recurrent incidents take place, motorist information systems are able to reroute drivers, which can be shown to be beneficial from both an individual and a system point of view (Chapter 12).

14.5 CONCLUSION

The impacts of both road-pricing and motorist information systems depend upon the behavioural responses of the travellers involved. Unfortunately, there has not been much empirical behavioural research carried out, which is mainly due to the lack of (rich) data sources available. The present chapter aimed to investigate one important empirical issue related to the likely effectiveness of road-pricing and motorist information systems: The factors that are influencing the work start time of workers. It was identified that flexibility in terms of work start time for a sufficiently large number of travellers is a necessary (but not sufficient) condition to render the implementation of road-pricing and motorist information systems successful.

The data, collected by means of telephone interviews, revealed that approximately 33% of peak-hour travellers have the opportunity to start working at flexible times. Furthermore, it was found that, among other things, the level of education and the net household income are significantly affecting the work start time flexibility. Workers with a relatively high level of education and a relatively high net household income are more likely to have a flexible work start time. This clearly has some implications for the debate on the regressiveness/progressiveness of road-pricing. It indicates that road-pricing is likely to be a regressive policy instrument as the relatively high income earners are more easily able to avoid paying the road-price.

Next, the level of satisfaction with the present work start time was analysed. The data revealed that 29% of the population of peak-hour workers is relatively dissatisfied with the present work start time. It was argued that the *return to the peak* phenomenon, often witnessed after the completion of new road infrastructure projects, is partly due to dissatisfied travellers returning to more desirable departure times.

Finally, the amount of the flexibility of work start times was investigated, both from the employers' and employees' point of view. The results indicated that employees do not always fully use the flexibility allowed for by the employer. Restrictions related to, among other things, family and work constrain the employees' work start time. Particularly, the work constraints imposed by employees onto themselves (the so-called employee-imposed constraints) are difficult to relax by means of government regulation as these are not imposed by the employer.

The data used in the analysis was based on respondents living in both congested and uncongested areas in The Netherlands. It would be very worthwhile to conduct a similar analysis with location-specific data of a highly congested area.

PART V

CONCLUSIONS

15 SUMMARY, CONCLUSIONS AND FUTURE RESEARCH DIRECTIONS

15.1 SUMMARY

Congestion affects most metropolitan areas around the world and has a strong negative impact on the economic activities. The traditional tool to tackle the congestion problem is to build more road infrastructure. However, the feasibility of this instrument is nowadays regarded to be rather low, as the social and environmental costs of enlarging the existing road infrastructure might exceed the economic benefits. As a consequence, transportation research is placing more emphasis on using existing road networks more efficiently. The research presented in this book analyses two instruments that have the objective to increase the road network efficiency: motorist information systems and road-pricing. These instruments have recently been gaining much interest, as there is a strong belief among academics and practitioners that motorist information systems and road-pricing are the main ingredients of a future government policy that aims at curbing levels of congestion in an effective and efficient manner.

In order to analyse the economic impacts of these two instruments, three main objectives were formulated: **(1)** the formulation of theoretical models in order to assess the economic impacts of motorist information systems and road-pricing, and the combination of both; **(2)** to build simulation models in order to investigate whether the theoretical results obtained from (1) also hold in a more realistic context; and **(3)** to empirically analyse the impact of motorist information systems and road- pricing, see also Section 1.3. These three objectives were divided into five intermediate objectives that were examined in Chapters 2 to 14.

The issues relevant to motorist information systems were explored in Chapter 2. The estimated beneficial effects prevailing in the literature were discussed together with the possible adverse effects of oversaturation, overreaction and concentration. Finally, attention was paid to the role played by Wardrop's principles when information is provided, and to the possible existence of positive and negative external effects due to motorist information systems.

Chapter 3 surveyed the arguments for and against implementing road-pricing. In the literature, road-pricing is the first-best solution for efficiently dealing with transport externalities, as it is in theory able to support socially optimal transport patterns. However, it was shown that the implementation of road-pricing will give rise to many problems, although, it might be expected to have some important advantages over other available measures for curbing levels of congestion.

The introductory part of the book was concluded by Chapter 4. Here, a stochastic route choice framework was proposed that treated link travel costs as random variables. This framework made it possible to model the impact of motorist information systems and road-pricing in a systematic manner. In this framework, informed travellers receive information on the realisations of the link travel costs; in contrast, uninformed travellers do not receive information on the realisations, and will therefore base their behaviour on expected link travel costs. The framework that followed these principles was referred to as stochastic network deterministic user equilibrium (SNDUE).

The SNDUE framework was further explored in Chapters 5 to 10. It should be stressed that in order to keep the analysis manageable a number of simplifying assumptions were made. First, the demand and cost structures were assumed to be linear over the relevant ranges considered. Second, the population of travellers was assumed to be homogeneous except for their respective willingness-to-pay for making a trip. Third, the choice of departure time was not modelled explicitly. It was assumed that travellers travel during either on-peak or off-peak hours.

In Chapter 5, the basic case was discussed: a one-link network with exogenous provision of information. It was shown that in such a simple context both the informed and uninformed travellers benefit from the motorist information system. Hence, implementing a motorist information system leads to a strict Pareto improvement in this model.

In Chapter 6, the analysis was extended to a two-link network. This made it possible to study beside *mode-split* effects, the impact of motorist information systems on *route-split* as well. It was shown that route-split effects are very significant at relatively low levels of market penetration. At higher levels of market penetration, Wardrop's first principle - the user equilibrium - ensures that travel costs on both links are identical in all possible states. As a result, the benefits due to information then exclusively arise from mode-split effects.

The acquisition of information was modelled in an endogenous manner in Chapter 7. It was assumed that a traveller will acquire information, only if the private benefits of being informed exceed the private costs of buying the information. Again, it was shown that information provision leads to a strict Pareto improvement. Furthermore, the analysis revealed that subsidising motorist information systems does not yield much additional efficiency gains. However, in combination with non-fluctuating (flat) tolling, subsidising motorist information might provide significant welfare effects.

In Chapter 8, it was acknowledged that route choice decisions are also based on the costs of uncertainty. This was done by including a term related to the uncertainty of the travel costs into the generalised cost function. Experiments revealed that in most circumstances, motorist information will yield a strict Pareto improvement.

The single origin-destination pair situation was abandoned in Chapter 9. Here, a simple network with two origin-destination pairs was investigated. The network was designed in such a manner that it is probably close to a worst case analysis for studying the welfare impacts of information provision for uninformed travellers. In spite of the network's design, it was found that information provision will always lead to a potential Pareto improvement, and in most cases even to a strict Pareto improvement.

The part of the book containing the theoretical models was completed with Chapter 10. In this chapter, the interaction between motorist information systems and various road-pricing regimes was studied. A distinction was made between no tolling, non-fluctuating (flat) tolling and fluctuating (fine) tolling. The analysis demonstrated that non-fluctuating tolling in combination with information provision to all travellers performed in practically all circumstances almost as good as the theoretically first-best option of fine tolling. Therefore, flat tolling (fluctuating within-the-day, but non-fluctuating between days) in combination with motorist information is an attractive option for policy makers.

To conclude the theoretical part, most theoretical models that were studied showed that information provision to travellers will lead to a potential (and sometimes even strict) Pareto improvement. Adverse impacts of information provision were also detected. However, the conditions under which this occurred seemed to be less realistic.

Next, the book turned to the simulation models. In Chapters 11 and 12, a relatively realistic behavioural mechanism was used for modelling traveller's behaviour. These models are rather complicated and hard to solve analytically. In order to overcome this problem, a simulation model was built. The simulation experiments with different types of information in Chapter 11 showed that the optimal level of market penetration is dependent on the quality of the information. Whereas a high penetration level is none of a problem when the quality of the information is good, it induces overreaction when the quality of the information is rather poor. Furthermore, the experiments revealed that information provision generates the external effects discussed in Chapters 2 and 5. Uninformed travellers are generally better off due to the information provision. Moreover, the efficiency of the transport network generally improves, particularly when the quality of the information is relatively good.

In Chapter 12, the simulation model was extended in order to analyse the occurrence of non-recurrent congestion. The experiments with this model first of all underlined the results obtained in Chapter 11. The interaction between: (1) the behavioural rules used for modelling driver behaviour, (2) the optimal level of market penetration, and (3) the quality of the information were found to be very significant. In addition, it was shown that there is more scope for motorist information systems when the stochasticity of the network is relatively large.

The empirical models were contained in Chapters 13 and 14. Chapter 13 contained an empirical analysis of driver's usage of two simple types of traffic information: radio traffic information and variable message sign information, respectively. Several interesting empirical results with respect to the complying propensity to traffic information were found. Particularly interesting were the results related to the willingness-to-pay for having in-vehicle dynamic traffic information. A large share of the population seemed unwilling to pay anything in order to obtain in-vehicle dynamic traffic information. Those with a relatively large willingness-to-pay appeared to be the male drivers on business trips.

Chapter 14 studied the flexibility of the work start time of employees and allowed for by employers. This empirical issue will strongly affect the likely effectiveness of road-pricing and motorist information systems. The data, collected by means of telephone interviews, revealed that approximately 33% of peak-hour travellers have the opportunity to start working at flexible times. Furthermore, it was found that, among other things, the level of education and the net household income are significantly affecting the work start time flexibility. Workers with a relatively high level of education and a relatively high net household income are more likely to have a flexible work start time. This provides a new argument for the hypothesis that road-pricing is likely to be a regressive policy instrument. Finally, it was found that restrictions that employees impose onto themselves (the so-called employee-imposed constraints) play an important role.

15.2 CONCLUSIONS

The book's first objective was to formulate theoretical models that provide insight into the economic impacts of motorist information systems, road-pricing and a combination of both. In Chapters 5 to 10, this objective was met, as several new theoretical insights were obtained with the models proposed in these chapters. In general, it can be concluded under the assumptions made that the implementation of a motorist information system is:

1. beneficial to the informed travellers;
2. beneficial to the uninformed travellers.

Particularly, the second observation is of great interest to both researchers and practitioners, as it has caused some controversy in the past. It has often been claimed that information to some travellers might negatively affect uninformed travellers. The theoretical results in this book are in general not in agreement with this statement; they show that in general information provision is beneficial for both the informed and the uninformed drivers. Nevertheless, some situations were detected where information provision had adverse impacts on system efficiency. However, the conditions under which this occurred seemed to be less realistic.

Consequently, the implementation of a motorist information system will lead to a potential (and sometimes even strict) Pareto improvement. However, due to the existence of the congestion externality, information provision will not direct the traffic flows towards the system optimum. The efficiency gains due to information provision, as a proportion of the maximum available efficiency gains (using the first-best instrument of road-pricing), are most likely between 0 and 40%. In order to accomplish a more efficient usage of the transport network, it is inevitable to implement some kind of road-pricing. The road-price serves two main purposes:

1. restricting road usage during congested periods by making road usage more expensive;
2. guiding drivers towards system optimal routes.

The combination of road-pricing and motorist information systems seems to be a viable policy alternative from both a practical and a theoretical perspective. Theoretically, it can be shown (Chapter 10) that the combination of flat tolling and information provision approximates first-best levels of welfare. This would imply that, in order to achieve a high level of efficiency in the transport network, a less sophisticated pricing system is sufficient, as long as it is accompanied by information provision. From a practical perspective, implementing both systems simultaneously might lead to synergy effects in the costs. Furthermore, in order to achieve the desired behavioural responses, travellers should be aware of the prevailing road-price; a motorist information system might be a suitable method to disseminate the necessary information on road-prices.

The results obtained with the applied models in this book underlined the findings acquired from the theoretical models and satisfied the book's second objective. Generally (as long as the quality of the information is good), both informed and uninformed drivers benefit from information provision. Furthermore, the implementation of a motorist information system will generate external effects to the road users. An additionally informed driver negatively affects the already informed ones, whereas an additionally informed driver positively influences the uninformed ones.

15.3 FUTURE RESEARCH DIRECTIONS

Building on the research in the present book, there are several important research questions that have to be addressed in the near future.

The theoretical framework that was proposed in Chapter 4, and further explored in Chapters 5 to 10, would gain much in realism if it were to be extended in the following four directions. First, in the models presented in this book it was assumed that travellers consist of a homogeneous group, differing only in terms of their willingness-to-pay for making a trip. However, as in reality the population is very diverse and heterogeneous, it is more realistic if the model would allow for this. For a start, it might be worthwhile to distinguish between two types of heterogeneity: heterogeneity in terms of the marginal value-of-income and in terms of the marginal value-of-risk (or uncertainty).

Second, the analysis has been conducted using relatively simple networks. Although the results obtained in Chapter 9 suggest that the conclusions obtained might be generalised to more complex networks, it is worthwhile to extend the analysis to more complex networks. In order to do so, Ran and Boyce (1994) have proposed to use the theory of variational inequalities (Nagurney, 1993).

Third, one of the most important degrees of freedom available to a traveller, the choice of departure time, has been modelled implicitly in this book. It was assumed that the so-called alternative mode - labelled as mode 0 in Chapter 4 - is a surrogate for alternative transport modes, suppressing the trip, and changing departure time (from on-peak to off-peak hours). However, it would be more elegant to model the choice of departure time explicitly. Ways to do so have been suggested by Ran and Boyce (1994).

Fourth, in realistic transport networks traffic signals play an important role, particulary in urban street networks. Consequently, the interactions between signal control and motorist information systems need attention, both from a theoretical and from a practical point of view. Smith and Ghali (1992) have already focused on the interaction between road-pricing and signal control.

It is probably not realistic to assume that all the extensions mentioned above can be treated in an analytical fashion. Therefore, it seems likely that simulation models will play an important role in further assessing the impact of motorist information systems and road-pricing. The simulation model used in this book in Chapters 11 and 12 might serve as a first step towards more sophisticated models. The traffic flow component in the simulation model would gain much in realism by applying some of the sophisticated methodologies developed and used in Nagel (1995).

The most important future research direction is perhaps related to the empirical analysis of driver's behaviour, and the way in which this behaviour can be influenced by means of motorist information systems and road-pricing. In Chapters 13 and 14, a modest attempt in this direction was made. However, it should be clear that much more has to be learned about driver's behaviour in order to make predictions of future developments with a reasonable degree of accuracy. An advanced empirical analysis of driver's behaviour will require sophisticated econometric techniques and rich data sources. Fortunately, the latter might become less of a problem as the market for in-vehicle navigation systems is booming in Japan (ITS, 1995, p. 37). Up-to-now, these systems have been providing static information. However, these devices will soon be capable of providing dynamic information on prevailing and (predicted) future levels of congestion as well, and could therefore serve as rich data sources for advanced empirical modelling.

The future research recommendations listed above indicate that still much has to be done in this challenging and fastly evolving field of research.

REFERENCES

Abelson, R.P. and A. Levy (1985) Decision making and decision theory, In G. Lindzey and E. Aronson (Eds.) *Handbook of Social Psychology*, **1**, 231-309, Random House, New York.

Adler, J.L., T.F. Golob and M.G. McNally (1993) A structural model with discrete-choice variables for predicting enroute behavior under ATIS, paper presented at the 73*rd* Annual Meeting of the Transportation Research Board, Washington, D.C., January 9-13, 1994.

ADVANCE (1990) Scope, feasibility and cost of a dynamic route guidance system demonstration, Final Report to Illinois Department of Transportation.

Al-Deek, H. and A. Kanafani (1993) Modeling the benefits of advanced traveler information systems in corridors with incidents, *Transportation Research*, **1C**, 303-324.

Allen, D. (1988) New telecommunications services: network externalities and critical mass, *Telecommunications Policy*, **12**, 257-271.

Allen, P.A. (1993) Driver response to parking guidance and information systems, *Traffic Engineering and Control*, **34**, 302-307.

Amemiya, T. (1973) Regression analysis when the dependent variable is truncated normal, *Econometrica*, **41**, 997-1016.

Arnott, R., A. de Palma and R. Lindsey (1990a) Economics of a bottleneck, *Journal of Urban Economics*, **27**, 11-30.

Arnott, R., A. de Palma and R. Lindsey (1990b) Departure time and route choice for the morning commute, *Transportation Research*, **24B**, 209-228.

Arnott, R., A. de Palma and R. Lindsey (1991) Does providing information to drivers reduce traffic congestion?, *Transportation Research*, **25A**, 309-318.

Arnott, R., A. de Palma and R. Lindsey (1992) Route choice with heterogeneous drivers and group-specific congestion costs, *Regional Science and Urban Economics*, **22**, 71-102.

Arnott, R., A. de Palma and R. Lindsey (1993) A structural model of peak-period congestion: A traffic bottleneck with elastic demand, *American Economic Review*, **83**, 161-179.

Arnott, R., A. de Palma and R. Lindsey (1994) The welfare effects of congestion tolls with heterogeneous commuters, *Journal of Transport Economics and Policy*, **28**, 139-161.

Arnott, R., A. de Palma and R. Lindsey (1996) Information and usage of free-access congestible facilities with stochastic capacity and demand, *International Economic Review*, **37**, 181-203.

Atkinson, T. and J. Stiglitz (1980) *Lectures on Public Economics*, McGraw-Hill, New York.

AVV (Adviesdienst Verkeer en Vervoer) (1995) *Verkeersgegevens; Jaarrapport 1994*, Dutch Ministry of Transport, Rotterdam.

Axhausen, K.W. and M. Alves (1994) Taming travel diaries: Some experiences with comparing travel diaries, In *Proceedings of the 26th UTSG Conference*, University of Leeds, Leeds, January 5-7, 1994.

BATT Consortium (1994) Context, design and results of the inter-urban survey, EC DRIVE II Programme, Commission of the European Community, Brussels.

Baumol, W.J. and W.E. Oates (1988) *The Theory of Environmental Policy*, 2*nd* Edition, Cambridge University Press, Cambridge.

Ben-Akiva, M., A. de Palma and P. Kanaroglou (1986) Dynamic model of peak period traffic congestion with elastic arrival rates, *Transportation Science*, **20**, 164-181.

Ben-Akiva, M., A. de Palma and I. Kaysi (1991) Dynamic network models and driver information systems, *Transportation Research*, **25A**, 251-266.

Blythe, P.T. and P.J. Hills (1992) Pricing And Monitoring Electronically of Automobiles: The DRIVE project PAMELA, *Traffic Engineering and Control*, **33**, 614-618.

Bonsall, P.W. (1992a) The influence of route guidance advice on route choice in urban networks, *Transportation*, **19**, 1-23.

Bonsall, P.W. (1992b) Research methods for the study of driver response to in-vehicle and roadside guidance-methods, In *Proceedings of 6th World Conference on Transport Research*, Lyon, France, 1992.

Bonsall, P.W. and M. Joint (1991) Evidence on drivers' reaction to in-vehicle route guidance advice, In *Proceedings of ISATA Conference*, Florence, May, 1991.

Bonsall, P.W. and T. Parry (1990) Drivers' requirements for route guidance, In *Proceedings of 3rd International Conference on Road Traffic Control*, IEE, London.

Bonsall, P.W. and T. Parry (1991) Using an interactive route choice simulator to investigate drivers' compliance with route guidance advice, paper presented at TRB Conference, Washington, D.C., January 1991.

Bonsall, P.W., L. Pickup, and A. Stathopoulos (1991) Measuring behavioural responses to road transport informatics, In *Proceedings of the DRIVE Conference*, 1457-1487, Brussels, February 4-6, 1991.

Borins, S.F. (1988) Electronic road pricing: An idea whose time may never come, *Transportation Research*, **22A**, 37-44.

Bovy, P.H.L. and E. Stern (1990) *Route Choice: Wayfinding in Transport Networks*, Kluwer Academic Publishers, Dordrecht.

Boyce, D.E. (1988) Route guidance systems for improving urban travel and location choices, *Transportation Research*, **22A**, 275-281.

Boyce, D.E., A.M. Kirson and J.L. Schofer (1991) Design and implementation of ADVANCE: The Illinois dynamic navigation and route guidance demonstration program, ADVANCE working paper series, No. 5 (August).

Boyce, D.E., L.J. LeBlanc and G.R.M. Jansen (1990) Design of a toll system for a congested intercity road network: The Dutch road pricing scheme, paper presented at the 37*th* North American Meeting of the Regional Science Association, Boston, Massachusetts, November 9-11, 1990.

Braess, D. (1968) Über ein Paradoxon der Verkehrsplanung, *Unternehmensforschung*, **12**, 258-268.

Brett, A.C. and J.R. Estlea (1989) Electronic route-guidance to electronic road-pricing: The next step?, In *Proceedings of the 17th PTRC Summer Annual Meeting*, 245-252, University of Sussex, UK.

Brown, M.B., R.C. Evans, H. Fox, P.J. Mackie and R. Sheldon (1993) The development of elasticities for a road pricing model, paper presented at 21*st* PTRC Summer Annual Meeting, Proceedings of Seminar F, 159-175.

Burtraw, D. (1991) Compensating losers when cost-effective environmental policies are adopted, *Resources*, **104**, 1-5.

Button, K.J. (1978) A note on the road pricing of commercial traffic, *Transportation Planning and Technology*, **4**, 175-178.

Button, K.J. (1992a) *Transport Economics*, 2*nd* edition, Edward Elgar, Aldershot.

Button, K.J. (1992b) Alternatives to road pricing, paper presented at the OECD/ECMT/NFP/GVF Conference on The Use of Economic Instruments in Urban Travel Management, Basel, June 18-19, 1992.

Capello, R. (1993) *Regional Economic Analysis of Telecommunications Network Externalities*, PhD dissertation, Free University, Amsterdam.

Caplice, C. and H.S. Mahmassani (1992) Aspects of commuting behavior: Preferred arrival time, use of information and switching propensity, *Transportation Research*, **26A**, 409-418.

Catling, I. and B. Harbord (1985) Electronic road pricing in Hong Kong: 2. The technology, *Traffic Engineering and Control*, **26**, 608-617.

CBS (1993) *De Mobiliteit van de Nederlandse Bevolking*, Dutch Central Agency of Statistics, Voorburg/Heerlen, The Netherlands.

Chang, G.-L., H.S. Mahmassani and R. Herman (1985) A macroparticle traffic simulation model to investigate peak-period commuter decision dynamics, *Transportation Research Record*, **1005**, 107-120.

Cohen, Y. (1987) Commuter welfare under peak period congestion tolls: Who gains and who loses?, *International Journal of Transport Economics*, **14**, 239-266.

Conquest, L., J. Spyridakis, M. Haselkorn and W. Barfield (1993) The effect of motorist information on commuter behavior: Classification of drivers into commuter groups, *Transportation Research*, **1C**, 183-201.

Daganzo, C.F. (1995) A Pareto optimum congestion reduction scheme, *Transportation Research*, **29B**, 139-154.

Daganzo, C. and Y. Sheffi (1977) On stochastic models of traffic assignment, *Transportation Science*, **11**, 253-274.

Dehoux, Ph. and Ph.L. Toint (1991) Some comments on dynamic traffic modelling in the presence of advanced driver information systems, In *Proceedings of the DRIVE Conference*, 964-980, Brussels, February 4-6, 1991.

de Palma, A. (1992) A game-theoretic approach to the analysis of simple congested networks, *American Economic Review: Papers and Proceedings*, **82**, 494-500.

de Palma, A. and P. Jehiel (1994) Queuing may be efficient in bottleneck models, working paper, 1994.9, Université de Genève, Genève, Switzerland.

de Palma, A. and C. Lefèvre (1983) Individual decision-making in dynamic collective systems, *Journal of Mathematical Sociology*, **9**, 103-124.

de Palma, A. and R. Lindsey (1994) The potential benefits and synergy of advanced driver information systems and road pricing, working paper, University of Genève, No. 1994.6.

Downs, A. (1962) The law of peak-hour expressway congestion, *Traffic Quarterly*, **16**, 393-409.

Dutch Ministry of Transport (1991a) *Studie Effecten Openstelling Ringweg Amsterdam: Verandering in Spreiding van Reis- en Werktijdstippen*, Directoraat-Generaal Rijkswaterstaat, Dienst Verkeerskunde, Ministry of Transport, Public Works and Water Management, Rotterdam.

Dutch Ministry of Transport (1991b) *Ringweg Amsterdam: Effecten van de openstelling*, Integraal Eindrapport Rijkswaterstaat, Dienst Verkeerskunde,

Dutch Ministry of Transport (1992) *Dynamic traffic management in The Netherlands*, Report of the Transportation and Traffic Research Division, Ministry of Transport, Public Works and Water Management, Rotterdam.

El Sanhouri, I.M. (1994) *Evaluating the Joint Implementation of Congestion Pricing and Driver Information Systems*, PhD dissertation, Massachusetts Institute of Technology.

Else, P.K. (1981) A reformulation of the theory of optimal congestion taxes, *Journal of Transport Economics and Policy*, **15**, 217-232.

Else, P.K. (1986) No entry for congestion taxes?, *Transportation Research*, **20A**, 99-107.

Emmerink, R.H.M. (1995) Drivers information in a two-link equilibrium model, *Tinbergen Institute Research Bulletin*, **7**, 53-67.

Emmerink, R.H.M., K.W. Axhausen, P. Nijkamp and P. Rietveld (1995a) Effects of information in road transport networks with recurrent congestion, *Transportation*, **22**, 21-53.

Emmerink, R.H.M., K.W. Axhausen, P. Nijkamp and P. Rietveld (1995b) The potential of information provision in a simulated road transport network with non-recurrent congestion, *Transportation Research*, **3C**, 293-309.

Emmerink, R.H.M., K.W. Axhausen, P. Nijkamp and P. Rietveld (1995c) Concentration, overreaction, market penetration and Wardrop's principles in an ATIS environment, *International Journal of Transport Economics*, **22**, 123-141.

Emmerink, R.H.M., P. Nijkamp and P. Rietveld (1995a) Is congestion-pricing a first-best strategy in transport policy? A critical review of arguments, *Environment and Planning B*, **22**, 581-602.

Emmerink, R.H.M., P. Nijkamp and P. Rietveld (1995b) The integration of road-pricing and motorist information systems, *Built Environment*, **21**, 236-245.

Emmerink, R.H.M., P. Nijkamp, P. Rietveld and K.W. Axhausen (1994) The economics of motorist information systems revisited, *Transport Reviews*, **14**, 363-388.

Emmerink, R.H.M., P. Nijkamp, P. Rietveld and J.N. van Ommeren (1996) Variable message signs and radio traffic information: An integrated empirical analysis of drivers' route choice behaviour, *Transportation Research*, **30A**, 135-153.

Emmerink, R.H.M. and P. van Beek (1997) Empirical analysis of the work schedule flexibility; Implications for road pricing and driver information systems, *Urban Studies*, **34**, 217-234.

Emmerink, R.H.M., E.T. Verhoef, P. Nijkamp and P. Rietveld (1996a) Information provision in road transport with elastic demand: A welfare economic approach, *Journal of Transport Economics and Policy*, **30**, 117-136.

Emmerink, R.H.M., E.T. Verhoef, P. Nijkamp and P. Rietveld (1996b) Endogenising demand for information in road transport, *Annals of Regional Science*, **30**, 201-222.

Emmerink, R.H.M., E.T. Verhoef, P. Nijkamp and P. Rietveld (1996c) Information in road networks with multiple origin-destination pairs. Forthcoming in *Regional Science and Urban Economics*.

Emmerink, R.H.M., E.T. Verhoef, P. Nijkamp and P. Rietveld (1996d) Information policy in road transport with elastic demand: Some welfare economic considerations. Forthcoming in *European Economic Review*.

Emmerink, R.H.M., E.T. Verhoef, P. Nijkamp and P. Rietveld (1996e) Information effects in transport networks with stochastic capacity and uncertainty costs. Forthcoming in *International Economic Review*.

Evans, Alan W. (1992) Road congestion: The diagrammatic analysis, *Journal of Political Economy*, **100**, 211-217.

Evans, Andrew W. (1992) Road congestion pricing: when is it a good policy?, *Journal of Transport Economics and Policy*, **26**, 213-243.

Fearon, J., M. Scott and M. Green (1994) Commercial vehicle responses to congestion charging, *Traffic Engineering and Control*, **35**, 52-55.

Fielding, G.J. and D.B. Klein (1993) How to franchise highways, *Journal of Transport Economics and Policy*, **27**, 113-130.

Fischer, M.M. (1993) Travel demand, In J. Polak and A. Heertje (Eds.) *European Transport Economics*, 6-32, Blackwell, Oxford.

FitzRoy, F. and I. Smith (1993) Priority over pricing: Lessons from Zürich on the redundancy of road pricing, *Journal of Transport Economics and Policy*, **27**, 209-214.

Foster, C. (1974) The regressiveness of road pricing, *International Journal of Transport Economics*, **1**, 133-141.

Frank, M. (1981) The Braess paradox, *Mathematical Programming*, **20**, 283-302.

Frey, B.S. and W.W. Pommerehne (1993) On the fairness of pricing - An empirical survey among the general population, *Journal of Economic Behavior and Organization*, **20**, 295-307.

Gärling, T. (1994) Behavioral assumptions overlooked in travel-choice modelling, In *Proceedings of the 7th International Conference on Travel Behavior*, Valle Nevado, Chile, June 13-16, 1994.

Ghali, M.O. and M.J. Smith (1992) Road pricing: A new model for assessing the many options, *Traffic Engineering and Control*, **33**, 156-157.

Giuliano, G. (1992) An assessment of the political acceptability of congestion pricing, *Transportation*, **19**, 335-358.

Glaister, S. (1981) *Fundamentals of Transport Economics*, Basil Blackwell Publisher, Oxford, England.

Gomez-Ibanez, J.A. and J.R. Meyer (1994) Private toll roads in the United States: Recent experiences and prospects, paper presented at the ESCR Seminar on Transport and Urban Development, London, April 18-19, 1994.

Goodwin, P.B. (1989) The rule of three: A possible solution to the political problems of competing objectives for road pricing, *Traffic Engineering and Control*, **30**, 495-497.

Goodwin, P.B. (1992) A review of new demand elasticities with special reference to short and long run effects of price changes, *Journal of Transport Economics and Policy*, **26**, 155-169.

Grieco, M. and P.M. Jones (1994) A change in the policy climate? Current European perspectives on road pricing, *Urban Studies*, **31**, 1517-1532.

Hall R.W. (1993) Non-recurrent congestion: How big is the problem? Are traveler information systems the solution?, *Transportation Research*, **1C**, 89-103.

Haltiwanger, J. and M. Waldman (1985) Rational expectations and the limits of rationality: An analysis of heterogeneity, *American Economic Review*, **75**, 326-340.

Hau, T.D. (1992) *Economic Fundamentals of Road Pricing: A Diagrammatic Analysis*, Transport Division, The World Bank (monograph).

Hendrickson, C. and G. Kocur (1981) Schedule delay and departure time decisions in a deterministic model, *Transportation Science*, **15**, 62-77.

Hicks, J.E., D.E. Boyce and A. Sen (1992) Static network equilibrium models and analyses for the design of dynamic route guidance systems, Technical Report in Support of the Design Phase of the ADVANCE Project, Illinois Department of Transportation.

Hicks, J.R. (1939) The foundation of welfare economics, *Economic Journal*, **49**, 696-712.

Hills, P. and Andrew W. Evans (1993) Road congestion pricing: A comment and a rejoinder, *Journal of Transport Economics and Policy*, **27**, 91-105.

Hols, M.C.A.B. (1992) Ruimtelijke effecten van rekening rijden, In P. Rietveld and E. Kreutzberger (Eds.) *Stedelijke Ontwikkeling, Infrastructuur en Transport in de Randstad*, 5-22, Delftse Universitaire Pers, Delft.

Horn, B., K. Hashiba and V. Feypell (1995) Fighting traffic congestion: An agenda for the future, *IATSS Research*, **19**, 6-15.

Horowitz, J.L. (1984) The stability of stochastic equilibrium in a two-link transportation network, *Transportation Research*, **18B**, 13-28.

Horowitz, J.L. (1985) Travel demand analysis: State of the art and research opportunities, *Transportation Research*, **19A**, 441-453.

Hounsell, N., M. McDonald and L. Breheret (1991) Models and strategies for dynamic route guidance, *Advanced Telematics in Road Transport*, 89-98, Elsevier Science Publishers, Brussels.

Iida, Y., T. Akiyama and T. Uchida (1992) Experimental analysis of dynamic route choice behavior, *Transportation Research*, **26B**, 17-32.

In 't Veld, R.J. (1991) Road pricing: A logical failure, In D.J. Kraan and R.J. in 't Veld (Eds.) *Environmental Protection: Public or Private Choice*, 111-121, Kluwer Academic, Dordrecht.

Ishtiaq, S. and N. Hounsell (1993) Short term forecasting of traffic congestion in urban areas, paper presented at the 25*th* UTSG Conference, University of Southampton, Southampton, January 6-8 1993.

ITS (1995) *Intelligent Transport Systems*, Issue no. 3, November 1995.

Jeffery, D.J. (1986) Route guidance and in-vehicle information systems, In P.W. Bonsall and M. Bell (Eds.) *Information Technology Applications in Transport*, 319-351, VNU Science Press, Utrecht.

Jeffery, D.J., K. Russam and D. Robertson (1987) Electronic route guidance by autoguide: The research background, *Traffic Engineering and Control*, **28**, 525-529.

Johansson, B. and L.-G. Mattsson (1995) *Road Pricing: Theory, Empirical Assessment and Policy*, Kluwer Academic Publishers, Boston.

Jones, P.M. (1991a) Gaining public support for road pricing through a package approach, *Traffic Engineering and Control*, **32**, 194-196.

Jones, P.M. (1991b) UK public attitudes to urban traffic problems and possible counter measures: A poll of polls, *Environment and Planning*, **9C**, 245-256.

Jones-Lee, M.W. (1990) The value of transport safety, *Oxford Review of Economic Policy*, **6**, 39-60.

Jou, R.-C. and H.S. Mahmassani (1994) Day-to-day dynamics of commuter travel behaviour in an urban environment: Departure time and route decisions, In *Proceedings of the 7th International Conference on Travel Behavior*, Valle Nevado, Chile, June 13-16, 1994.

Judge, G.G., W.E. Griffiths, R.C. Hill, H. Lütkepohl and T.-C. Lee (1985) *The Theory and Practice of Econometrics*, 2*nd* Edition, John Wiley & Sons, New York.

Kaldor, N. (1939) Welfare propositions of economics and interpersonal comparisons of utility, *Economic Journal*, **49**, 549-552.

Kanafani, A. and H. Al-Deek (1991) A simple model for route guidance benefits, *Transportation Research*, **25B**, 191-201.

Keller, H., H. Hampe and J. Schmickl (1983) Forschung, Straßenbau und Straßenverkehrstechnik: Störungen im Fahrt- und Verkehrsablauf auf Schnellstraßen, Heft 396, herausgegeben vom Bundesminister für Verkehr, Abteilung Straßenbau, Bonn-Bad Godesberg.

Khattak, A.J., J.L. Schofer and F.S. Koppelman (1993) Commuters' enroute diversion and return decisions: Analysis and implications for advanced traveler information systems, *Transportation Research*, **27A**, 101-111.

King, G.F. (1986) Driver performance in highway navigation tasks, *Transportation Research Record*, **1093**, 1-11.

King, G.F. and T.M. Mast (1987) Excess travel: causes, extent and consequences, *Transportation Research Record*, **1111**, 126-134.

Knight, F.H. (1924) Some fallacies in the interpretation of social cost, *Quarterly Journal of Economics*, **38**, 582-606.

Koh, W.T.H. and D.K.C. Lee (1994) The vehicle quota system in Singapore: An assessment, *Transportation Research*, **28A**, 31-47.

Koutsopoulos, H.N. and H. Xu (1993) An information discounting routing strategy for advanced traveler information systems, *Transportation Research*, **1C**, 249-264.

Kulash, D. (1974) Income distributional consequences of roadway pricing, *Transport Research Forum*, **15** (1) 153-159.

Lave, C. (1994) The demand curve under road pricing and the problem of political feasibility, *Transportation Research*, **28A**, 83-91.

Layard, R. (1977) The distributional effects of congestion taxes, *Economica*, **44**, 297-304.

Lindley, J.A. (1986) Quantification of urban freeway congestion and analysis of remedial measures, Federal Highway Administration RD-87/052, Washington, D.C.

Lindley, J.A. (1987) Urban freeway congestion: Quantification of the problem and effectiveness of potential solutions, *ITE Journal*, January, 27-32.

Lindley, J.A. (1989) Urban freeway congestion problems and solutions: An update, *ITE Journal*, December, 21-23.

Lindsey, R. (1995) Optimal departure scheduling for the morning rush hour when capacity is uncertain, paper presented at the 7*th* World Conference on Transport Research, Sydney, Australia, July 16-21, 1995.

Lindveld, Ch.D.R., E.P. Kroes and H. de Ruiter (1992) DYNA; Real time congestion prediction, In P.M. Blok (Ed.) *Colloquium Vervoersplanologisch Speurwerk - 1992 - Innovatie in Verkeer en Vervoer*, 875-893, C.V.S., Delft.

Lotan, T. and H.N. Koutsopoulos (1993) Models for route choice behavior in the presence of information using concepts from fuzzy set theory and approximate reasoning, *Transportation*, **20**, 129-155.

Maddala, G.S. (1983) *Limited-Dependent and Qualitative Variables in Econometrics*, Cambridge University Press, Cambridge.

Madden, J.F. (1981) Why women work closer to home, *Urban Studies*, **18**, 181-194.

Mahmassani, H.S. and G.-L. Chang (1985) Dynamic aspects of departure-time choice in a commuting system: Theoretical framework and experimental analysis, *Transportation Research Record*, **1037**, 88-101.

Mahmassani, H.S. and G.-L. Chang (1987) On boundedly rational user equilibrium in transportation systems, *Transportation Science*, **21**, 89-99.

Mahmassani, H.S. and P.S.-T. Chen (1991) Comparative assessment of origin-based and en route real-time information under alternative user behavior rules, *Transportation Research Record*, **1306**, 69-81.

Mahmassani, H.S. and R. Herman (1990) Interactive experiments for the study of tripmaker behaviour dynamics in congested commuting systems, In P.M. Jones (Ed.) *Developments in Dynamic and Activity-Based Approaches to Travel Analysis*, 272-298, Avebury, Aldershot.

Mahmassani, H.S. and R. Jayakrishnan (1988) Dynamic analysis of lane closure strategies, *Journal of Transportation Engineering*, **114**, 476-496.

Mahmassani, H.S. and R. Jayakrishnan, (1991) System performance and user response under real-time information in a congested traffic corridor, *Transportation Research*, **25A**, 293-308.

Mahmassani, H.S. and S. Peeta (1993) Network performance under system optimal and user equilibrium dynamic assignments: Implications for ATIS, paper presented at the 72*nd* Annual Meeting of the Transportation Research Board, Washington, D.C., January 10-14, 1993.

Mahmassani, H.S. and C.-C. Tong (1986) Availability of information and dynamics of departure time choice: Experimental investigation, *Transportation Research Record*, **1085**, 33-49.

Mannering, F., S.-G. Kim, W. Barfield and L. Ng (1994) Statistical analysis of commuters' route, mode, and departure time flexibility, *Transportation Research*, **2C**, 35-47.

May, A.D. (1992) Road pricing: An international perspective, *Transportation*, **19**, 313-333.

Meland, S. and J.W. Polak (1993) Impact of the Trondheim toll ring on travel behaviour: Some preliminary findings, paper presented at 21*st* PTRC Summer Annual Meeting, Proceedings of Seminar F, 103-115.

Milne, D. (1993) The network effects of alternative road user charging systems, paper presented at the 21*st* PTRC Summer Annual Meeting, Proceedings of Seminar F, 91-102.

Mirchandani, P. and H. Soroush (1987) Generalized traffic equilibrium with probabilistic travel times and perceptions, *Transportation Science*, **21**, 133-152.

Mishan, E.J. (1971) The postwar literature on externalities: An interpretative essay, *Journal of Economic Literature*, **9**, 1-28.

Mogridge, M.J.H. (1990) *Travel in Towns: Jam Yesterday, Jam Today and Jam Tomorrow?*, The Macmillan Press Ltd., London and Basingstoke.

Mohring, H. and M. Harwitz (1962) *Highway Benefits*, Northwestern University Press, Evanston, Illinois.

Nagel, K. (1995) *High-Speed Microsimulations of Traffic Flow*, PhD dissertation, University of Cologne, Germany.

Nagurney, A. (1993) *Network Economics: A Variational Inequality Approach*, Kluwer Academic Publishers, Norwell, MA.

Nash, C.A. and P.K. Else (1982) Notes and comments: A reformulation of the theory of optimal congestion taxes, *Journal of Transport Economics and Policy*, **16**, 295-304.

Newbery, D.M.G. (1990) Pricing and congestion: economic principles relevant to road pricing, *Oxford Review of Economic Policy*, **6**, 22-38.

NHTV/BGC (1992) *Statistisch Overzicht: Woon-/Werkverkeer en gebruik voorzieningen*, Breda, The Netherlands.

Nijkamp, P., G. Pepping and D. Banister (1996) *Telematics and Transport Behaviour*, Springer-Verlag, Berlin.

Noland, R.B. (1995) Commuter responses to travel time uncertainty under congested conditions: Expected costs, information, and flexible scheduling, paper presented at the Meeting of the Western Regional Science Association, San Diego, California, February 1995.

Noland, R.B., K.A. Small, P.M. Koskenoja and X. Chu (1994) A discrete choice simulation model of urban highway congestion incorporating travel reliability, paper presented at Annual Meeting of the Regional Science Association International, Niagara Falls, Ontario, Canada, November 1994.

Noland, R.B. and K.A. Small (1995) Travel-time uncertainty, departure time choice and the cost of the morning commute, paper presented at the 74*th* Annual Meeting of the Transportation Research Board.

Nowlan, D.M. (1993) Optimal pricing of urban trips with budget restrictions and distributional concerns, *Journal of Transport Economics and Policy*, **27**, 253-276.

Oldridge, B. (1990) Make yourself scarce, *New Civil Engineer*, 1 March, 46-47.

Ortúzar, J. de Dios, and L.G. Willumsen (1994) *Modelling Transport* (2*nd* Ed.) John Wiley & Sons, Chichester.

Oum, T.H., W.G. Waters II and J.-S. Yong (1992) Concepts of price elasticities of transport demand and recent empirical estimates, *Journal of Transport Economics and Policy*, **26**, 139-154.

Phang, S.-Y. (1993) Singapore's motor vehicle policy: Review of recent changes and a suggested alternative, *Transportation Research*, **27A**, 329-336.

Pigou, A.C. (1920) *Wealth and Welfare*, Macmillan, London.

Polak, J.W. and P.M. Jones (1993) Using stated preference methods to examine traveller preferences and responses, working paper Transport Studies Unit, University of Oxford, TSU 773. Forthcoming in M.E.H. Lee-Gosselin and P.R. Stopher (Eds.) *Understanding Travel Behaviour in an Era of Change*, Pergamon Press, New York.

Polak, J.W., P.M. Jones, P. Vythoulkas, R. Sheldon and D. Wofinden (1993) Travellers' choice of time of travel under road pricing, TSU Ref. 766, paper presented at the 21*st* PTRC Summer Annual Meeting.

Poole, R.W. (1992) Introducing congestion pricing on a new toll road, *Transportation*, **19**, 383-396.

PROMETHEUS (1989) Functions or how to achieve PROMETHEUS objectives, issued by PROMETHEUS Office, Stuttgart, Germany.

Quinct, E. (1990) *Economics of Natural Resources and the Environment*, Hemel Hempstead, Harvester Wheatsheaf.

Rakha, H., M. Van Aerde, E.R. Case and A. Ugge (1989) Evaluating the benefits and interactions of route guidance and traffic control strategies using simulation, In *Proceedings of Vehicle Navigation and Information Systems (VNIS) Conference*, 296-303.

Ran, B. and D.E. Boyce (1994) *Dynamic Urban Transportation Network Models: Theory and Implications for Intelligent Vehicle-Highway Systems*, Springer-Verlag, Berlin.

Ran, B., D.E. Boyce and L.J. LeBlanc (1993) Dynamic congestion pricing models in an ATMIS environment, paper presented at 40*th* North American Meeting of the Regional Science Association, Houston, Texas.

Richardson, H.W. (1974) A note on the distributional effects of road pricing, *Journal of Transport Economics and Policy*, **8**, 82-85.

Rienstra, S.A., J.M. Vleugel and P. Nijkamp (1996) Options for sustainable passenger transport; an assessment of policy choices. Forthcoming in *Transportation, Planning and Technology*.

Rouwendal, J. and P. Rietveld (1994) Changes in commuting distances of Dutch households, *Urban Studies*, **31**, 1545-1557.

Rumar, K. (1990) Driver requirements and road traffic informatics, *Transportation*, **17**, 215-229.

Seale, K. (1993) Attitudes of politicians in London to road pricing, PTRC 21*st* Summer Annual Meeting, Proceedings of Seminar F, 117-128.

Second Structure Plan (1988) *Beheerste mobiliteit, bereikbaarheid, leefbaarheid, Deel A: Beleidsvoornemen*, Tweede Kamer, 1988-1989, 20922, 1-2.

Segal, D. and T.L. Steinmeier (1980) The incidence of congestion and congestion tolls, *Journal of Urban Economics*, **7**, 42-62.

Sharp, C. (1966) Congestion and welfare - an examination of the case for a congestion tax, *The Economic Journal*, **76**, 806-817.

Shefer, D. and P. Rietveld (1994) Congestion and safety on highways; towards an analytical model, Tinbergen Institute Discussion Paper, TI 94-92.

Sheffi, Y. (1985) *Urban Transportation Networks: Equilibrium Analysis with Mathematical Programming Methods*, Prentice-Hall, Englewood Cliffs, N.J.

Sheffi, Y. and W.B. Powell (1981) A comparison of stochastic and deterministic traffic assignment over congested networks, *Transportation Research*, **15B**, 53-64.

Sheldon, R., M. Scott and P. Jones (1993) London congestion charging: Exploratory social research among London residents, paper presented at the 21*st* PTRC Summer Annual Meeting, Proceedings of Seminar F, 129-145.

Shladover, S.E. (1993) Potential contributions of Intelligent Vehicle/Highway Systems (IVHS) to reducing transportation's greenhouse gas production, *Transportation Research*, **27A**, 207-216.

Simon, H. (1955) A behavioral model of rational choice, *Quarterly Journal of Economics*, **69**, 99-118.

Simon, H.A. (1990) Invariants of human behavior, *Annual Review of Psychology*, **41**, 1-19.

Sloof, R. (1995) A comment on drivers information in a two-link equilibrium model, *Tinbergen Institute Research Bulletin*, **7**, 69-72.

Small, K.A. (1982) The scheduling of consumer activities: Work trips, *American Economic Review*, **72**, 467-479.

Small, K.A. (1983) The incidence of congestion tolls on urban highways, *Journal of Urban Economics*, **13**, 90-111.

Small, K.A. (1992a) *Urban Transport Economics*, Chur, Harwood.

Small, K.A. (1992b) Using the revenues from congestion pricing, *Transportation*, **19**, 359-381.

Smeed, R.J. (1964) The traffic problem in towns: A review of possible long term solutions, *Town Planning Review*, **35**, 133-158.

Smith, M.J. and M.O. Ghali (1992) Two new models for assessing Urban Traffic Control and road pricing strategies, *Traffic Engineering and Control*, **33**, 245-249.

Spyridakis, J., W. Barfield, L. Conquest, M. Haselkorn and C. Isakson (1991) Surveying commuter behavior: Designing motorist information systems, *Transportation Research*, **25A**, 17-30.

Stergiou, B. and A. Stathopoulos (1989) Traffic models and Road Transport Informatics (RTI) systems, *Traffic Engineering and Control*, **30**, 580-586.

Stern, E., E. Holm and M. van Maarseveen (1990) Choice behaviour in congested situations: The effects of information, Report on Research Structure Formulated with NECTAR Twining Grant.

Stoelhorst, H.J. and A.J. Zandbergen (1990) The development of a road pricing system in The Netherlands, *Traffic Engineering and Control*, **31**, 66-71.

Supernak, J. (1992) Temporal utility profiles of activities and travel: Uncertainty and decision making, *Transportation Research*, **26B**, 61-76.

Thomson, J.M. (1977) *Great Cities and Their Traffic*, Gollancz, London.

Transportation Research (1986) Special issue: Road pricing, K.J. Button (Ed.) *Transportation Research*, **20A** (2).

Tsuji, H., R. Takahashi, H. Kawashima and Y. Yamamoto (1985) A stochastic approach for estimating the effectiveness of a route guidance system and its related parameters, *Transportation Science*, **19**, 333-351.

van Berkum, E.C. and P.H.J. van der Mede (1993) *The Impact of Traffic Information*, PhD dissertation, Universiteitsdrukkerij, Delft.

van der Mede, P.H.J. and E.C. van Berkum (1991) BEAST: A Behavioral Approach to Simulating Travellers, In *Proceedings of the 6th International Conference on Travel Behaviour*, Quebec.

van der Zijpp, N.J. and P.H.L Bovy (1994) Modelling driver information acquisition behaviour in route forecasting, Paper presented at the 1*st* Erasmus-Network Conference on Transportation and Traffic Engineering, Kerkrade, The Netherlands, September 4-6, 1994.

van Vuren, T. and D.P. Watling (1991) A multiple user class assignment model for route guidance, *Transportation Research Record*, **1306**, 22-31.

van Vuren, T. and D. van Vliet (1992) *Route Choice and Signal Control*, Avebury, Aldershot.

Varian, H.R. (1992) *Microeconomic Analysis*, Third Edition, Norton & Company, New York.

Vaughn, K., M. Abdel-Aty, R. Kitamura, P. Jovanis and H. Yang (1992) Experimental analysis and modeling of sequential route choice behavior under ATIS in a simplistic traffic network, Research Report UCD-ITS-RR-92-16, September 1992, Institute of Transportation Studies, University of California, Davis.

Verhoef, E.T. (1994a) External effects and social costs of road transport, *Transportation Research*, **28A**, 273-287.

Verhoef, E.T. (1994b) Efficiency and equity in externalities, *Environment and Planning*, **26A**, 361-382.

Verhoef, E.T. (1995) The demand curve under road pricing and the problem of political feasibility: A comment, *Transportation Research*, **29A**, 459-464.

Verhoef, E.T., R.H.M. Emmerink, P. Nijkamp and P. Rietveld (1996) Information provision, flat- and fine congestion tolling and the efficiency of road usage, *Regional Science and Urban Economics*, **26**, 505-529.

Verhoef, E.T., P. Nijkamp and P. Rietveld (1995a) Second-best regulation of road transport externalities, *Journal of Transport Economics and Policy*, **29**, 147-167.

Verhoef, E.T., P. Nijkamp en P. Rietveld (1995b) The economics of regulatory parking policies, *Transportation Research*, **29A**, 141-156.

Verhoef, E.T., P. Nijkamp and P. Rietveld (1996) Second-best congestion pricing: The case of an untolled alternative. Forthcoming in *Journal of Urban Economics*.

Verhoef, E.T. and J.C.J.M. van den Bergh (1994) Sustainable transport: Modelling interactions between transport, infrastructure, economy and environment in a spatial price equilibrium setting, Tinbergen Institute Discussion Paper, TI 94-63.

Vickrey, W. (1955) Some implications of marginal cost pricing for public utilities, *American Economic Review*, **45**, 605-620.

Vickrey, W.S. (1969) Congestion theory and transport investment, *American Economic Review*, **59**, Papers and Proceedings, 251-261.

Vickrey, W.S. (1971) Responsive pricing of public utility services, *The Bell Journal of Economics and Management Science*, **2**, 337-346.

Visser, J.G.S.N. (1992) Toltunnels in de randstad, *Infrastructuur, Transport en Logistiek*, **16**, Delftse Universitaire Pers, Delft.

von Winterfeldt, D. and W. Edwards (1986) *Decision Analysis and Behavioral Research*, Cambridge University Press, Cambridge.

Walters, A.A. (1961) The theory and measurement of private and social cost of highway congestion, *Econometrica*, **29**, 676-699.

Wardrop, J. (1952) Some theoretical aspects of road traffic research, In *Proceedings of the Institution of Civil Engineers*, **1** (Part II), 325-378.

Watling, D. (1994) Urban traffic network models and dynamic driver information systems, *Transport Reviews*, **14**, 219-246.

Watling, D. and T. Van Vuren (1993) The modelling of dynamic route guidance systems, *Transportation Research*, **1C**, 159-182.

Westerman, M. (1995) *Real-Time Traffic Data Collection for Transportation Telematics*, PhD dissertation, Delft Institute of Technology, Delft.

Wheeler, J.O. (1969) Some effects of occupational status on work trips, *Journal of Regional Science*, **19**, 69-78.

White, M.J. (1986) Sex differences in urban commuting patterns, *American Economic Review*, 368-372.

Yang, H., R. Kitamura, P.P. Jovanis, K.M. Vaughn and A. Abdel-Aty (1993) Exploration of route choice behavior with advanced traveler information using neural network concepts, *Transportation*, **20**, 199-223.

SUBJECT INDEX

AUTHOR INDEX